Artificial Intelligence in Engineering

Artificial Intelligence in Engineering

Edited by

Graham Winstanley
Information Technology Research Institute
Brighton Polytechnic
UK

JOHN WILEY & SONS
Chichester · New York · Brisbane · Toronto · Singapore

Other Wiley Editorial Offices

John Wiley & Sons, Inc., 605 Third Avenue,
New York, NY 10158-0012, USA

Jacaranda Wiley Ltd, G.P.O. Box 859, Brisbane,
Queensland 4001, Australia

John Wiley & Sons (Canada) Ltd, 22 Worcester Road,
Rexdale, Ontario M9W 1L1, Canada

John Wiley & Sons (SEA) Pte Ltd, 37 Jalan Pemimpin #05-04,
Block B, Union Industrial Building, Singapore 2057

Library of Congress Cataloging-in-Publication Data:

Artificial intelligence in engineering / edited by Graham Winstanley.
        p.   cm.
    Includes bibliographical references.
    ISBN 0 471 92603 5
      1. Engineering—Data processing.   2. Artificial intelligence.
    3. Expert systems (Computer science)   I. Winstanley, Graham.
    TA345.A78   1990
    620'.00285'63—dc20   90-31728
                                                            CIP

British Library Cataloguing in Publication Data:

Artificial intelligence in engineering.
    1. Engineering. Applications of artificial intelligence
    I. Winstanley, Graham
    62.00286563

    ISBN 0 471 92603 5

Printed in Great Britain by Courier International Ltd, Tiptree, Essex

Contents

List of Contributors

Mr N. J. Avis
Department of Medical Physics and
   Clinical Engineering
University of Sheffield and Health
   Authority
The Royal Hallamshire Hospital
Glossop Road
Sheffield S10 2JS

Dr D. Bose
Department of Mathematical Sciences
Brighton Polytechnic
Lewes Road
Brighton

Dr S. W. Ellacott
Information Technology Research Institute
Brighton Polytechnic
Lewes Road
Brighton
East Sussex BN2 4AT

H. A. Heathfield
Information Technology Research Institute
Brighton Polytechnic
Lewes Road
Brighton
East Sussex BN2 4AT

Mr A. B. Hunter
Department of Computing
Imperial College of Science and
   Technology
South Kensington
London SW7 2BZ

Dr T. Katz
Department of Electrical and Electronic
   Engineering
Brighton Polytechnic
Lewes Road
Brighton
East Sussex BN2 4GJ

Mr J. M. Kellet
Brighton Polytechnic
Lewes Road
Brighton
East Sussex BN2 4GJ

Mr P. W. Kuczora
Information Technology Research Institute
Brighton Polytechnic
Lewes Road
Brighton
East Sussex BN2 4AT

Dr G. Marshall
AI Unit (S 5090)
British Airways
PO Box 10
Halton Cross
Middlesex TW6 2JA

Dr F.N. Teskey
IT Planning Division
National Westminster Bank plc
PO Box 114
Goodmans Fields
74 Alie Street
London E1 8HL

Dr R. Thomas
Department of Electrical and Electronic
   Engineering
Brighton Polytechnic
Lewes Road
Brighton
East Sussex BN2 4GJ

Dr G. Winstanley
Information Technology Research Institute
Brighton Polytechnic
Lewes Road
Brighton
East Sussex BN2 4AT

List of Trademarks

704, Intel 80386, iPSC and Direct Connect Routing are trademarks of Intel Corporation.

ACRONYM is a trademark of Stanford Research Institute (SRI).

Allegro Common Lisp is a trademark of Coral Software Corporation.

ALPHA is a trademark of Fujitsu.

APOLLO and PRISM are trademarks of Apollo, Inc.

Apple is a registered trademark of Apple Computer, Inc.

AUTOVIEW is a trademark of BRSL (British Robotic Systems Labs.).

The ART System, which includes Inference ART, Art-LISP and ART-IM, are registered trademarks of Inference Corporation.

Connection Machine, CM Lisp and NEXUS are trademarks of Thinking Machines, Inc.

CRAY and X-MP are trademarks of Cray Research Corporation.

Crystal is a trademark of Intelligent Environments Ltd.

dBASE and dBASE III are trademarks of Ashton Tate.

Distributed Array Processor (DAP) is a trademark of ICL and Active Memory Technology (AMT).

Ethernet, Loops and Interlisp-D are trademarks of Xerox Corporation.

Explorer, MicroExplorer and NuBus are trademarks of Texas Instruments, Inc.

GOLDWORKS, Golden Common Lisp, GCLISP and Common Concurrent Lisp are trademarks of Gold Hill Computers, Inc.

Hewlett Packard and Hewlett Packard 9000 are trademarks of Hewlett Packard, Inc.

IBM, IBM-PC, PS/2, IBM 6150 and Personal Computer AT are registered trademarks of International Business Machines, Inc.

The KEE system, which includes KEEconnection, KEE/C, KEEpictures and SimKit are trademarks of IntelliCorp.

KES, KES PS, KES HT and KES Bayes are trademarks of Software A&E.

KEYSIGHT, CONSIGHT and INSPECTOR are trademarks of General Motors, Inc.

Knowledge Craft is a registered trademark of Carnegie Group, Inc.

LAMBDA is a trademark of Lisp Machines, Inc.

Lotus and 1-2-3 are trademarks of Lotus Development Corporation.

Macintosh is a trademark licensed to Apple Computer, Inc.

MPP is a trademark of Goodyear.

MSDOS and Microsoft Windows are trademarks of Microsoft Corporation.

PUMA is a trademark of Unimation, Inc.

SEQUENT and BALANCE are trademarks of Sequent, Inc.

SMALLTALK-80 is a trademark of Xerox Corporation and ParcPlace Systems.

Sun Workstation and SPARC are trademarks of Sun Microsystems, Inc.

Symbolics is a trademark of Symbolics, Inc.

T-Series is a trademark of Floating Point Systems.

Transputer is a trademark of Inmos.

UNIX is a trademark of AT&T Bell Laboratories.

VAX, VAX-11, VMS, MicroVax, PDP and DEC are trademarks of Digital Equipment Corporation.

Introduction

This book is intended to convey some of the important issues associated with the introduction and successful use of artificial intelligence (AI) in engineering. Engineers are well used to applying high technology to the solution of complex problems, and are therefore ideal candidates for the utilisation of this relatively new and exciting innovation. It is possible to employ techniques of AI in (almost) any branch of any of the engineering fields, but one of the major problems faced by the potential user is an accurate and indeed useful definition of artificial intelligence itself.

It is generally considered that the discipline of Artificial Intelligence began in the Summer of 1956, at the now-famous conference held at the Dartmouth College in Hanover, New Hampshire, USA. The conference was to debate the possibilities of producing computer programs with the ability to accurately simulate human reasoning and behaviour. One of the 'core concepts' at the heart of that conference was the proposition that any aspect of human intelligence could theoretically be simulated by machine. This conjecture relies largely on the possibility of accurately describing every feature of intelligence.

The major organiser of that fateful conference in the discipline of Artificial Intelligence was John McCarthy, who was then an assistant professor of mathematics at Dartmouth. McCarthy is also believed by many to have first coined the term 'artificial intelligence'. A working AI system, existing at around that time, was known as The Logic Theorist, developed by Newell and Simon, and was effectively a computer program capable of proving theorems in mathematical logic taken directly from *Principia Mathematica* (a now 'classic' text by Whitehead and Russell, written in 1910). This text, and in particular the acknowledged expertise of the authors, was very highly regarded, and the impact of a computer program able to prove theorems taken directly from this definitive work was immense. It was regarded as a viable demonstration of the possibility of computer-based 'intelligence', and had a great influence on the direction of AI research immediately after 1956. The Logic Theorist can now be categorised into a particular branch of AI in which intelligent behaviour is simulated using methods of automated mathematical logic. AI research is not restricted to this narrow domain of formal mathematics and logic, but it is interesting to note that the current trend is towards representation and reasoning based on formal mathematical methods, able to provide the kind of validation and verification possibilities required in large and highly sophisticated systems.

The Dartmouth conference provided a focus for the work and ideas of a new generation of computer scientists, influenced by the powerful notion that computers may eventually simulate most (if not all) human faculties. These people had been influenced by the principles of major research figures such as John Von Neumann and Alan Turing. Some quite significant computer developments had already been established long before the summer of 1956. In

1936 Alan Turing developed the principles of a universal machine capable of carrying out any conceivable calculation. He suggested that a test of the intelligence of such a machine would be if it were possible (or not) to distinguish its answers from those derived by a human being. This hypothetical test of a machine's intelligence, measured directly against the human capacity, became known in 1963 as 'The Turing Test'. Von Neumann is considered by many to be the 'founding father' of present-day computer architectures, but his ideas of sequential processing and stored programs were significantly influenced by Turing's work. It is, of course, possible to trace the origins of AI as far back in the history of science and philosophy as might be desired for reasons of curiosity. For example, the study of thought and reasoning can easily be identified with ancient philosophers and logicians such as Aristotle in the Middle Ages, and the Renaissance thinkers such as Descartes, Locke, Leibnitz and Kant, and, more recently, with George Boole, whose major work was concerned with mathematical logic. He is well known for his development of the Boolean algebra, which is familiar to engineers, but he was also keenly interested in the mechanisms of human thought and reasoning. Boole's interests can be identified in the statement which forms part of his published work in 1854, 'An Investigation of the Laws of Thought on Which are Founded the Mathematical Theories of Logic and Probabilities':

> The laws we have to examine are the laws of one of the most important of our mental faculties. The mathematics we have to construct are the mathematics of the human intellect.

The separate and identifiable discipline which has emerged since the Dartmouth conference is, by its very nature, multi-faceted. The term 'artificial intelligence' covers an extremely broad area of research and development. In its broadest sense, the field of academic study which represents AI is concerned with the investigation and simulation of human intelligence. In order to understand what that entails, it is tempting to consider the meaning of intelligence itself. However, an accurate and widely agreed definition of human (or natural) intelligence is notoriously difficult to obtain; all but the most committed psychologist would resist any attempt to define intelligence, because it is such a vague term. It has elements of:

- reasoned, and in some way directed behaviour,
- organised and dynamic memory,
- the ability to communicate effectively,
- the ability to adapt in the light of experience.

The four basic points above encompass some of the more obvious attributes associated with human 'intelligent-like' behaviour. The ability to remember in a structured fasion is important. Facts must be stored in an organised way, with interrelationships well-formed. The ability to adapt, or learn, in the light of experience is highly important, and possibly crucial, in intelligence. The ability to learn, and possibly the rate at which the learning process takes place, is a frequent indicator of intelligence. Insight is yet another extremely vague term, but nevertheless has meaning because we can all identify with it.

In the inevitable evolution of artificial intelligence, the discipline has undergone a significant degree of differentiation. It may no longer be possible to be expert in AI, but rather in one of the speciality areas commonly associated with the term. Such areas of special study, independent development and implementation include:

- natural language processing,
- machine vision,
- planning,
- robotics,
- expert systems,
- theorem proving,
- game playing,
- machine learning.

All of the above, apart perhaps from game playing, are evident in engineering applications of AI. Machine vision, robotics and planning are very closely related disciplines in the sense that the general field of robotics can easily be identified to have elements of planning and vision. Natural language processing, although not discussed in this book, is subject to much active research in the AI community, because it holds the promise of the 'ultimate' user interface, and is a key element in understanding. It can therefore be recognised as a future generic AI system component. Machine learning is another active and fundamental research topic, and is set to become a central issue of major importance.

Although it may be true that game-playing techniques are not obviously evident in engineering systems, the development of automatic systems able to play games has provided researchers with realistic and complex situations on which to base their work. Sometimes quite simple games provide a stimulating academic challenge when attempting to design an automatic system able to understand the opponent's moves and plan an optimal strategy for eventual success. The principles of games of this kind are also easily understood, and therefore the success or otherwise of the system is immediately obvious. One of the basic principles in automated game playing is to plan on minimising the maximum damage that the opponent could cause. This successful principle is termed the 'minimax' strategy, and it essentially involves a 'search' for the worst-case move an opponent can make, and chooses its own move in an attempt to minimise the effects. The basic concepts resulting from this line of research have been:

- search,
- optimisation,

and have had a major impact on the research and development of AI systems, including those suitable for engineering application. By creating a 'search tree', it is possible for a computer system to search every possible move that an opponent could make, and thereby plan an 'optimum move'. However, when playing a game such as chess, the number of search pathways required in the implementation of a complete search are enormous. In order to provide a realistic (that is, successful) search, the notion of search space minimisation based on rules of thumb, or 'heuristics', became popular and was the basis for the 'General Problem Solver' (GPS), introduced by Ernst and Newell in 1969. Success in 'pruning' the search space was found to rely on expert knowledge of the situation at hand. For example, in order to produce a successful heuristic search for optimal moves in chess, those heuristics must be defined by an expert game player (or at least, someone who knows the rules well). This particular problem introduces the notion of 'domain-specific knowledge'.

In the 1960s, a system known as DENDRAL arrived on the scene. This program was designed to perform the kind of chemical analysis function otherwise required of an expert experienced chemist. The reason for its development (apart from the academic evolution of AI) was the remote analysis of Martian soil. NASA was planning an unmanned mission to Mars, and they needed expert chemical analysis functionality on-site. At the end of the project, the eventual DENDRAL system was physically too large to go on the mission, but the project had a significant impact on future directions of AI, or at least, a certain part of AI.

Today there is much more diversity in the range of research in AI. There is a trend to provide mathematical 'soundness' to representation schema through the use of conceptual structures, formal logic and structured design processes. This was an inevitable consequence of AI systems 'graduating' from the academic research laboratories into industry and commerce. In situations where AI tools are actually implemented and expected to perform a useful role in dealing with real-life situations, the amount of information and knowledge required is both massive and ever-increasing. The data processing world has long been aware of the problems of large database management, but with AI systems, which rely to a great extent on very detailed and accurate information, the problems are mangnified.

Chapter 1 deals with the problems associated with complexity in real-life situations, and introduces systems concepts and techniques designed for dealing with complexity. The discussion hinges on the definition of the problem as a bounded system, and the application of the generic concepts of models, languages and methodologies. In essence, the chapter attempts to demonstrate that the systems movement has a philosophy, which shows how models give rise to methodologies, comparable with those of AI and software engineering. The systems philosophy is able to elucidate commonalities with these apparently disparate disciplines and can provide the means to approach the solution to complex engineering problems with AI techniques.

Chapter 2 represent an introduction to the important issues of representation and reasoning in AI systems. The methods used to represent domain knowledge are central to such systems. However, there are many different forms of knowledge and the representation of that elusive quality presents all-too-familiar problems. The initial sections of chapter 2 deal in some detail with domain models, and therefore follows quite naturally from the previous chapter's discussion of conceptual models. The discussion in Chapter 2 on representation includes relational data models, semantic networks and logic, and the problems of searching such models. An introductory section on automated reasoning concludes the chapter.

Chapter 3 examines the various types of tools that are available to the developer of knowledge-based systems. In particular, it concentrates on implementational and practical issues relating to their use. The large hybrid development systems such as KEE and ART are contrasted with the less sophisticated expert system shells such as KES and Crystal. Examples are given to provide the user with an insight into how these tools may be utilised in practice. The chapter also introduces the concept of symbolic computer languages and demonstrates how they can best be used in the development of knowledge-based systems.

Chapter 4 deals with a subject which cannot be underestimated: the knowlege engineering process. The elicitation and computer application of a domain expert's knowledge is not a trivial task. It is widely accepted as the 'bottleneck' in the development of knowledge-based systems applications, and the eventual success of such an application depends to a large extent on the quality and thoroughness of the knowledge engineering process. This chapter examines the problems of elicitation and shows various methods of dealing with the situation, from interview techniques to structured and formalised methodologies. Examples are given of tried and tested techniques, and several systems designed to assist the knowledge engineer are described.

Chapter 5 is the first of three applications chapters and deals with engineering design manufacture and testing. This chapter presents very much a practical viewpoint on the tasks required of computer-based systems destined for engineering applications, and discusses the various issues pertinent to the choice of development/implementation strategy. As such, it represents a logical extension of earlier material (Chapter 3) on environments, shells and languages. Knowledge representation is again the subject of discussion in this chapter, as is domain modelling, but the text also deals with operational considerations. Consideration is given to the implications of AI in design and manufacture, but the chapter uses an extended example in automatic test to give a practical demonstration of the processes involved in designing and developing an engineering knowledge-based system.

Chapter 6 gives an overview of the development of AI planning techniques and their application to planning and project management as a whole within engineering domains, such as construction and manufacture. This chapter traces the development of 'traditional' strategic AI planning techniques, beginning with the concept of search and continues by discussing means-end and non-linear hierarchical planners. Detailed examples of each type are included from within the literature. The topic of backtracking and truth maintenance is discussed in relation to planning systems, and then later, more sophisticated techniques such as case-based, opportunistic and meta-planners are introduced. The discussion then moves to wider approaches than 'pure AI', such as AI-levered systems, and non-AI techniques in tactical planning as applied to uncertain domains. The chapter concludes with an appraisal of likely future developments in the use of AI-related techniques for engineering project management.

Chapter 7 examines the impact of AI in industrial computer vision. The chapter looks at the actual tasks involved in machine vision and how they have been emulated in dedicated computer systems. A historical perspective is given, and examples of existing systems are included. Care has been exercised in providing a general coverage of machine vision, existing as a discipline which can benefit from the application of AI techniques, but is not reliant on this approach for much of its low-level functionality. The case is made, however, in model-based recognition in robot guidance applications, where more flexible computer vision systems will eventually be developed with the ability to operate reliably and repeatedly within completely unstructured environments, and locate objects within three-dimensional space.

Chapter 8 extends the book's treatment of AI by examining developments in reasoning. Much of the attention of AI has been at the symbolic level. This is the level at which the representation and manipulation of information, or knowledge, is in the form of tangible concepts (such as dog, car, etc.). Symbolic AI requires the manipulation of such symbols in order to satisfy given goals. From this, AI can be interpreted to mean the formalisms for representing and manipulating symbolic-level knowledge. It is now apparent that an increasing number of those involved in AI are aware of the shortcomings of current symbolic reasoning technology. Much current research, both in academia and industry, is directed towards developing more appropriate formalisms, with supporting technology, for engineering applications. Chapter 8 is intended to introduce some of the significant issues in this area, including the management of uncertainty, time and space. Much of this work remains in the laboratory, but it is likely that the new methodologies introduced in this chapter will have widespread ramifications in AI systems of the future.

Chapter 9 has the title 'Mathematical Foundations of Artificial Intelligence'. The language of mathematical analysis is particularly well-suited to engineering problems, but in AI the need is to model non-numeric data in the form of interrelated information structures, rules, etc. Discrete mathematics has been used successfully to deal with these kinds of problems, and the chapter helps to develop an understanding of the range of problems faced by the AI system builder, and the underlying mathematical concepts at the heart of many AI systems. Included in the chapter are the topics of set theory, relations, graph theory, and the mathematics of reasoning. Much use is made of example material to emphasise the points made. Although much (if not all) of the material in this chapter is general to AI and its related disciplines, the fundamental formalisms of AI are relevant to engineering systems.

Chapter 10 illustrates the types of processing platforms available for running AI applications, and is essentially organised into two parts. The first part examines the suitability of the conventional Von Neumann computer for the processing of AI applications, and identifies the areas appropriate for modification in the drive for increased computational efficiency. The second part of the chapter examines the role of VLSI and the development of parallel processing architectures aimed specifically at AI systems. Case studies of commercial and research systems are used throughout the text to illustrate how design considerations are implemented in practice. The chapter concludes with an introduction to the connectionist approach to AI, in which is likely to have a profound effect on future systems architectures and their performance.

In this book, an attempt has been made to include the important theoretical issues of representation and reasoning, as well as some of the more application-related material. The reader may, in fact, take several 'paths' through the text, depending on their level of familiarisation/expertise and the intended use of the material. For a thorough treatment of AI for engineers, which incorporates a good deal of fairly introductory material, the sequence should be Chapters 1 to 7 inclusive. This includes some basic systems theory, an introductory treatment of representation and reasoning, some important introductory material of a survey nature on expert system shells and languages, a treatment of the important problem of knowledge engineering, and three specific applications chapters. This order should provide a good introduction to the potential and realities of AI, and should give the reader a good 'feel' for how these techniques have been used, and will in the future be used in 'real' engineering situations. Each chapter is provided with a bibliography, chosen to reflect definitive work to supplement the material in the chapter, and should be treated as necessary reading material in any serious study. The reader who wishes to take the study further would wish to read Chapters 8 and 9 for an advanced treatment of some of the more formal aspects of AI.

For a reader already familiar with the concepts introduced in the early stages of the book, and who may be interested in the application of AI techniques in engineering, the three applications chapters (5, 6 and 7) should provide sufficient information (along with the references included in every chapter) to ensure an appreciation of the problems and potential of real-life engineering applications. In some ways, the example material included in these chapters are a better indication of the realities of applied AI than the more theoretical treatment found in many AI texts. Specific reference is made at key places in each chapter, which permits the fuller exploration of important concepts which may be 'skipped over' in one chapter, but is discussed in much more detail in another. One of the most common occurrences of this within the book is a reference to the mathematical treatment of fundamental AI concepts to be found in Chapter 9.

The third 'pathway' through the book is to concentrate on the more advanced material to be found in Chapters 8, 9 and 10. Chapter 8 can be read after gaining a good understanding of the introductory material in Chapter 2, but could also stand on its own as a good treatment of reasoning in logic. Chapter 9 covers some of the important mathematical fundamentals required in a deep understanding of AI, and can be regarded as a similar standard to Chapter 8. Chapter 10 is different. It deals with the hardware aspects of artificial intelligence systems technology, and can be read and assimilated on its own if desired. However, no matter what the level of competence of the reader, or their specific goals, this chapter is recommended. It discusses the functionality of AI languages in a way which would be impossible without including the underlying mechanisms. In many ways, the hardware must 'catch up' with the available software, and only when near-real-time operation becomes a reality will embedded knowledge-based systems, for example, become commonplace. The engineer undertaking an appraisal of AI as it may be applied to a particular situation must be aware of the significant hardware considerations required, and will be all too aware of the consequences of making an uneducated decision.

Finally, there are topics considered important in AI but not dealt with specifically in this book. One of these subject areas is natural language processing, and another is artificial neural networks (also referred to as connectionist systems). In order to provide an appropriate level of treatment within the text it was decided to concentrate on what may be regarded as fundamental (or core) issues which are sufficient, when collected together, to gain a good overall appreciation of the subject. Neural networks are becoming popular in the research and development field, and it is certainly true that they have a future in many fields. Although it was a deliberate decision not to include a specific connectionist chapter in this book, the subject relevance is discussed in several places in the text, and reference is made to permit and encourage further study. In particular, Chapter 10 includes a detailed discussion of their structure and function in the context of computer hardware architectures.

1 The Engineering Domain and Systems

G. Winstanley, UK

Brighton Polytechnic, UK

1.1 INTRODUCTION

The definition of an engineer in the Collins English Dictionary states that an engineer is 'one who constructs, designs, or is in charge of engines, military works, or works of public utility (roads, docks, etc.)'. This would appear to be a somewhat limited definition in a modern society blessed with Genetic Engineers, Software Engineers, Knowledge Engineers and Sales Engineers. However, it is possible to sympathise with such an over-simplistic definition if the actual intellectual processes involved in each of these apparently disparate disciplines are held under closer scrutiny. In each case, the engineer is faced with a problem, or problem situation, for which he or she has specific responsibilities to effect a solution (via the engineering process). In effect, engineering is problem-solving, but this is too wide a definition; mathematicians and philosophers involve themselves deeply in problem solving, but they would hardly be included in the fellowship of engineering. There is something more, and it relates back to the dictionary definition of engineer: 'constructs, designs ...'. The engineer will be involved in the design of something to rectify a perceived problem or area of concern, and will be responsible, at one level or another, for specifying, designing, constructing, testing and evaluating a system which may be used beneficially in satisfying a given set of demands, relating to the initial area of concern (the problem situation). It is the practical aspects of problem-solving, which lead to verifiable solutions, that characterise the discipline, i.e. the process from initial specification through design to production and evaluation (and beyond, within the 'product lifecycle').

Due to the practical problem-solving nature of engineering and its traditional emphasis on problems with a highly deterministic nature, an inexorable increase in the number and complexity of tools and techniques made available to assist in the process has been apparent. In effect, a cycle of development exists, both intellectually and technologically. This was referred to by Sir Ieuan Maddock (1983) as 'The Technological Explosion', in which every branch of technology has been seen to advance at great speed. One of the key factors in the 'speeding up' of technological change has been the phenomenon of interdiscipline

Artificial Intelligence in Engineering. Edited by Graham Winstanley
ⓒ1991 by John Wiley & Sons Ltd.

interaction. This has been facilitated by several relatively recent enabling technologies, which include:

- Computer Technology (programmability),
- Communications Technology (accessibility),
- Systems Technology (approachability and maintainability),
- Artificial Intelligence Technology (possibility).

Undoubtedly, other branches of technology have had a massive impact on individual engineering disciplines, but those factors shown above can be viewed as 'universal' in relevance and applicability. The availability of equipment which can be made (engineered) to perform a particular task within a problem situation, via the design of a non-hardware based program, has had an impact on the world at large, of which we are all consciously or subconsciously aware. This technology is itself responsible for our digital communications capabilities, which are perhaps even more significant to the future direction of technological integration and use in the world. It will affect working practices, business operation, technical innovation and everyday life in general, by virtue of the ease of accessibility to specialised information and knowledge. Traditional boundaries, both national and international will disappear, and information will become a rich and available commodity. Systems technology provides a means of dealing with the inevitability of increasing complexity with concepts, models and, above all, methodologies for dealing with complex situations. Artificial intelligence (AI) is the common concern within this book, and in its short history has been shown to provide concepts, methods and specific techniques to deal with situations not usually associated with the engineering professions, i.e. the less well-defined problems characterised by uncertainty and incomplete information. Examples of this are in project planning and management, which are complex, multiple-discipline situations having highly uncertain components—people. Although knowledge-based systems can be used, and have been used, in situations in which strict algorithmic solutions exist, they have a much more important role in situations dominated by heuristics, and in which beliefs and underlying concepts may be challenged and subsequently modified during the process of system building, via the mechanism of knowledge engineering. Michie talks about 'Knowledge Refinement' (Michie, 1983), where human knowledge is enriched and refined due to a 'back-translation' process from a possibly-simplistic knowledge base and inference process to the system user. This means that, in effect, the very processes involved in making knowledge and reasoning strategies explicit renders that knowledge appropriate to the learning situation.

The fusion of the four technological areas of Computing, Communications, Systems and Artificial Intelligence has had a discernible impact in many spheres of expertise, but will have consequences significant, not only for engineering, but for the world at large.

1.2 THE SYSTEMS PHILOSOPHY

It has already been stated that engineering is a problem-solving discipline. Likewise, systems engineering is firmly established in providing the means to alleviate problem situations. However, there is much more to the field of systems than structured engineering. It is a holistic philosophy which attempts to rationalise the whole of our natural environment, from the behaviour of subatomic particles to the nature of the universe. Indeed, many general systems principles have been established through research into the nature of man's physical environment. For example, research into the nature of atoms and molecules has resulted in the concepts of:

- subatomic particles, neutrons and protons (entities),
- energy in terms of a 'nuclear force',
- physical (or notional) boundaries,
- relations between entities, and
- constraints—energy is somehow constrained by nuclear force in the formation of the nucleus, and individual atoms are constrained in the formation of molecules.

These concepts are useful in understanding the physical existence and behaviour of material. In this context, relations can be readily associated with constraints. For example, protons all have charges of the same polarity, resulting in a repulsion effect. However, that behaviour is constrained by the relation of nuclear force that binds them together. In fact, all systems appear to contain forces of opposition existing in a situation of 'balance', or equilibrium. These opposing forces, or constraints, have been referred to as 'polarities', and the general tendency for systems to exhibit a continual balancing of polar extremes has been termed 'equilibration' (Downing Bowler, 1981). The notion of an imposed boundary in the analysis of the system is another important factor which has implications in the wider context. However, there are no 'hard and fast' rules for defining the correct boundary in complex systems, since any artificially-imposed limits to observation will inevitably exclude relationships with the 'outside world', and will therefore render the observed situation incomplete. Interrelationships exist between all entities within the physical universe. Entities, or systems, can be said to exist only in relation to other entities or systems, otherwise there is no way to describe it in any situation—it would be a 'non-entity'.

In developing interrelationships into yet another important systems concept; the hierarchy, it is instructive to consider the position of man in a wider context. Figure 1.1 is an extreme diagrammatic approximation to the complex hierarchy which imposes constraints on our nature and behaviour. The 'top level' in Figure 1.1 is shown as the world ecosystem in which a population of human beings exist in balanced harmony with other living organisms. An ecosystem has been defined as 'any unit of nature in which energy and materials are exchanged between the biotic community and the abiotic community' (Gerking, 1974), where a biotic community is defined as 'a natural assemblage of organisms living under uniform habitat conditions'. It is instructive at this stage to ponder the concepts considered earlier at the atomic and subatomic levels. It is in fact the ecosystem level within the hierarchy of Figure 1.1 which demonstrates most forcibly the concept of interrelationships and the inevitable effects on one part of the overall system (as bounded in Figure 1.1) due to change caused at some lower level.

Below the ecosystem level exists a population (or populations if the hierarchy were to be expanded) and, below that, a community. There is no doubt that populations and communities impose severe constraints (some would say restraints) on the behaviour of the individual. Within this level of the hierarchy, concepts such as psychosocial interaction, personal identity and personal meaning play an important part in the complex interrelationships of the community. A direct relationship is apparent with the 'higher'

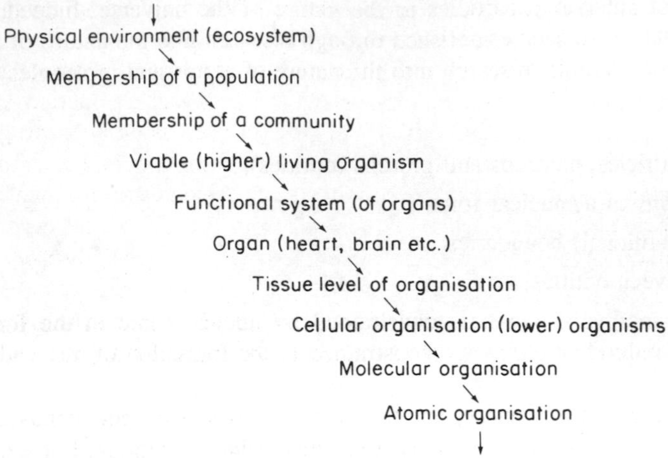

```
                            ↓
          Physical environment (ecosystem)
                             ╲
            Membership of a population
                              ╲
               Membership of a community
                                ╲
               Viable (higher) living organism
                                  ╲
                  Functional system (of organs)
                                    ╲
                    Organ (heart, brain etc.)
                                      ╲
                      Tissue level of organisation
                                        ╲
                        Cellular organisation (lower) organisms
                                          ╲
                          Molecular organisation
                                            ╲
                            Atomic organisation
                                              ↓
```

Figure 1.1   An approximate physical hierarchy relating to man

ecosystem. The human being is expressed in Figure 1.1 as a viable (higher) living organism around which a meaningful boundary can be established. This classification is qualified as 'higher' in order to distinguish it from the cellular level of viability. Interrelationships between the living organism and the community are both physical and psychosocial. Below this level, organs are 'organised' into organ systems, tissue is an organisation of individual cells, and so on.

Figure 1.1 certainly indicates the levels of organisational complexity relating to man and his place in the world, but it does very little to highlight the interrelationships which could provide an indication of cause and effect. It does, however, serve to illustrate one of the most important systems concepts: the emergent properties of higher (or new) levels of organisation. The choice of subject with which to illustrate the hierarchical properties of systems in Figure 1.1 is rather relevant by virtue of the history of systems thinking. In general, some of the most important developments in systems ideas can be attributed to the 'orgasmic biologists', who, at around the turn of the century, were concerned with studying the 'complete' organism rather than the component parts in isolation. By considering organisms at a particular level of complexity, they were able to identify properties and behaviour that were meaningless at lower levels in the hierarchy. In effect, as each higher level is considered, emergent properties or 'novelty' (Downing Bowler, 1981) is apparent. This is easy to visualise in Figure 1.1: the human heart comprises large amounts of muscle (the myocardium), nerve tissue, etc., which exist at the tissue level of organisation. A heart has properties which are 'novel' in relation to the individual components, i.e. it is a very efficient pump which is highly controllable. However, when the heart is included in the circulatory system, that system has the emergent properties of nutrient distribution, body temperature distribution, etc. in addition to facilitating disease defence and blood gas exchange. These properties and behaviour patterns do not exist at lower levels of organisational complexity. Hierarchy theory has developed in recent years, and much of the interest appears to centre on the biological hierarchy from cell to species (Simon, 1962).

The notion of 'wholeness' is central to the general theory of systems as proposed by L. Von Bertalanffy (1940). The basic tenet of his argument was that the concepts identified by the early biologists could be universally applied to any sphere of organised complexity (Checkland, 1981 uses this argument).

It is important at this stage in the discussion to attempt to define the term 'system' and to document a concensus of opinion on what it is about a situation which merits the title. Fortunately, general systems theory can be expressed as an attempt to develop useful generalisations about systems. It is based on the assumption that a commonality exists between all systems, and properties not common to all provide a useful means to differentiate between systems. The word 'system' is derived from the Greek and quite literally means 'to cause to stand together', but the term would appear to have been widely interpreted. For example, we are accustomed to hearing about computer systems, communication systems and security systems, but what about storage systems, or gambling systems? From the previous discussions on the basic philosophy of systems thinking, the following concepts are pertinent:

Wholeness	The whole system, not just the individual components, or entities, even in situations in which individual entities provide the focus of attention. The concept may appear obvious in some situations, but not so in others.
Functionality	The system must have a function. At the most abstract level, it must affect the external environment through a process of change. This concept ties in with the concept of control, i.e. the control of continual change towards a state of equilibrium (but never actually achieving it!).
Relationships	(or constraints). The relationships which exist between individual systems, subsystems or entities. The result of entities existing in an ordered relationship is usually referred to as a hierarchical structure. Figure 1.2 shows a basic inverted tree structure representing part of a domestic television system. The 'root' of the tree is the overall system, and it is seen to be composed of subsystems, components and 'primitive components'. It should be noted that each subsystem can be considered a system in its own right. In other words, the structure has a recursive nature.
Boundaries	The imposition of boundaries around systems in such a way that the resulting system, i.e. that which lies within the 'artificial barrier', is largely impervious to external influences. This means, in effect, that control over the system is achieved (mainly) from within the boundary, and therefore, for some systems, viability is ensured.
Emergent properties	The properties and behaviour of a system related to its level of organised complexity. This has important consequences since the emergent properties of a system to be designed can be used in its definition. The familiar saying 'The whole is greater than the sum of its individual parts' (an argument attributed to Aristotle) is highly relevant and correct when associated with systems thinking.
Control	In Downing Bowler's definition (1981) of equilibrium, a physical system would have the tendency towards a state of equilibrium.

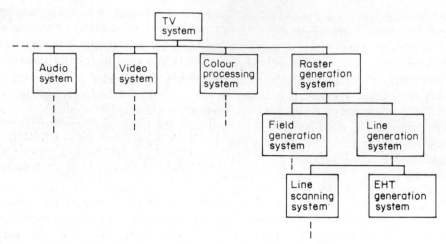

Figure 1.2 A hierarchical structure of part of a domestic television system

In the general sense, all systems will tend towards a situation of maximum entropy without control.

Communication   The regulatory function of control can only be facilitated by the availability of relevant information. That information relates to the external environmental influences on the behaviour of systems or subsystems. It may entail the communication of changing weather patterns in the clothing we wear to maintain body warmth, or the information communicated may take the form of an error signal in a piece of electronic equipment or management situation.

In order to use these potentially useful generic concepts, which are pertinent to highly complex situations, there are tools and techniques relevant to the systems scientist (or engineer):

- models,
- languages, and
- methodologies.

Models are used in the effective representation of bounded situations, Systems (Modelling) Languages provide the means for building and manipulating models and Methodologies give us the means for actually carrying out activities such as analysis and design in a structured fashion.

1.3 SYSTEMS CLASSIFICATION

Before attempting to use these 'philosophically-interesting' concepts, it is instructive to define the term 'system' in a real-world sense, and to identify important classes of systems along with their important or relevant attributes. As a first step, in consideration of the previous discussion, it is obvious that a system is a finite set of elements or entities which have a meaning or function. Within the boundary which encapsulates the individual entities, relationships exist between them. Downing Bowler (1981) talks of 'closed systems' and 'open systems'. The classification can be summarised as below:

Closed Systems The model is assumed to be isolated from the external environment and normally has a well-defined input set which maps explicitly onto the output. This classification represents an approach to systems which is traditionally favoured by scientists and follows from the reductionist approach.

Open Systems This systems model allows for dynamic interaction with the external environment. They are characterised by their non-predictable deterministic output. The inherent interaction with the external environment make it impractical to predict the outcome of a given set of input conditions, even when some of the basic elements are well-defined.

Checkland (1981) outlined a more detailed and potentially more useful classification:

Natural Systems As depicted in Figure 1.1. systems forming part of the physical universe.

Designed Systems Designed and described by man. Examples include computers, ships, etc. The definition includes the more abstract systems of mathematics, language and philosophy. The systems hierarchy of Figure 1.2 is an example of such a structure

Human Activity Systems This is a very important class of systems, applicable to situations involving human activity, such as political systems, industrial systems involving man–machine interaction, etc. The management hierarchy shown in Figure 1.3 is an example of one such human activity system.

Social and Cultural Systems A wider context for human activity systems, where interpersonal relationships exist between people (the individual entities of the system) alongside influencing factors from natural systems. This particular class of system is very important as a consideration in situations characterised by human activity systems.

Figure 1.4 illustrates the basic idea of a system. Individual entities, shown as nodes, have relationships indicated by the arrowed-links between them. The system boundary is explicit, but in establishing a boundary, some relationships are inevitably excluded. These are shown as broken lines in Figure 1.4. The diagram also illustrates the relationship between a system and its external environment. Two points emerge from the analysis:

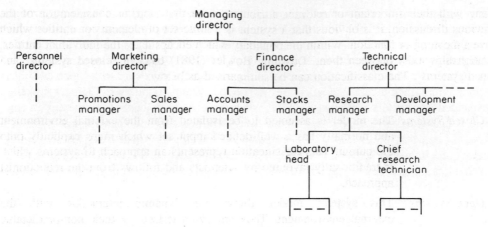

Figure 1.3 A hierarchical (line) management structure

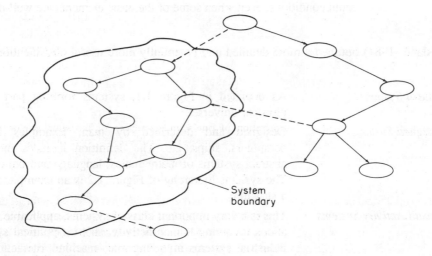

Figure 1.4 A basic system

- To have meaning, the system must have a function within a 'higher' system of greater complexity. Its relationships with the outside world are important in establishing its very existence.
- The function of any system is to influence its external environment by a process of change or transformation. For example, an electricity generating plant may exist as a system, but its relationships with the outside world are obvious; its input is fuel (or energy-bearing material) and its main output is energy in the form of electricity. The transformation process is a physical one involving chemical or nuclear processes.

If the diagram of Figure 1.4 were to be applied to the physical hierarchy of Figure 1.1, say for example, at the organ level, then the entities within the system boundary would perform activities such as mechanical movement of material, take-up and distribution of nutrients, organised discharge of waste, etc. (the digestive system as an example). The external influencing factors would include strict mechanical constraints imposed through the attachment of tissue, the fine chemical balance of surrounding fluid, innervation, etc. These influences represent interrelationships that exist within the system boundary at the next highest level of complexity as shown in Figure 1.1, i.e. the functional system (of organs) level. If the next lower level of Figure 1.1, i.e. the tissue level, were to be included in a system diagram, then the individual entities within the system boundary of Figure 1.4 would be expanded into systems (of tissues) in their own right. At each level of complexity in any system (other than, perhaps, the subatomic level), entities can be expanded into systems, of systems, etc. until the required level of detail or 'resolution' is reached. This concept is actually quite important in the initial system design phase, where an overall 'picture' of the problem situation can be obtained by deliberately restricting the apparent complexity in this way.

1.4 MODELS

Models are essential in the analysis and design of complex systems. they provide a means to describe the behaviour of a situation, or system to a given level of accuracy. They can be quantitative or qualitative, as the situation demands but, more importantly, they must serve to answer a particular question reliably. A definition has been given that refers in these terms to quantitative and qualitative representations (Chestnut, 1965), but a much broader definition was outlined by Wilson (1984) as:

> 'A model is the explicit interpretation of one's understanding of the situation, or merely one's ideas about that situation. It can be expressed in mathematics, symbols or words, but is essentially a description of entities and the relationships between them. It may be prescriptive or illustrative, but above all, it must be useful.'

One may get the idea that this definition is so broad that it caters for all interpretations. However, it was arrived at through the application of systems concepts in the highly complex field of management. Again, it is useful to classify models in terms of their applicability. A widely used classification was introduced by Ackoff (1962), in which he identified basic types: iconic, analogic, analytic. The iconic class of models involved scaled or unscaled replicas, which may be used to demonstrate physical behaviour in proportion to the 'real situation. The analogic type would be used in a similar fashion, i.e. to exhibit proportional behaviour, but more from a position of analogy than physical similarity. The analytic type uses mathematics or logic. Wilson (1984) adds a fourth type to this classification, the conceptual model. This is defined as a pictorial or symbolic representation of the qualitative aspects of the model. These types of models can be used almost intuitively in the initial stages of systems development.

The iconic class of model is intuitively obvious, and its role in analysis and design is readily apparent. However, the distinction between analogic and analytic is not so obvious. For example, the flow of electric current through the human body is often represented by an electrical network made up of capacitors and resistors. The electrical circuit is designed to simulate the complexities of tissue layers and boundary conditions. The circuit representation is certainly an analogy, but the underlying model is mathematical in nature. An analogue computer, which may be used to simulate the complex behaviour of a passenger train suspension system, is another example. The model is based on differential equations, even though an analogy is apparent. The simple answer is that the two types are commonly interrelated. In both of the examples given, the analogic model would be used either in conjunction with the analytic, or immediately afterwards. The distinction may be blurred to some extent, but, in general, analogic models incorporate some form of physical or physically based representation, whereas analytic models are based on pure mathematical or logical relationships which represent the physical laws controlling the system's behaviour.

The level of abstraction expressed in these classifications may be subject to further 'refinement'. In particular, the analytic class of model deserves further consideration. This is particularly the case in the engineering domain, where mathematical analysis could be considered the 'cornerstone' of the science. The conceptual model is even more important in terms of the significance of refinement. Conceptual models are produced in real-life situations every day. The old adage 'it was designed on the back of a matchbox' encompasses the true definition of a conceptual model, i.e. an almost intuitive representation of a situation to a maximum level of abstraction consistent with associated meaning. The whole problem of the choice, design, refinement and implementation of conceptual models will be covered later in the discussion of systems characterised by human activity of one kind or another.

A classification of analytic models was outlined by Wilson (1984). The classification distinguishes between deterministic and non-deterministic models, which are characterised by temporal dependency (dynamic) and models which are not subject to such constraints (static). A deterministic system is defined as having determined relationships governed by physical laws, and a non-deterministic model is based on statistical distributions. Mathematics of some kind is used as a modelling language in analytic systems, and the particular modelling tool most commonly used in each of the model class types are shown as elements. Examples of the scheme are now described.

1.4.1 Steady-state, Deterministic

Situations describable in terms of algebraic equations. Simple examples in engineering include 'passive' electrical networks. In electrical power transmission, a three-phase arrangement is common. A three-phase electrical system consists of three equal voltages, differing by 120° and caused by the rotation of three equally-spaced coils in an electrical generator. A passive three-phase load on the power generated may be represented, as in Figure 1.5, as a 'delta' configuration (Figure 1.5(a)) or a 'star' configuration (Figure 1.5(b)) of electrical resistances. By using network theorems, it is possible to show that the two circuits of Figures 1.5(a) and 1.5(b) are equivalent if their respective input, output and transfer resistances are equal. Moreover, it is possible to obtain star-to-delta transforms as below (applied to the respective resistances of Figures 1.5(a) and (b)):

$$R_a = \frac{R_1R_2 + R_1R_3 + R_2R_3}{R_3} \qquad R_1 = \frac{R_aR_b}{R_a + R_b + R_c}$$

$$R_b = \frac{R_1R_2 + R_1R_3 + R_2R_3}{R_2} \qquad R_2 = \frac{R_aR_c}{R_a + R_b + R_c}$$

$$R_c = \frac{R_1R_2 + R_1R_3 + R_2R_3}{R_1} \qquad R_3 = \frac{R_bR_c}{R_a + R_b + R_c}.$$

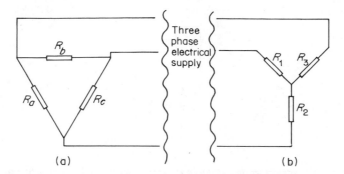

Figure 1.5 A model of a passive electrical load

Perhaps one of the most important outcomes of this kind of analysis is that the obtained transforms serve to identify two mnemonic rules which can be used to provide any necessary transformations:

For star to delta: Any resistance of the delta circuit is equal to the sum of the products of all possible pairs of star resistances divided by the opposite resistance of the star circuit.

For the delta to star: Any resistance of the star circuit is equal to the product of the two adjacent delta resistances divided by the sum of the three delta resistances.

1.4.2 Steady-state, Non-deterministic

Systems under this category would be characterised by a degree of uncertainty about the physical laws or mechanisms which govern their behaviour. In such situations, assumptions must be made about the dependence of system components on overall behaviour, and the nature of interrelationships. Statistics are used commonly in these situations to establish the degree and nature of relationships between individual entities (variables). Regression analysis is a useful tool which can be used to relate two variables. However, in Wilson's classification, probablistic relationships are included in this class of model. Probability models are very important in Artifical Intelligence (AI) systems, since they provide a structured means of dealing with relative uncertainty. Examples in engineering include communications systems, electronic device design (physics), etc., and decision theory at the company management level.

1.4.3 Dynamic, Deterministic

Systems of this kind are commonly described by differential equations. The analytical model is often associated with a convenient representation of the system with an input, output and a central 'transformation process' (an analogic model). The transformation process represents the temporal nature of the system. As an example, consider the simple spring damping system of Figure 1.6. The diagram shows a mass (m), suspended by a spring (spring constant (ideal spring) K) within a containing vessel. The friction existing due to the mass/spring contact is denoted in the diagram by a friction coefficient (f). The force acting upon the mass is denoted as $r(t)$, and the distance travelled by the mass is included (s).

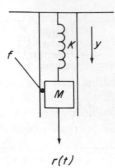

r(t)

Figure 1.6 A simple spring–mass damping system where $r(t)$ is the force; K, the spring constant; M the mass of the piston; y the volocity; and f, the friction constant

The behaviour of this simple mechanical system is defined by Newton's second law of motion. Therefore the following second-order differential equation is found:

$$M \frac{d^2 s(t)}{dt^2} + f \frac{ds(t)}{dt} + Ks(t) = r(t)$$

and since velocity (v) is the rate of change of distance, ds/dt, the above equation becomes:

$$M \frac{dv(t)}{d(t)} + fv(t) + K \int_0^t v(t) \, dt = r(t).$$

This analytic model can be used directly to provide an analogic model in terms of an electrical network composed of resistance (R), capacitance (C) and inductance (L). The analogic model of Figure 1.7 shows an RLC network in a circuit configuration necessary to describe its behaviour in terms of Kirchhoff's current law. A constant current source is shown, generating an electrical current ($i(t)$), and the output is measured in terms of an electrical potential difference ($v(t)$). The following equation describes its dynamic behaviour:

$$\frac{v(t)}{R} + C \frac{dv(t)}{dt} + \frac{1}{L} \int_0^t v(t) dt = i(t).$$

It is quite obvious that the two equations are equivalent, and therefore the electrical circuit analogic model can indeed be used in representing and simulating the behaviour of the 'real' mechanical system.

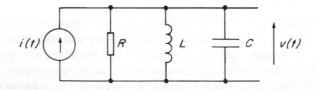

Figure 1.7 An electrical network analogue of a mechanical damper where C is the capacitance, L, the inductance, R the resistance, $i(t)$, the electrical current flowing from a constance current source and $v(t)$ the electrical potential difference measured at the output

1.4.4 Dynamic, Non-deterministic

In Wilson's definition, this is characterised by discrete event simulation. Systems of this nature are limited by a level of uncertainty about the actual mechanisms involved in the system's transfer process (transfer function). An example of this is the situation which may exist within the marketing department of a commercial organisation. In such a situation, actions are frequently determined through a process of correlating discrete series of past events such as sales figures, results of previous market surveys, results of seasonal trend analysis, etc. in order to maximise the effects of future commercial behaviour. Control system models are frequently used to represent situations of this kind, in which control is seen to operate in a 'feedforward' arrangement. A good description is given by Anderson (1976) on forecasting methods, and a treatment of the subject in the context of control is given by Box and Jenkins (1970).

Discrete event simulation is important in a large number of systems employing AI. If it is accepted that Artificial Intelligence finds much use in non-deterministic situations, then it must also be realised that most real-life situations, in which AI solutions are in demand, are characterised by their temporal nature. This is particularly true in non-deterministic situations which involve human activity. Examples of situations in which discrete event simulation is appropriate include:

- planning (robotic task planning and project planning),
- project management and control,
- CAD-based simulation (electrical, mechanical, etc.) ,
- biological and biotechnological , and
- communications network design.

Models within this class of system include elements of time and order. Temporal representation and reasoning are important concepts in such situations, and future developments in this vital area of AI will have a significant impact in future systems. Reference should be made to Chapter 8 for a detailed discussion of temporal reasoning.

1.4.5 Control Models

In general, control theory and its use of a specific kind of model has been extensively adopted in all branches of engineering. Dorf (1974) states that: 'A control system is an interconnection of components forming a system configuration which will provide a desired system response'. This quotation has some interesting semantic similarities with the definition of a system given earlier: the notion of a SYSTEM being composed of individual entities having well-established interrelationships–interconnections. As far as a control model can be used, i.e. in an analogic situation, the term 'desired system response' says it all. The basis for control engineering lies in feedback theory and has its foundation in linear system analysis, where linearity is defined, in control engineering, in terms of the system excitation and response. As an example, if $x_1(t)$ is the excitation of a system and $y_1(t)$ is the corresponding response, and if $x_2(t)$ and $y_2(t)$ are also excitation/response pairs of the same system, then an excitation of the form $x_1(t) + y_2(t)$ must result in the response $y_1(t) + y_2(t)$, i.e. conforming to the Superposition Theorem.

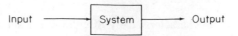

Figure 1.8 A simple control system model

Perhaps the simplest model of a control system is that shown in Figure 1.8. This model is an extreme abstraction and fails to represent some key concepts of systems:

- communication,
- relationships between individual components, and
- control to facilitate equilibration.

A better model is shown in Figure 1.9. This provides a higher level of detail in identifying individual components. The component at the 'heart' of the model is labelled the 'transformation process'. This is the actual process, existing within all systems, of transforming an input condition (problem situation) into an output condition (solution). The box is also assumed to contain a 'controller', which is able to make use of the output from the comparator—the error signal. The output condition is said to be 'sampled' at the sample point, and communication of this information is facilitated by the 'sampled information processor' box, which in practical terms serves to scale the output information to a suitable form for comparison and control. Without this feedback loop, such a system would, in Downing Bowler's terms (Bowler, 1981), tend towards a situation of maximum entropy.

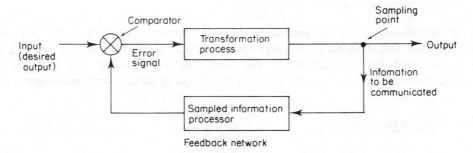

Figure 1.9 A basic negative feedback system model

Figure 1.10 shows the same modelling technique applied to a somewhat over-simplified and 'lumped' impression of the national economy. Even though this model may appear to be an over-simplification, the elements of control and the necessary information feedback is easy to identify and possibly agree with. The model serves to represent, in very simple terms, a highly complex situation with many important interrelationships. The model of Figure 1.9 also demonstrates the usefulness of modelling at an initially abstract, or 'conceptual' level, i.e. it provides a useful vehicle for more detailed model refinement and analysis.

The control system diagram in Figure 1.10 is very much a conceptual model. It is:

- *Qualitative*. It represents the analyst's understanding of the gross behaviour of the system

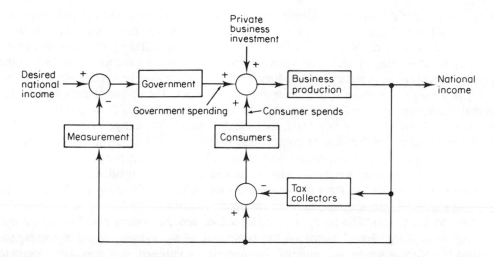

Figure 1.10 A basic feedback control system model of the national economy. After Dorf (1974)

in an abstract sense, i.e the specific form of model used cannot be 'mapped' directly onto the real world, it can only be validated through logic.

- *A tool to assist in the thought process.* The very act of attempting to produce a suitable conceptual model tends to 'concentrate the mind' on the various components of the system, how they interrelate and the implications of interaction at the wider system level, i.e. at the system boundary.

Models within this classification are well known to engineers, since they are qualitative and serve to assist in the intellectual process of problem analysis. The actual form they might take and the use to which they are put tends to be the subject of personal taste, even though some model forms are well established and widely used (the control diagram for example). In some spheres of engineering, the term conceptual model may not be used. In such cases, schematic diagrams often serve the same purpose of representational abstraction. In software engineering, a model frequently used is the 'abstract model' in which a program module can be represented by specifying its interface procedures via operations on mathematical objects such as sets, trees, graphs, etc. (Lamb, 1988).

1.4.6 Conceptual Models

The model of Figure 1.10 represents a system made up of components which relate to human activities. For example, tax collection is an activity carried out by a group of people, business production is dependent on human endeavour and it is almost certain that we can all identify with the consumer. In Checkland's terms (1981), this particular class of system is defined as the 'Human Activity System'. Human activity systems are essentially intellectual constructs which describe some purposeful human activity. They are intellectual in the sense that an accurate model of human activity would be extremely complex, and conceptual models used to represent them are necessarily limited to the minimum activities consistent with meaning. As an example of a conceptual model applied to a human activity system, consider the situation existing within a construction firm wishing to construct a certain building. Figure 1.11 shows a conceptual model which describes, albeit in fairly abstract terms, a series of activities which form a possible building site acquisition and development planning system. The system input is the need to produce a building to the client's specification, in order to meet a long-term goal, which may be to construct a building suitable for a given purpose with a predetermined life span. The output of the system is the actual acquisition and development of the site, i.e. 'begin to build'. There are components of such a system which are important, individually identifiable and generic in nature. The actual expertise on the business of construction is embodied within the three 'know about' components. These are vital in determining the actual choice of site, the manpower required to carry out the various and diverse tasks associated with any relatively large building, and also the resources necessary to complete the job. Although these have links to various activities shown in the diagram, they are inherently linked to external events or influences external to the system. The business of construction, and the current practises within that domain have a significant impact on the behaviour of the system. In other words, the system boundary is somewhat 'artificial', but necessarily imposed. It is important though to be constantly aware of cross-boundary influences. One other important aspect of a system model, even at the conceptual level, is the inclusion of a control component. In Figure 1.11, this is shown as an 'extra component' with the rather bland label 'activity control', with an input 'monitoring information' and an output 'control output'. If the model was to be extended into a more realistic and potentially useful representation, the actual source and

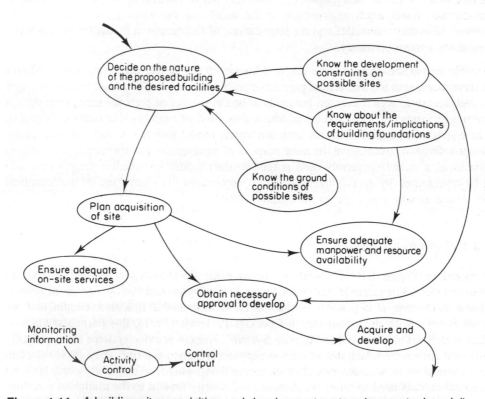

Figure 1.11 A building-site acquisition and development system (conceptual model)

form of such information would be made explicit, and the control action would be seen to affect individual system components in a much more detailed fashion.

In AI systems, the concept of 'know about' is a central feature. The concept of acquiring and explicitly representing knowledge, in such a way that that knowledge can be used in the transformation process of a computer-based system, is vitally important. It is also important to adequately assess the limitations imposed on any resulting AI system, by virtue of the choice of system boundary, i.e. what external influences are excluded from the system's behaviour by deliberately not providing a link, or association, or interrelationship path? In effect, it is apparent that:

systems thinking is highly relevant to the area of knowledge-based systems.

1.4.7 An Approach to the Building of Models

Conceptual models of all kinds are intuitive in nature. This is perhaps even more so with human activity systems, in which the primary aim of a 'first-level' model is to represent and structure the thought process, with a view to focus attention on the important issues of the current problem situation. However, certain fundamental systems concepts have been identified, and can be applied to the actual process of model-building, in order to ensure that a 'meaningful' model is the result of the process. In essence, these are the fundamental system concepts of:

- boundary choice,
- system component identification,
- interrelationships between components within the system,
- effects across the system boundary,
- communication and
- control.

The conceptual model of Figure 1.11 corresponds to all of these issues. The boundary choice can be justified in terms of the functionality of the system and the readily available information on cross-boundary influences. Communication is apparent in the interrelationships, which are explicit, and a control element has been added, even though it seems to be almost an afterthought. It is useful at the conceptual model stage, to produce an extreme abstraction which represents the analyst's personal view of the world, i.e. a model of the problem situation with the minimum number of elements consistent with meaning. This results in a system model composed of a very small number of highly general components which can be easily justified and reasoned with. The model of Figure 1.11 can be regarded as an initial model, or a model to the 'first-level of resolution'. Once this has been produced and validated, each element within the system can be represented as subsystems. For example, in the system of Figure 1.11, the component labelled 'plan acquisition of site' can be expanded into a full-blown planning system with an independent set of 'lower level' activities with interrelationships and dependencies. The component labelled 'obtain necessary approval to develop' has been expanded in Figure 1.12 to include extra detail in terms of specific activities to obtain legal permissions and rights. It is apparent that one element of the system

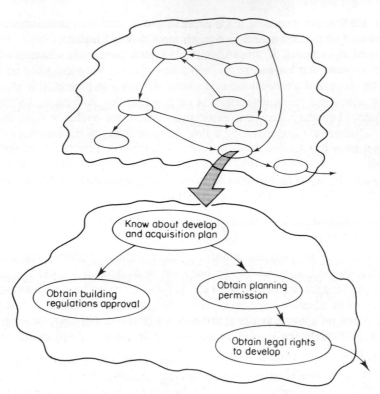

Figure 1.12 Incremental development of a 'first-level' conceptual model

of Figure 1.11 has been expanded to the 'second-level of resolution', i.e. more detail has been added to just one of the original system components.

It is important to establish the way that a conceptual model (or a systems model) comes into being. In effect, a specification must come first: a definition of the functionality of the system in terms of its input (the problem), its transformation process (functionality within the system boundary) and its output (the goal). Wilson talks about a 'root definition' (Wilson, 1984) which goes beyond a statement of the objectives of a system to a level of description which incorporates a point of view, making the activities and behaviour of the system meaningful. In the system of Figure 1.11, the root definition may be:

> 'A system to achieve the acquisition and efficient development of a building site to the satisfaction of the client within the structural and legal constraints of the national building regulations'.

The resultant system may be somewhat different if the following root definition is made:

> 'A system to acquire and develop building land to meet the short-term needs of the market place with the maximum financial return on investment.'

Unfortunately, there is no absolutely strict formal methodology for producing good root definitions. It depends so much on the personal world-view of the individual analyst. There are, however, guidelines by which good models can be built. These guidelines rely heavily on a formalisation of general systems concepts, and have been collectively introduced as the 'formal systems model' (Checkland, 1981). The basic principles, which can be used as a 'check-list' in model validation, are:

- A statement of objectives and purpose. The objectives should be incorporated into the root definition and should be clearly identifiable.

- Well-defined connectivity. The links between the various components of the system which facilitate the transmission of effects. In a human activity system, these are made up of activities, connectivity manifesting itself primarily as logical dependence, i.e. one activity must precede another in the transformation process.

- Performance evaluation. The measure of performance (error signal), which signals progress in striving to meet the stated objectives.

- The incorporation of a monitoring facility and control mechanism. A specific mechanism to effect control via continual monitoring. This usually manifests itself in a control system with monitoring inputs and a control action output.

- The incorporation of a decision-taking process. The decision taking process is responsible for taking regulatory action.

- A system boundary should be imposed. The boundary should be well defined as the area within which the decision-taking process operates directly. The boundary encloses the system, but it is apparent that the system exists and functions within the 'wider system'. The decision-taking process may influence the wider system, but it should not directly affect it.

- The utilisation of resources. For a human activity system, activities are assumed to operate on resources, which are at the disposal of the decision-taking process.

- A hierarchical structure should be apparent in the system. System components can be represented as systems in their own right (subsystems). It should be possible to represent the system at various levels of abstraction, or levels of resolution.

It should be noted that the guidelines above are not totally independent. Some of the points are inherently interrelated. It is impossible to completely separate monitoring, control, decision-taking and measures of performance. They are, however, slightly different in their nature within the system, and attempts should be made to validate conceptual models using all eight points in the check-list.

1.5 METHODOLOGY

A methodology, in systems terms, provides the means for carrying out analysis and design in a structured fashion. In this sense, we talk of 'problem-solving methodologies'.

It is very important to clarify the meaning of problem and methodology in much the same way that the terms model and modelling language have been discussed. The significance of the term 'problem' is quite profound and can be applied to situations in which the desired conclusion, or goal, is well defined. In Checkland's terms (1981), these particular problems are termed 'hard', i.e. a problem which can be formulated as the search for an efficient means of achieving a defined goal. The type of problem frequently encountered by engineers fall firmly into this category. For example, if a planned new road is required to cross a river, then the goal is apparent: ensure that motor vehicles eventually travelling on that road are able to continue their journey without leaving the road, i.e. the road should be continuous from one bank of the river to the other. The engineer knows what to achieve, but not necessarily how to achieve it. Conceptual models may be constructed of bridges of various kinds, or perhaps tunnels.

At the other end of the 'problem spectrum' is the so-called 'soft' problem. Soft problems have less structure than their 'hard' counterparts, in that there is no well-defined endpoint; no known desired goal. As an extreme example, take the problem: 'what do we do about the national drugs abuse situation?'. It may be apparent that there is an obvious desired goal: eradicate the problem of drugs abuse, but the problem is essentially ill defined. Unlike the bridge situation, which had an easily identifiable 'what' component (what to do? ... get across the river), leaving only the 'how' to do it, the drugs problem lacks a well-defined 'what' element. Indeed, there may be many perceptions and firm proposals put forward in the determination of what to do before subsequently considering how to do it. As a further example, consider an active industrial company. There will be a line management structure, and the various managers concerned in the business enterprise will have their own individual and possibly different perceptions of the aims of the organisation. In particular, a manager will have personal perceptions of the aims of the organisation at his/her level in the line management structure. Even at the higher, strategic levels, the various people involved will 'see' it differently. One perception may be that the primary aim of the company is the satisfaction of a particular market need, and another may be the desire to make the most efficient use of production resources. Soft problems are characteristic of 'human activity systems' and are prime candidates for the application of AI techniques. The German word 'weltanschauung' conveys the concept of multiple perceptions of the same problem. Translated it means 'world view', i.e. the impression of the world which has meaning to the individual. It shapes the individual's observation of events in the problem situation and affects his behaviour. The weltanschauung is continually modified by experience and relevant external influences.

In general, soft problems can be identified as having two distinct elements which are recognisable as systems in their own right: a system made up of activities (what to do), having relationships which are essentially logical dependencies, and a social system in which the components are people doing the activities (how to do it). In this latter case, the relationship linking the components are interpersonal. Here, a consideration of the range of problems and the different situations in which the term 'problem' may be used, leads to the conclusion that 'problem' may be too narrow a concept, and 'problem situation' would appear to broaden the scope of the term in such a way that the meaning can be extended to cover situations not immediately perceived as problems in the strict sense, but rather a concern; a reason for the system's transformation process. Within this context, the term methodology is taken to mean a structured set of guidelines which, when followed, will lead the analyst to a solution or solution situation. It will determine whether a particular technique is suitable in the process, and therefore the terms methodology and technique should not be confused.

1.5.1 Systems Engineering Methodologies

The term 'systems engineering' has been widely used to cover such topics as systems analysis, systems design and management of all kinds. In fact, the term may well be much abused; being used in situations characterised by the general problem-solving process and in no way conforming to the systems ideas outlined previously. However, one thing is certain: systems engineering is synonymous with good engineering, and is synonymous with common-sense in a large number of cases. That common-sense has merely to be articulated and formalised in such a way that general, formal-like principles emerge, which can be used in situations of a complex nature.

Sporn (1964) has a definition of an engineer couched in terms of the translation of scientific knowledge into tools, resources, energy and labour in order to directly benefit mankind. The definition firmly identifies the engineer as a key figure in the material progress of the human race. The definition is meant to convey the particular role of the engineer in the applied side of science, as having the skill to appreciate fundamentally new branches of scientific knowledge, seeing the potential for development, and taking the idea from a purely abstract and formative stage, right up to useful products able to take part in the technological evolution of man. The definition also has some important consequences for systems thinking. It mentions a broad spectrum of engineering enterprise, incorporating tools, resources and energy as one would expect, but it also includes labour. In fact, this particular definition of an engineer, perhaps with some poetic license, can be seen to encompass the whole business of technology (product) specification, design and production, and also the provision and management of a highly-skilled labour force.

The historical aspects of engineering, from simple to highly-complex systems demands a more structured, formalised approach. An example of the scope of engineering effort could perhaps be the road transportation problem. A road bridge may be regarded as a relatively complex entity, but it merely forms part of a much larger local, regional and national road network (transport system?). A pair of radio transceivers are potentially useful communications devices, but when incorporated into a communications network supported by a cellular infrastructure, computer control and satellite facilities, a whole new set of organisational problems emerge. Of course, this is one of the basic tenets of systems thinking, i.e. emergent properties of each new level of sophistication (complexity). This particular phenomenon is significant in its own right when considering, not merely the problems which must be addressed by the systems engineer, but also the emergent properties of the wider system, existing as a direct consequence of the availability and use of new technology. In many ways, the technological evolution of mankind, which is undergoing something of an 'exponential explosion' has had effects on the social and cultural aspects of everyday life. These effects are changing the way we live and go about our business, and the rate of such changes can surely only increase with time.

Engineering is all about solving problems. The discipline of problem understanding, a command of technological tools and technique, and perhaps more importantly, an overriding objective to research, develop and produce practical solutions. Checkland (1981) traces the development of science as a human activity from 624 BC (the natural philosophers of Iona) to the present day, and explores the emergence of the systems movement in the history of science and engineering. He makes the point that the exposition of methodology for the total engineering of complex equipment appeared around the 1950s. The driving forces behind the emergence of a systemic approach to engineering can be listed:

- the rapid increase in technological discovery and development, strongly influenced by the military situation of that era, leading to an almost overwhelming increase in system complexity;
- the appearance of technological tools (e.g. the computer) had the emergent properties of further extending the scope and complexity of technology—this is particularly apparent today with the availability of sophisticated CAD/CAM systems;
- market forces resulted in a high level of competition, leading to the need for optimisation in terms of system reliability and efficiency, development and production efficiency and effective market research.

All of the above points remain relevant today. The design process represents a significant investment for industry and must be seen in the wider context of the company organisation and business plan. Williams (1961) introduces an explicit design methodology, in which the design process is represented by a hierarchical arrangement of feedback loops. The control system model used to cater for interrelating factors is configured into several feedback control systems, which include:

- analysis of possible ventures,
- production scheme optimisation.

Modules existing within the control system include:

- evaluation of possible yields and product distribution,
- sales and pricing optimisation,
- overall costs and profit optimisation,
- design optimisation.

As an exercise in model-building, it would be possible to extend each of the system modules into systems in their own right, either as a more complex control system or as an alternative hierarchical conceptual system model (as in Figure 1.12).

Engineering design can be considered to be the activity of making decisions about the structure of a system on the basis of an initially abstract specification. The process includes an identification of the 'structure' of the outline system, and the components necessary to implement it. This process continues with a recursive decomposition of the outline structure into subsystems, thereby producing a hierarchy of specifications and designs. In this way, the designer creates a hierarchy of specifications and designs, but the user/client interface rests generally with the requirements specification for the system (or product). Of course, each design must be expressed in an appropriate language or notation. Lewin (1986) talks of four major engineering design levels:

Behavioural This is also defined as the 'information flow' level, or 'algorithmic'. Such systems are described in terms of interconnections of processes specifying required behaviour. They are, in essence, implementation independent. Suitable design tools include formal specifications methods, directed graphs, algorithmic validation and theorem provers.

Functional   The 'architecture level'. Systems are partitioned into subsystem components. Suitable logical processes are detailed. Suitable design tools include system design languages, systems and logic simulation (at the functional level), and hardware description languages.

Structural   The 'logic' level. This includes a description of actual hardware/software realisations to be used in implementing subsystem functions. Suitable design tools include logic diagrams (and simulation), test generation and circuit simulation (system simulation at the detailed structural level).

Physical   The 'circuits' level. This level is based on a transformation of a structural representation into a physical realisation. Suitable tools include layout languages, interconnection tools and design rule checkers.

The view of design as a problem-solving process can also be seen to progress as a 'piecewise refinement' or 'transformational development'. Each design stage, or level (apart from the last), produces a partial solution in the form of an abstract model of the eventual system. In this way, as the design progresses from a position of maximum abstraction to a point of maximum detail, the transformational model becomes more and more refined to the point of physical implementation.

1.6 LIFECYCLE CONCEPTS

A useful tool in the planning and organisation of a product is the product lifecycle model, which indicates not only the essential steps in the development process, but also gives an indication of sequence and iteration. The diagram of Figure 1.13 demonstrates the concept. It shows some of the major activities within a development project, but there are some essential, interrelating concepts missing. These include planning, customer service, maintenance, etc.

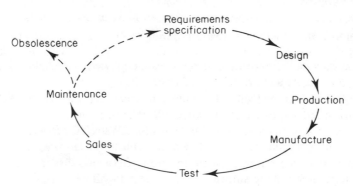

Figure 1.13 A product lifecycle

Planning is implied in the whole lifecycle model, but perhaps more so in the specification and design stage. However, we have already discussed the need for a systemic (and systematic) approach. The planning concept (and therefore the lifecycle model) should include:

- requirements and constraints—applied to the product,
- an assessment of the activities and resources required to realise the product development,
- an assessment of the resources available in-house and those required to be procured,
- an evaluation of the organisational constraints on design, production, test, manufacture and marketing,
- an assessment of the implementation details, i.e. the client interface; installation, customer service and maintenance,
- an evaluation of lifecycle length (time),
- a full and accurate cost/benefit analysis.

1.6.1 Software Systems Engineering

The design of a software system has become a more recent candidate for the introduction of formal methods. Software engineering has become an established discipline quite apart from computer programming or the more traditional academic field of computing in general. The development of the software engineering specialisation is fairly recent, around the 1960s. It was realised that, as software systems were becoming larger and more complex, a set of difficulties were emerging which almost mirrored the problems accepted somewhat earlier in engineering. These were especially:

- The software systems were becoming more and more complex.
- Development time was becoming extremely long, labour intensive and expensive. Large teams of programmers needed to be managed in such a way that the systems were realised within the projected timescale, and that cooperation between individuals was optimised to the benefit of the overall system quality.
- In many cases, the eventual software systems delivered to the customer were found not to provide the functionality expected (or wanted).
- The software systems were not able to 'evolve' to meet changing needs.
- Quality assurance was difficult to establish during the development process.

Therefore, it was inevitable that the emerging software engineering discipline should develop along the lines of systems and systems engineering.

Software engineers talk in terms of systems, lifetimes, specifications, methodologies, control, hierarchies, modularisation and, above all, models at various levels of abstraction. The most famous lifetime software model is called the 'Waterfall Model' and is shown, in diagrammatic form, in Figure 1.14. The model identifies with a 'waterfall effect', where the life of a system is characterised by clean breaks from one distinct phase to another. The modularisation of the entire software system lifecycle allows for planning, formalised design to cater for complexity, and quality assurance. Indeed, the waterfall model is said to be 'document-driven', with the following documents included in various stages of the lifecycle:

- feasibility study report and requirements specification,
- user reference manual and system test plan methodology,

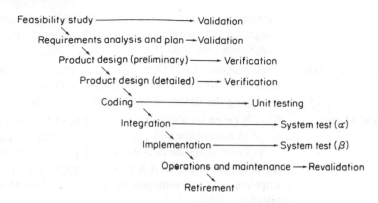

Figure 1.14 A representation of the waterfall model of software development

- project plan and configuration management plan,
- schedule and budget,
- training plan,
- security plan,
- module decomposition and module dependencies,
- integration test plan,
- module specifications,
- module implementation plans and unit test plans,
- code walkthrough reports,
- release notes and problem reports.

The waterfall model came under attack in the early 1980s, mainly because of the common observation that problems relating to early stages in the process are not discovered until the later stages. Another common observation, that is pertinent to all engineering disciplines, is that we often try to design products that have never been designed before. This often means that the actual functionality of at least some interrelating components are not known in detail in the early stages. As far as software engineering is concerned, the waterfall model has enjoyed a good deal of acceptance because it provides a detailed lifecycle model and a structured methodology for dealing with complex systems. Parnas and Clements (1986) suggests that software design documents should be written as though the waterfall model had been strictly adhered to, even though that did not occur. Some of the inherent difficulties with the model have been largely overcome by adding extensions to cater for prototyping and incremental development, parallel developments, program families, accommodation of evolutionary changes, formal software verification and validation, and risk analysis.

In recent times, new software lifecycle models have been proposed to cater for different classes of software, some of which have been found not to map well onto the waterfall model. The following examples are some of the more well known models:

The Evolutionary
Development Model   This model is ideally suited to the development of fourth generation language applications. It consists of expanding

increments of an operational software product, with the directions of evolution being determined by operational experience (McCracken and Jackson, 1982).

The Transform Model   This assumes the existence of a capability to automatically convert a formal software product specification into a program satisfying the specification (Balzer *et al*., 1983).

The Spiral Model   This is based on a risk-driven spiral model, which incorporates a radial dimension which represents cumulative costs incurred in accomplishing the development steps in a traditional software development process. It has been suggested as a candidate for improving the software process model situation as outlined here (Boehm, 1988).

The situation of development processes for AI systems is discussed more fully elsewhere in this book, but it is important to provide some analysis of software engineering methodologies as they have been applied to the development of expert systems, in the light of systems concepts (see Chapter 3 for a practical discussion of development using commercial expert systems environments, and Chapter 4 for a discussion on knowledge engineering). Software engineering, at its purest level, has much in common with systems. It is concerned with the wholeness of the proposed system, from conception through to formalised design and eventual implementation. It addresses the concern of quality assurance and maintainability. It also takes account of the management of the project and its expected lifetime. In other forms of engineering, the same concepts apply, although the actual mechanics of the lifecycle approach to design are not as inherently integrated as in software engineering, i.e. the various components of the 'whole system' may be physically different, relying on differing design, production and test technologies. Computer-aided design has provided an enabling technology for highly complex design and the emergent properties of this technology have had an enormous impact on the whole engineering field. More recently, computer-integrated manufacture has provided the basis for systemic manufacture.

1.7 AI AND SYSTEMS

In order to rationalise the various independently-perceived concepts of engineering and systems when considering AI, it is instructive to identify possible meanings for AI; what is the nature and what are the processes involved in such a grandly named idea? An accurate and widely agreed definition of human (or natural) intelligence is notoriously difficult to obtain. All but the most committed cognitive scientist would resist any attempt to define a root definition for intelligence, because it is such a vague term. It has elements of:

- reasoned and in some way directed behaviour,
- organised and dynamic memory,
- the ability to communicate effectively, and
- the ability to adapt in the light of experience.

The four basic points above encompass some of the more obvious attributes associated with human intelligent-like behaviour. The ability to remember in a structured fashion is important. Facts must be stored in an organised way, with interrelationships well formed. The ability to adapt, or learn, in the light of experience is highly important and possibly crucial in intelligence. The ability to learn and possibly the rate at which the learning process takes place is a frequent indicator. The ability to communicate is important in the development of intellect, but it is dangerous to apply this to the human case. In terms of an automatic system, however, it is easy to see how communication at all levels would be vital.

Gardner (1985) talks of 'insight'; yet another extremely vague term, but which nevertheless has meaning to us because we can all identify with it. A definition of insight has been bravely attempted by Mandler and Mandler (1964) as 'the appearance of a complete solution with reference to the whole layout of the field'. In fact, Gardner himself states that 'what characterises higher learning or "intelligent" processes, wherever found, is the capacity to grasp the basic fundamental relations in a situation'.

In addition to the quest for a reliable definition for intelligence itself, there is also a need to rationalise the scope of AI as it is perceived today. A useful definition has been put forward by Charniak and McDermott (1985) as:

> Artificial intelligence is the study of mental faculties through the use of computational models.

It is not correct to define a rule-based expert system as AI; it merely represents a subcomponent of AI. Figure 1.15 indicates the various widely-accepted components of AI, but it is not, however, the whole picture. It is possible to construct an input/output model for AI (Parsaye and Chignell, 1988) which has the above components indicated as outputs whilst the inputs are essentially:

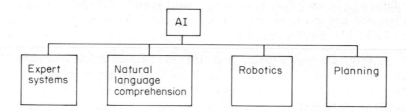

Figure 1.15 A hierarchy of AI

- psychology,
- mathematical logic,
- computer science,
- linguistics and
- philosophy.

It is important to add to this list, another major discipline which has had a significant impact on AI: systems.

Systems thinking is characterised by communication, control, structure (hierarchical) in addition to an awareness of boundary constraints and emergent properties. In some ways, the two sets of concepts can be mapped onto each other, if the concept of intelligence were to be applied to the problem-solving process.

Intelligent problem solving involves the conscious or subconscious modelling of the situation, moulded by experience and the individual perceptions (*weltanschauung*) of the situation. The problem is modelled intuitively in such a way that an artificial boundary is imposed, necessary for intellectual analysis and resolution. Communication is important at all levels, and the reasoning process itself makes use of intellectual and available mechanistic transformation processes, suitable for matching a desired goal with a system output state.

This argument allows a close parallel with knowledge-based systems. Such systems are usually equipped with a highly structured database, holding an internal model of the 'world' in terms of interrelated facts. A knowledge base, often in the form of rules, is present and can be automatically invoked by an inference mechanism in order to facilitate the automatic and reasoned use of those facts. the database holds a system model and the rules. When rules are used in conjunction with data in a controlled fashion, this permits the employment of heuristics in cases where traditional algorithmic problem-solving have always suffered from severe limitations. In some ways, this facility introduces the *weltanschauung* into the problem-solving equation. In carrying out the knowledge engineering function at the knowledge acquisition phase, the knowledge engineer is unwittingly programming a particular expert's world view of his domain into the system for as long as it is used. This is in no way a limitation in its use, because the actual purposes of such systems normally includes:

- replacement of a highly experienced expert,
- use of expertise in training,
- use of expertise on-line and/or in real time (embedded),
- use of consultative expertise (as opposed to stand-alone).

There are, however, exceptions to this generality, and perhaps the most obvious of these is the medical expert system. Although this particular domain has seen some of the earliest 'successful' systems, such as Mycin, Internist, Protean etc., their general acceptability is limited by several factors which are a mixture of inherent operating functionality, limits on their embedded knowledge depth and, perhaps much more significant, the actual agreement between experienced practitioners on correct diagnostic procedures, indicators and suitable follow-up action. In effect, the knowledge of one, albeit highly thought-of and eminent medic, will not necessarily be accepted by his peers. The significance of the individual *weltanschauung* can never be underrated in any AI project involving the medical domain.

The engineering domain is somewhat different. Unless a particular system is to operate as an expert at the forefront of science, taxonomic knowledge is usually well documented and respected. It is usual in such cases to think in terms of 'hybrid systems' which include algorithmic functions to cater for deterministic problems or subproblems, and specific heuristics which can be used reliably to either reduce the computational effort by employing well known or tried-and-trusted 'rules of thumb', or can be used to cater for those situations traditionally dealt with well in the light of experience, but with no formal basis.

Engineering has been defined in terms of:

- problem analysis/definition (modelling),

- problem solution (language, technique, methodology),
- organisational and process management (complex system design and production control).

Taking all these points into consideration, and summarising to what may be an extreme, the schematic of Figure 1.16 is an attempt to 'equate' the three disciplines of AI, systems and engineering.

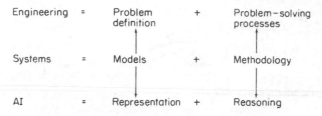

Figure 1.16 The synergy between engineering, systems and AI

1.8 CONCLUSIONS

The need for a systemic approach to all engineering disciplines has been apparent in the last several decades along with the development of engineering as a professional and vital sphere of enterprise. The inevitable increase in complexity of engineering and engineered systems has lead to an analysis and design philosophy centred on systemic principles. It is no longer feasible to consider the design of a complex system, with all the short- and long-term financial burdens that that entails, without taking into account implications at the individual, company, national and international levels. It is certainly possible to model a company as a complex feedback control system in which a slight change in one part of the organisation can be seen to have a significant, sometimes drastic effect on another. Such an approach has been demonstrated by Dorf (1974) using a management control structure as an example. The message of this chapter is that a systemic approach is required for the analysis or design of a complex system, using appropriate models and modelling techniques, and systems methodologies. This chapter has demonstrated that systems engineering is good engineering, whether applied to software engineering, electronic engineering, sales engineering, etc.

AI techniques can be applied to any type of problem. It has been defined as 'the task of engineering a technology of thought' (Tennant, 1986). The technology is, however, found to be much more useful in problems characterised by large amounts of data, sometimes poorly structured and quite often incomplete. Such problems are traditionally approached by experts who have, through a long period of 'focused' education and committed experience, acquired a structured memory capacity which can be used in the deductive process. A successful lawyer will have the 'edge' on his opposite number if he can remember more points of law and more cases of precedent. The reasoning process in most problem-solving disciplines can be resolved into a number of rules by which large amounts of data can be used to solve problems or reach goals.

Figure 1.17 shows a small part of what could be a specific document hierarchy. The root object of the hierarchy is labelled 'feasibility study', which has the subcomponents of 'engineering report', 'cost breakdown' and 'structural design'. Figure 1.17 is actually an associative network, one of the most common modelling structure found in use in expert systems. There are three important concepts in the diagram:

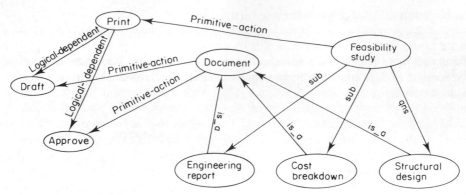

Figure 1.17 An associative network

Actions   In fact, `primitive_actions` are shown in Figure 1.17. This represents the 'lowest level' actions. For example, a higher-level (more abstract action) could be `produce_feasibility study`, which may involve the more primitive actions of print, copy and bind.

Sub-components   In the example shown in Figure 1.17, engineering report, cost breakdown and structural report are all interrelated to feasibility study by the labelled `sub_component` links.

'Is_a' relationships This is a powerful concept appropriate to large and complex information structures lying at the heart of expert systems. It allows for the inheritance of attributes 'higher up the hierarchy'. For example, engineering report, cost breakdown and structural design are all documents, and as such all share the primitive actions of draft and approve. They all inherit these necessary actions from the higher concept of document.

This allows for any explicit representation of a logical dependence between actions. In the example of Figure 1.17, print is logically dependent on draft and approve.

The mixing of concepts in this way permits the representation of actions on objects and also the order in which those actions should be carried out. The same structure could be used to build up a model of a much more complex system, with actions, resources, logical dependency and time-dependent states (Marshall and Boardman, 1987).

One of the most important features of expert systems is their ability to make reasoned decisions about a course of actions, choice of object or path of reasoning. Systems research has demonstrated that a viable approach to the solution of complex problem situations involves the ordered development of an accurate world model. One of the most successful methodologies in AI is the use of rules to control the reasoning process, and to direct the acquisition of information from the domain model. In rule-based expert systems, such rules may be employed to achieve a desired goal by sequentially achieving subgoals in a 'backward-chaining' fashion, or its operation may be 'driven by data' in a 'forward-chaining' process. In the case of Figure 1.17, links may be made to exist between concepts (nodes) which may or may not exist in real life, and must therefore be decided at run time on the basis of inference.

If the model of AI as shown in Figure 1.15 is used to classify the technology, its systemic nature can be appreciated. One of the long-term 'dreams' of some of the early AI protagonists was the emergence of an anthropomorphic facsimile: an artificial person, able to move, carry out complex tasks automatically, but above all to think. Figure 1.15 has some of the attributes necessary for such a creation. It has computer vision, not only able to form pictures, but also to reason about their content and meaning. It has natural language interface capabilities, available to communicate in a very human-like way. It has the ability to plan its actions and to coordinate its mobility through the vehicle of robotics. Not shown explicitly in Figure 1.15 is the ability to learn from experience, although such a feature could be well employed in all the AI 'boxes' shown. Machine learning is a very active research topic, and rightly so, with rich rewards for the eventual winner. It involves concept modification and association-building in the light of experience during system use. However, just as intelligence itself is extremely difficult to define accurately, so learning is a notoriously vague concept. It is possible for a knowledge-based system to cause some changes in its internal data structures in an attempt to modify 'current beliefs', and certain learning algorithms have been found to be successful in certain circumstances (genetic algorithms, neural networks, etc.), true intellectual development through learning remains essentially elusive.

The link between systems engineering and the design of AI systems for use in engineering is a close one. Its association is demonstrated in Figure 1.18.

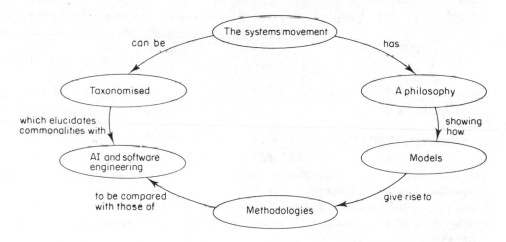

Figure 1.18 The importance of systems thinking in engineering, and the role of AI

REFERENCES

Ackoff, R. L. (1962) *Scientific Method: Optimising Applied Research Decisions*, Wiley, New York.

Anderson, O. D. (1976) *Time Series Analysis and Forecasting*, Butterworths, London.

Balzer, R., Cheatham, T. E. and Green, C. (1983) Software Technology in the 1990s: Using a New Paradigm, *Computer* (Nov.), 39–45.

Bertalanffy, L. Von (1940) The Organism Considered as a Physical System, Reprinted in: (1968) *General Systems Thinking*, Braziller, New York.

Boehm, B. W. (1988) A Spiral Model of Software Development and Enhancement, *Computer*, (May), 61–72.

Box, G. E. P and Jenkins, G. M. (1970) *Time Series Analysis: Forecasting and Control*, Holden-Day, San Francisco.

Charniak, E. and McDermott, D. (1985) *Artificial Intelligence*, Addison-Wesley, New York.

Checkland, P. B. (1981) *Systems Thinking, Systems Practice*, Wiley, Chichester.

Chestnut, H. (1965) *Systems Engineering Tools*, Wiley, New York.

Dorf, R.C. (1974) *Modem Control Systems*, Addison-Wesley, Reading, MA.

Downing Bowler, T. (1981) *General Systems Thinking: Its Scope and Applicability*, North Holland, Amsterdam.

Gardner, H. (1985) *The Minds New Science, A History of the Cognitive Revolution*, Basic Books, New York.

Gerking, S. D. (1974) *Biological Systems*, Saunders, New York.

Lamb, D. A. (1988) *Software Engineering: Planning for Change*, Prentice Hall, N.J.

Lewin, D. (1986) 'Systems, Models and Engineering Design', Based on his Professorial inauguration lecture 'A System is a System Is a System', unpublished.

Maddock, I. (1983) The Future of Work, In: *Intelligent Systems: The Unprecedented Opportunity*, Eds by J. E. Hayes and D. Michie, Ellis Horwood, Chichester.

Mandler, J. M. and Mandler, G. (1964) *Thinking From Association to Gestalt*, Wiley, New York.

Marshal, G. and Boardman J. T. (1987) *A Methodological Modelling Scheme for Project Management*, Proc. IEE Colloquium on Expert Planning Systems, Digest No. 1987/59, Peter Peregrines, London.

McCracken, D. D. and Jackson, M. A. (1982) Life Cycle Concept Considered Harmful, *ACM Software Engineering Notes*, (April), 29–32.

Michie, D. (1983) A Prototype Knowledge Refinery, In: *Intelligent Systems: The Unprecedented Opportunity*, Eds J. E. Hayes and D. Michie, Ellis Horwood, Chichester.

Parnas, D. L. and Clements, P. C. (1986) How and Why to Fake it, *IEEE Trans. Software Engineering*, **12** (2), 251–257.

Parsaye, K. and Chignell, M. (1988) *Expert Systems for Experts*, Wiley, Chichester.

Simon, H. (1962) The Architecture of Complexity, Reprinted in *The Sciences of the Artificial*, MIT Press, Cambridge, MA.

Sporn, P. (1964) *Foundations of Engineering*, Pergamon Press, Oxford.

Tennant, H. (1986) The redesign of thought, *Texas Instruments Engineering Journal* **3** (1), Jan–Feb, 9–11.

Williams, T. J. (1961) *Systems Engineering for the Process Industries*, McGraw Hill, NJ.

Wilson, B. (1984) *Systems: Concepts, Methodologies and Applications*, Wiley, Chichester.

G. Winstanley
Information Technology Research Institute
Brighton Polytechnic
Lewes Road
Brighton
East Sussex BN2 4AT

2 Representation and Reasoning

F. N. Teskey

IT Planning Division,
NatWest Bank,
London, UK

2.1 INTRODUCTION TO DOMAIN MODELS

One of the main features of artificial intelligence (AI) systems is that the knowledge used by them is represented explicitly, rather than being implicit in the program code. In conventional programs it is only the data that is represented explicitly. For example in finite element programs the properties of the elements are represented explicitly as data values, whilst the program code is used to represent the relationships between the elements.

The methods used to represent domain knowledge are central to AI. In addition to representing knowledge we need to be able to apply it to specific problems; this entails developing methods of reasoning about the knowledge representation. A domain model will therefore consist of two parts: the knowledge representation language and the reasoning methods used. It could be argued that the reasoning methods are themselves knowledge and so should be represented explicitly within the model. This type of knowledge is called meta knowledge, as distinct from domain knowledge, and some systems do indeed provide explicit representations of part of the meta knowledge.

In this chapter we will look at various methods of representing and reasoning about domain knowledge. In addition to looking at the general theory of knowledge representation, we will also look, in some detail, at a specific knowledge representation environment.

There is a wide range of knowledge representation used for various types of application, and, before discussing these in detail, it is worth taking a general look at the different types of knowledge we wish to represent. A detailed classification of knowledge types is outside the scope of this book, but we can make an initial philosophical classification of three types of knowledge: theory, rules and examples; and two types of reasoning: induction and deduction. The relation between these types of knowledge is shown in Figure 2.1.

This classification can be explained by looking at how it applies to a particular domain. Take, for example, our knowledge about the weather. An example of the most simple type of knowledge, is that concerning specific examples of weather, i.e. it is fine today, there was a red sky last night, etc. From this example of knowledge, it is possible to induce

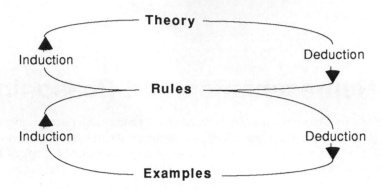

Figure 2.1   Types of knowledge

rules about the weather that can be applied in more general circumstances. This type of knowledge is often captured in folk lore, such as 'red sky at night is a shepherd's delight', or in more scientific rules, such as high pressure in summer means fine weather. From these specific rules, it may be possible to induce a general theory of weather. From the rules about atmospheric pressure, we could induce a theory of weather based on circulation of air masses. This example illustrates two of the techniques used in AI: reasoning from first principles (theory) and reasoning from heuristics (rules). It should be clear that the knowledge at each level is increasingly complicated, and it is for this reason that most of the work in AI has concentrated on rules and examples. The one exception to this is in the domain of mathematics, where the basic theory can be well described, and here there has been some success in using AI techniques (Lenat, 1982).

In addition to using induction to move from specific knowledge to more general knowledge, it is also possible to use deduction to move from the general to the more specific. Continuing the meteorological illustration, once a general theory of weather is established, we can use that to deduce specific rules about how the weather will behave; for example about the circulation of winds in a depression. Once again, from these rules, it is possible to deduce examples of specific weather patterns, such as wind speed and direction on a particular day. The standard form of these rules, which will be discussed again later in this chapter, can be represented as:

> Rule:
>   *if*   evidence
>   *then* hypothesis

In common with many of the current applications of AI, the engineering applications considered here will use mainly rules and examples. At present the methods for reasoning from first principles are very much in the development stage and are not widely used in practical applications. However, they are discussed in some more detail in Chapter 8. The types of knowledge representation to be considered in this chapter are grouped under the following headings:

- Relational Models,
- Semantic Networks,

- Inference Networks,
- Predicate Logic.

2.2 RELATIONAL DATA MODELS

The relational data model, at its simplest, is based on the concept of a relation table consisting of a unique key item and a number of related attributes. The model has been developed and extended over a number of years and is now gaining in popularity. Though it was not developed as an AI tool, it provides a number of facilities that are useful for AI work, but also has a number of limitations in this area. The relevance of the model to AI is discussed below.

2.2.1 Relational Theory

The relational model is usually considered to be a method for structuring databases rather than knowledge bases. Yet, from the preceding discussion, it can be seen that example data represents just one aspect of the larger concept of knowledge. Indeed as we move to more complex types of relational databases, then the models become more suitable for representing the more complex types of knowledge. This can be seen in the similarity of the Binary Relational Model (BRM) to semantic nets; the former was developed by the DBMS (Database Management Systems) community, the latter by the AI community, yet both produce similar models.

The basic concept of the relational model is simple; data is stored in tables, each row in the table contains data about a particular entity and each column represents attributes of the entities. A typical example would be a table representing engineering components, their name, size, weight etc. as shown in Figure 2.2.

Components	Name	Size (in mm)	Weight (in gms)
	Bolt	5	10
	Nut	7	4
	Screw	5	15
	...	...	...

Figure 2.2   A simple relational model

This model was first proposed by Codd (1970) and has been developed by himself and others since then. There are now numerous texts describing the benefits of relational databases; one of the most authoritative texts being the two-volume work by Date (1975). No attempt will be made to duplicate this work, but an assumption is made that the reader is familiar with the main concepts of the relational model. From the point of view of this treatment of the subject, the important point is not the undoubted success of the relational model in specific applications, but rather its application and influence on more general problems of knowledge representations.

From the point of view of knowledge representation, the most interesting areas of relational database theory are those of key attributes and functional dependencies. In simple terms, a key is a set of attributes that can be used to determine an entity uniquely, for example in an inventory, the attributes model and serial number will determine an item uniquely. Though the concept was introduced for database management systems, it has major importance for other types of knowledge representation. When we come to represent entities in a knowledge base, it is desirable for each entity to have a basic set of attributes which define it uniquely. These basic attributes behave very like relational database keys and should obey similar rules. For example, all non-key attributes should depend only on the key attributes (the Third Normal Form of Relational Theory defined by Codd). There was a tendency in the early days of AI to develop knowledge representations in an '*ad hoc*' fashion, because it was claimed that this was the only way to capture the rich variety of types of knowledge. The belief is now held that for engineering AI applications, there is a need for a more structured approach, and this should draw on the existing expertise in relational database modelling.

One of the problems of relational database theory is the lack of any tools for explicitly modelling the functional relationships between entities in different relationships. Whilst this is not essential for simple relational database systems it becomes more important as we consider fourth and fifth normal forms for relational databases, and similar structures for knowledge based systems. One conclusion that can be drawn from this is that large complex relations should be split up into smaller, simpler ones. In the extreme case, this leads to the concept of binary relations: relations with only two attributes. Before considering this in more detail, some of the other limitations of relational models will be discussed.

2.2.2 Limitations of Relational Models

The relational model has been very successful in dealing with well structured data, but it is less suited to dealing with the more unstructured information that is often used in AI. Before looking at the knowledge representations for AI, it is useful to see exactly what the limitations of the relational model are in dealing with this type of knowledge. Five major problem areas can be identified:

Modification   One of the basic aims of relational databases is to allow easy modification of the data; this is achieved by converting the data to normal form, based on the implicit functional dependencies in the data. This approach works well, so long as these functional dependencies are static, but if, for example, a functional dependency develops between two non-key attributes, then the process of normalising the relations would require that the data be unloaded from the relation, split into two and reloaded into two separate relations. In AI systems, these types of change are likely to be quite common as the knowledge base grows, and so the relational approach may no longer be suitable.

Semantics   Much of the processing of data in the relational model is achieved by use of the 'join operator'; this allows information in two separate relations to be combined on the basis of common data values. The join is a syntactic operation with no method of imposing semantic constraints. Thus, it is possible to perform joins which are relationally correct but have no semantic meaning. This problem is due to the current emphasis on using tuple rather than domain calculus for relational theory (Lacroix and Pirotte,1977).

Inference The relational model is very good at reorganising data using the *project* and *join* operators, but it is not able to infer new information from its existing database. As a simple example of this, consider the following two relations:

```
capital_of (city, country )
is_in ( X, Y )
```

with the specific instances:

```
capital_of (Paris, France )
is_in (France, Europe )
```

It seems easy to infer that Paris is in Europe, assuming that a capital city is in the country of which it is the capital, and that if X is in Y and Y is in Z then X is in Z. Representing and processing this type of knowledge is central to AI, but is difficult to perform in the standard relational model.

Meta knowledge This term is applied to the whole class of information that describes how the main data is stored and processed in a system. Some of the more powerful relational database management systems do provide a series of system tables describing the logical structure of the data. As we have seen the data required for AI is likely to have a complex structure. The Meta data will be correspondingly complex, and so more powerful methods will be required.

User Interface Since relational databases will usually have a fairly simple and static structure, the user interface can be correspondingly simple, usually requiring nothing more than form filling or query by example. Once again, as the structure of the data becomes more complex and variable then alternative interface methods are necessary.

To summarise, the relational model is well suited to representing knowledge within a fixed framework but knowledge representations for AI require a much more dynamic model to allow the knowledge to be accessed and processed in new and unforeseen ways. Nonetheless, the relational model has provided a basis for some types of AI representations, for example OPS5's tuples, and when semantic attributes are applied to the model it can be transformed into a simple semantic net.

2.2.3 Binary Relations

The Binary Relational Model (BRM) has been proposed as a method of knowledge representation, which combines the benefits of the standard relational model with the more complex processing required for AI applications (Lavington and Azmoodeh, 1982). The simplicity of the model means that it is much easier to deal with functional dependencies, meta data, etc., than in the standard relational model.

The basic concepts of the BRM are similar to the relational model, except that relations are restricted to having only two attributes. Information is stored in simple tables, such as in Figure 2.3.

EMPLOYEE	EARNS	SALARY
Jones		£15000
Smith		£10000
Edwards		£12000
Jones		£18000

Figure 2.3 Sample Binary Relation

The meta knowledge implicit in this relation is the fact that an employee earns a salary; this can be made explicit as in Figure 2.4.

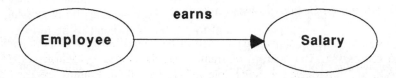

Figure 2.4 A simple BRM

The important point is that this simple model can be extended to more complex models, as illustrated in Figure 2.5. The knowledge modelled here should be obvious from the diagram; this is indeed one case where a picture is worth a thousand words.

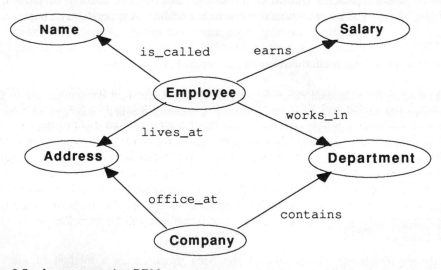

Figure 2.5 A more complex BRM

There are, however, some points that are worth emphasising. First, entities such as 'address' and 'department' are attributes of more than one entity, since both employees and companies have addresses. In a conventional relational database, these would be represented as separate columns in separate relations, though sharing a common name. In the BRM, these attributes are, as can be seen from the diagram, represented by the same entity. Second, the use of meta data to represent the structure of the model itself is much easier in the BRM than in standard relational models. It is possible to build entities to represent relations and their properties, such as 1–1, 1–many, etc.

2.3 SEMANTIC NETWORK MODELS

Though the semantic network model and the BRM have been developed to meet different needs, there are many similarities between the two models. The similarities and differences of the two models are discussed in the following section.

2.3.1 Overview of Semantic Networks

The theory of semantic nets was developed in the late 1960's to support work on natural language processing. The concept of frames was developed by Minsky (1975) from this early work, and has played a major role in AI since then. Frames are used to represent concepts and the relations between concepts. In practise it is useful to identify two types of frame:

- generic or prototype frames; and
- specific or instance frames.

A generic frame represents knowledge about a general concept, and can be thought of as providing a prototype for information about actual objects. A generic frame has:

- a type name; and
- one or more default properties/slots.

The slots in a frame correspond to the attributes of an entity in the BRM. The slots will have a type associated with them specifying whether the attribute is a raw data value, a link to another frame or a procedure. For example, in a model of an engineering system, we could have a number of generic frames representing various types of components such as nut, bolt, screw, etc. Each of these would have a number of slots such as size, weight, etc. This example will be developed further in the following section. Reference should be made to Chapter 3 for examples of commercially available systems based on the frame paradigm.

A specific frame represents knowledge about a specific object. In most systems every specific frame will be an instance of a generic frame and will be modelled on that frame. These frames will have:

- an instance name which will be unique to the individual object represented;
- a frame type which will identify the generic frame used as a prototype for this specific frame; and
- a number of property/slot values inherited from the generic frame.

Continuing the previous example, the model of our engineering system could have a number of specific frames representing individual components such as nut_27, bolt_36, etc. The slots in each of these would contain the actual size, weight, etc. of the individual objects.

This type of model lends itself very readily to a graphical representation, in the same way as the binary relational model. The nodes in the graph represent concepts of interest, and their properties and the arcs represent relations between concepts. One of the significant differences from the relational model, is that the semantic model identifies two basic relations between frames. The first is the relation between a specific frame and its prototype frame; we will refer to this relation as the 'inst' (instance) relation. (It is in fact the mathematical set membership relation, though there is considerable confusion in the AI literature about what it should be called.) The second relation is a relation between one prototype frame and a more general one. For example, in addition to having generic frames representing the concept of nuts, bolts, etc., we could have frame representations of more general concepts such as engineering components, physical objects, etc. We will refer to this relation as the 'is_a' relation. (It is in fact the mathematical subset relation, though once again there is considerable confusion about terminology.) Just as instance frames inherit properties from their generic frame, often referred to as their parent frame, so generic frames can inherit properties from their parents. An example of such a structure is shown in Figure 2.6).

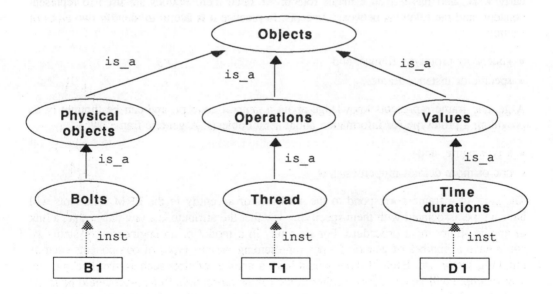

Figure 2.6 A simple semantic net

In this model, as in most semantic nets, it is customary to have a top level node representing all objects. Below that, there are a number of different types of object each with their own set of properties. Thus 'Physical Object' is a type of object with its own set of properties such as size, weight, location, etc. In addition to physical objects, abstract concepts, such as operations on engineering components can be represented; these operations would have properties such as duration, actor, agent, etc. At the next level of the network are frames representing more specific concepts. For example, the class of bolts is a subclass of the class of physical objects, the class of operations of putting a thread on a bolt is a subclass of the class of operations, and so on. It is important to realise that these are still generic concepts rather than specific instances; we have used a different notation in the network to distinguish between generic and specific frames; generic frames are represented by ovals and specific frames by rectangles in the network. At the lowest level in the network we have the representations of specific instances of generic objects. Thus, B1 is the representation of a specific bolt, T1 is the representation of a specific operation, and so on. In passing, it can be noted that further relations could be developed to represent a fact such as the operation, T1, of putting a thread on bolt B1 took a time D1.

2.3.2 Other Frame Relations

The basic semantic net model provides two types of frame relations—'`inst`' and '`is_a`'. These relations are used to represent property inheritance. There are, however, very many other possible relations between frames that one may wish to model. For example, generalising the example in the previous section, we can see that any operation will require an actor and an agent, and will have a duration, etc. All this knowledge can be stored in the generic frame, by providing slots for each of these attributes and specifying that they link to a particular type of generic frame. Thus, we would create a slot called 'duration' in the frame 'operation', and specify that the slot should be filled by a pointer to a frame of type 'time duration'. At the instance level, the slots on each specific operation frame would then be linked to another specific frame containing the time duration. This is illustrated in Figure 2.7.

It is at this point that the semantic net model again differs from the BRM. The semantic net model uses pointers to frame values rather than raw data values, as in the BRM. The advantage of this approach is that it is possible to impose more structure on the frame values, for example rules about likely time durations, and this can assist the process of making inferences from the model.

In summary, the frame base consists of two parts: the generic plane and the specific plane. The generic plane maps out the scope for the model and, as such, all possible interrelationships between concepts are described. The specific plane describes a particular manifestation of the model. To build, or instantiate, the specific model from the generic, the system must know how to reason about information in the generic plane. Apart from using the property inheritance of the '`is_a`' and '`inst`' links, the main method of achieving this is by the use of procedural attachments.

2.3.3 Procedural Attachments

Much of the power of semantic nets, stems from the ability to attach procedures to frame slots. When a semantic model is being constructed, the designer may not know the actual values to place in specific slots of specific frames, but he may have a general set of rules that could be applied to each specific case, in order to generate the actual values. Continuing with the example used earlier, the duration of each individual operation in our engineering system may not be known, but we may have procedures that calculate the duration of each

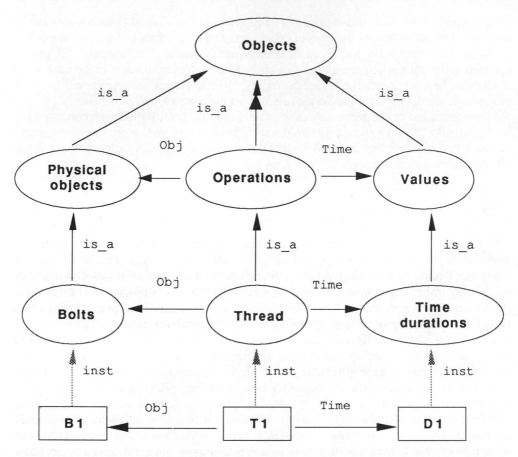

Figure 2.7 A semantic net with additional relations

operation, given the type of operation, the size of the component, etc. This knowledge can be encoded as a procedure and attached to the 'duration slot' of the generic frame 'operation'. As specific frames representing individual operations are created, they can invoke the procedure and calculate their own duration. Since this may be a lengthy process, the values will usually be calculated only when they are required, rather than when the frame is created; this gives rise to the so-called 'If Needed' values.

A second, and perhaps more important, type of procedural attachment is the 'If Changed' values (refer to Chapter 3 for a practical treatment of this concept, using commercially available systems). It is possible that in a semantic model, the introduction or change of some high-level concept could have major consequences in the rest of the model. For example, the introduction of faster machine tools could affect the calculation of the duration of some of the operations using that tool. In this case, if a value (say, the speed of the machine tool) is changed, then procedures should be run to change various parts of the model. These procedures can be attached to 'If Changed' values in the slots of the relevant frames. Since the activation of these procedures is, to a large extent, outside the control of the programmer, they are often referred to as 'demons'.

2.3.4 Object-oriented Programming

We have seen that much of the power of semantic nets results from the ability to attach procedures to individual frames. The logical extension of this is an object-oriented programming system (OOPS). Object-oriented programming is based on the concept that each item of data should contain within itself methods to specify how it can be processed. This was first formulated in the Smalltalk system in the early 1970's and subsequently developed (Goldstein and Bobrow, 1980). As an example of an object, consider the data representing a person—age, height, etc.—the 'person' object should be able to respond to messages such as:

```
Add one year to your age.
Print out your weight.
```

but would ignore messages such as:

```
Add your age to your weight.
```

The philosophy behind the object oriented approach is that it provides an effective means of maintaining data integrity. Obviously it would be very inefficient for every item of data to contain all such procedures, and so an object-oriented programming system will consist of generic objects which contain procedures describing how specific objects will respond to specific messages. The procedures, or 'methods', in each object can respond to messages by:

- updating local variables,
- displaying variables,
- sending messages to objects.

It can be shown that frame-based systems are a special case of object-oriented systems; there is one main type of object—the frame—and it responds to two simple messages: 'update slot' and 'display slot'. Chapter 3 contains a discussion of the object-oriented programming paradigm, and includes relevant examples of its use.

2.3.5 Some Limitations of Semantic Nets

Semantic nets were originally developed to try to model 'common-sense knowledge', by providing a system to facilitate the inheritance of properties. This has proved very useful in well structured domains, where such inheritance can be clearly defined, but in other areas, and most notably in the area of common-sense knowledge, the method has had limited success. The main reason for this, is that in such cases there is no strict inheritance of properties. In 99 cases out of 100, a property may indeed be inherited by its subclasses, but the one remaining case will prove difficult to deal with. The classic example is the rule:

> All birds can fly ...
> ... *except* penguins *and* ostriches *and* dead birds *and* ...

This type of reasoning is referred to as non-monotonic reasoning—the truth value of a proposition does not increase monotonically as more evidence is accumulated, but in some cases the truth value may change from 'true' to 'false' as more information is added. This can be dealt with to a limited degree by providing specific slot values which override the inherited ones. It is now recognised that this is a serious limitation of semantic nets (Brachman, 1983) and alternative methods are needed to deal with the problem. Various methods of dealing with the problem of non-monotonic reasoning are discussed more fully in Chapter 8.

Another problem with semantic nets is that of multiple inheritance. In many cases we may wish a class of objects to inherit properties from a number of different parent classes. For example, a group of engineering components, such as turbine blades, might inherit one group of properties from the engine in which they are installed and another set of properties from the material from which they are made. So long as the properties are distinct, there is no problem, but as soon as there is a possibility that an object may inherit the same property from more than one source, then there is the potential for inconsistencies.

Finally, we have seen that semantic nets can be used to represent some knowledge about time durations but they do not provide any method for reasoning about time. Methods of temporal reasoning are discussed in Chapter 8.

2.4 USING LOGIC FOR REPRESENTATION

2.4.1 Types of Logic

The title of this section is, perhaps, a little misleading, in that it suggests that there is only one type of logic to consider. Certainly there is one major logic: the classical logic that was developed by the early Greek philosophers. One of the impacts of trying to apply this logic to the problems of knowledge representation in AI is that a number of its weaknesses have been exposed; we have already seen this in the problem of non-monotonic logic. Indeed, as a result of these problems, AI has 'spawned' a whole range of different types of logic, and some of these are discussed in Chapter 8. In this section, we will consider the application of the two main types of classical logic: propositional logic and predicate logic. This will be based on the mathematical work in Chapter 10, but here the emphasis is on applying the logic to the problem of knowledge representation. No attempt will be made here to provide a rigorous mathematical definition of predicate and propositional logic, but rather to outline some of the concepts of logic that are relevant to AI.

2.4.2 Propositional Logic

In the previous chapter, a proposition was defined as a statement which is either true or false. Simple examples of propositions might be:

α 'It is raining today.'
β 'John is a man.'

 γ 'Mary is a woman.'
 δ 'John is married to Mary.'

Simple propositions can be combined using the logical connectives:

 and $\quad\wedge$
 or $\qquad\vee$
 not $\quad\sim$
 implies $\;\Rightarrow$

At its simplest, this would yield propositions such as:

 $\sim\gamma\quad$ 'Mary is not a woman.'
 $\beta\wedge\gamma$ 'John is a man and Mary is a woman.'

We would expect the truth or falsity of these statements to be easy to establish, and it should be clear that they can be used to represent simple facts about the world. It must be emphasised, however, that these propositions are only tokens, it is up to us to decide how to interpret them and to decide whether they represent correct or incorrect knowledge. This is particularly so when we consider propositions involving the implies operator ($\Rightarrow$). For example, consider the proposition:

 $\delta\wedge\beta\Rightarrow\gamma$ John is married to Mary
 and John is a man
 implies Mary is a woman.

The interpretation of this seems logically true. However, as far as propositional logic is concerned, the proposition $\delta\wedge\beta\Rightarrow\gamma$ is just another proposition which may be true or false.

 This problem of interpretation leads to one of the limitations of propositional logic, namely that there is no way to relate different propositions that have related interpretations. Consider the two propositions

 β 'Bill is a man.'
 γ 'Charles is a man.'

Though the interpretations of β and γ are very similar, there is no logical connection between them. This severely limits the type of analysis that can be undertaken with these propositions.

 A second limitation of propositional logic is that there is no way of representing general concepts such as:

 'All men are mortal.'

At best, we can produce a set of propositions, one for each man, stating that that particular man is mortal; we then run into the previous problem of not being able to relate these similar propositions.

2.4.3 Predicate Logic

Predicate logic is an attempt to overcome some of the limitations of the simple propositional logic. It is based on classes of objects, rather than individual objects. A predicate is a statement about a class of objects; for each object in the class the statement is either true or false. Some examples of simple predicates are given below:

rain(x)　　　'It was raining on day x this month.'
is_wheel(y)　'Component y in this car is a wheel.'
sub_part(y,z)　'Component y is a sub-part of component z.'

As can be seen from these examples, predicates are represented in 'functional notation', with the parameters of the function referring to the members of the underlying class. A predicate can be regarded as a logical function in one or more variables; the function takes the value true or false depending on the value of the variables. Predicates can be combined using the same logical connectives as in propositional logic, but in addition there are two further operations—the universal ($\forall$: for all) and existential ($\exists$: there exists) quantifiers.

The universal quantifier specifies that a given predicate is true for all possible values of the variable. For example, if we have a proposition:

temperature (day, minimum, maximum)

which specifies that the temperature on the specified day was never less than the minimum and never more than the maximum, then we can express simple propositions such as 'The temperature on 1 April is between 0 °C and 10 °C' by the proposition:

temperature (1 April, 0 °C, 10 °C)

However, this can be taken further, since we know for example, that every day the temperature is between −100 °C and 100 °C. This can represented by using the universal quantifier as follows:

$\forall$ day : temperature (day, −100 °C, 100 °C)

The existential quantifier specifies that there exists at least one value of a variable which makes the predicate true. Continuing with the previous example, if we know that there are days when the temperature never rises above 0 °C, then we can represent that fact using the existential quantifier as follows:

$\exists$ day : temperature (day, −100 °C, 0 °C)

The use of predicate logic for knowledge representation and processing has proved very fruitful. It is at the basis of the Prolog system, which is discussed more fully in Chapter 8 (and also in Chapter 3, from a different point of view).

2.4.4 Prolog

Prolog was developed as a method of programming computers using logic rather than conventional programming languages. It has been widely used by the AI community for a large range of tasks and there are now numerous textbooks on the language (Clocksin and Mellish, 1981; Bratko, 1986). We will not seek to provide a full description of the language and its uses here, but rather to highlight those areas that are relevant to knowledge representation. In particular, attention will be focused on Prolog data structures and rules.

Prolog, like most computer languages, provides facilities for representing various types of data. The main difference is that Prolog allows its programmers to create their own data types by means of 'functors'. This is illustrated in Figure 2.8.

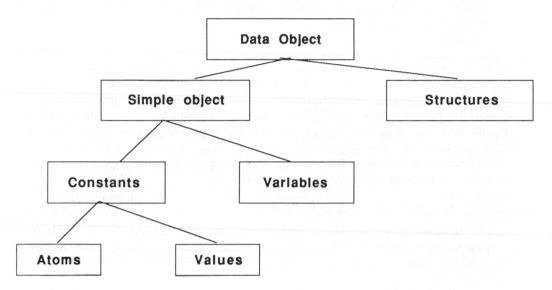

Figure 2.8 Prolog data structures

At the lowest level, Prolog 'atoms' are used to represent objects. They have no internal structure, and any interpretation that we may place on them is due purely to our own naming convention. At the syntax level, atoms are represented by alpha-numeric strings, starting with a lower case letter. As with all programming languages, it is wise to choose a sensible naming convention, so for example, we might represent an object such as a particular bolt by the atom '`bolt_1`'. Values are used to represent the attributes of entities and may be either integers, reals or quoted character strings. These two types—atoms and values—make up the set of Prolog constants.

Variables in Prolog correspond to universally quantified variables in predicate logic. At the syntactic level they are represented by alpha-numeric strings, starting with an upper case letter. Thus '`Bolt`' could represent a Prolog variable which we might *expect* to take values such as `bolt_1`, `bolt_2`, etc. Variables and constants together make up the class of simple objects in Prolog.

The most important type of data object in Prolog is the structure. Prolog data structures are used to represent logical predicates; the name of the predicate is represented by a functor and its parameters by components, thus:

```
functor ( component_1, component_2, ...)
```

A component is either a simple object or another structure. Examples of structures might be:

```
date (1,may,1983)
point(1,3)
line (point(1,3), point(3,5))
```

In these cases, the predicate represented by the functor can be interpreted as specifying whether or not a particular relation holds between the objects represented by the components of the functor. A necessary condition for this condition to be true may be that the components themselves have a certain type. Thus

$$line(P1,P2) \Rightarrow \text{P1 and P2 are endpoints of points of a line}$$
$$\Rightarrow \text{P1 and P2 are points with (X,Y) coordinates}$$

It is this richness of data types that gives Prolog its power to represent a wide range of types of knowledge. In fact, Prolog can be used to represent two main types of knowledge: facts and rules. So far we have looked at how Prolog data structures can be used to represent facts. To represent rules, use can be made of Prolog program structures.

The representation of rules in Prolog is achieved by means of the language's declarative semantics. This means that specification of tasks is achieved by defining the result of the task, rather than by defining a set of operational procedures to achieve those results. Rules are represented by a head, or goal, and a body, or set of conditions; the rule declares that the goal is satisfied if all the conditions of the body are satisfied. For example, if we wished to specify that an engineering part of a particular type, say a widget, is to be made from either a wi30x and a wi31y, say, or two wi40z then we could represent these by Prolog rules of the form:

```
widget(Sub_part_1, Sub_part_2) :-
      wi30x( Sub_part_1 ) , wi31y ( Sub_part_2 )
widget(Sub_part_1, Sub_part_2) :-
      wi40z( Sub_part_1 ), wi40z( Sub_part_1 )
```

The predicate `widget(Sub_part_1, Sub_part_2)` is true if the variables `Sub_part_1` and `Sub_part_2` represent correct sub-parts of a widget. Similarly the predicate `wi30x` is true if its variables represents an object of this type, and so on. The way in which these rules can be used will be discussed in the following sections.

2.5 SEARCHING WORLD MODELS

So far in this chapter the discussion has concentrated on how to represent knowledge rather than how to process it. There are, however, two areas of knowledge processing that require discussion: searching world models and using logic for reasoning. These are discussed in this and the following section.

2.5.1 Pattern Matching

One of the basic processes in AI is pattern matching. There is no benefit in developing methods of representing large complex volumes of knowledge, if we have no way of retrieving and processing it. This implies that we need some method of matching and retrieving information stored in the knowledge base. In standard database management systems there are only a small number of ways of matching data, but with knowledge bases, the situation is much more complex. In databases, there are usually only two types of simple data—numeric or character— with corresponding types of matching: lexical equality and lexical ordering for characters and numeric equality, and numeric ordering for numbers. In knowledge representation we are dealing with complex structures, and the methods for matching them are correspondingly more complex.

There are a number of possibilities that can occur when matching structured objects:

- strict equality—the two tokens to be matched in fact refer to one and the same object, for example, the two objects 'the author of this chapter' and 'F.N. Teskey';
- functional equality—when fully evaluated in a suitable environment the two objects are strictly equal, for example, 'the average temperature today' and '10 °C';
- structural equality—the two objects have the same structure, and where the structures are fully defined, they are functionally equal, for example, '1/4/89' and 'a day in 1989'.

There is a further level of matching, on top of these types of equality. The knowledge base will usually contain a hierarchy of specific instances of objects and generic classes of objects. There is a need to link objects to classes, and so it is possible to apply the same types of equality test to check whether or not a specific object matches the prototype object of a class; linking '1/4/89' to 'the set of all days in 1989' is an example of this.

2.5.2 Depth or Breadth First

Because of the complexity of matching, searching knowledge bases takes on an added complexity. Methods of searching a database management system are based on the assumption that the search request will retrieve a single, well defined item of data. In contrast, requests from a knowledge base are often 'fuzzy', and could return numbers of different, and possibly conflicting, pieces of information. We have only to consider an apparently simple knowledge-based system, such as a natural language interface, to understand the ambiguities that can arise. A further effect of the complexity of matching, is that it is necessary to compare very many more objects than would be the case in a simple database, where index methods can be used to limit the search space.

Thus, it is necessary to consider carefully the methods for searching the knowledge base and to seek ways to limit the search space. The two main considerations are:

- in what order should the hierarchy be searched;
- how many matching solutions should be looked for.

The former gives rise to the main distinction in search strategies between depth first and breadth first, and the latter raises the issue of backtracking to find alternative solutions. This is dealt with in more detail in Section 2.6.

The implementation of breadth-first and depth-first searching should be clear from Figure 2.9. It may be possible, with varying degrees of certainty, to eliminate parts of the tree from the search path, depending on the organisation of the tree and the type of search being

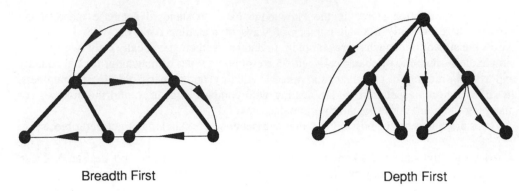

Breadth First   Depth First

Figure 2.9   Search strategies

performed. The more specific the search, the easier it will be to limit the search space. This method of limiting the search space is called 'tree pruning'. A full discussion of this is outside the scope of this chapter, but for more details see Mellish (1985).

2.5.3 Forward and Backward Chaining

We have seen how knowledge can be represented in the form of rules:

> Rule:
>> *If*   evidence
>> *Then* hypothesis

It has been shown that the hypothesis of one rule can be used as the evidence for another. Thus, it is possible to build up a hierarchy of evidence, rules and hypothesis. This is illustrated by the rule tree in Figure 2.10. There are two ways of searching this tree:

(1) Start with a goal hypothesis which requires proving, and from that identify the evidence that is needed to support this hypothesis. After this, check to see if this evidence is indeed true in the outside world, and if it is, then it can be concluded that the original goal hypothesis is also true. This is known as backward chaining, since this is effectively working from the hypothesis back to the evidence.

(2) Input all the available evidence to the system and from that let the system deduce which hypothesis may or may not be true. This is known as forward chaining, since this process involves working from the data towards the goal.

2.6 USING LOGIC FOR REASONING

It has been demonstrated, in Section 2.4 , how logic can be used for representing knowledge, but the great strength of logic in AI is that it can also be used for reasoning about knowledge. Because knowledge has been represented as standard logical formulae, so standard logical techniques of theorem proving can be used to ascertain the truth of any hypothesis with

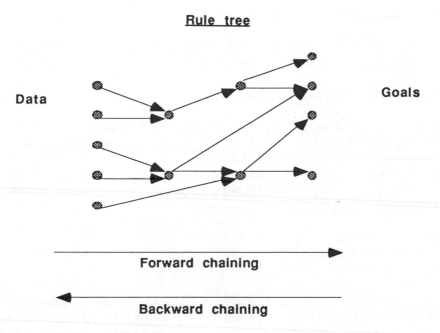

Figure 2.10 Searching a rule tree

respect to the available evidence. Traditionally, logic has relied on mental effort to prove theorems, often relying heavily on intuition and trial and error. For AI, the need exists to consider methods of automating this process.

2.6.1 Automated Theorem-proving

The concept of automated theorem-proving is simple: namely to automate the mental processes used by logicians to prove theorems. Since there is a well defined set of logical inference rules, it might be thought that the process of automating them would be simple. However, the problem is that there is a potentially infinite number of methods of representing any logical predicate or inference rule. Any attempt to apply these exhaustively would produce a combinatorial explosion of possible proofs to investigate. Most, if not all, automated theorem proving systems address this problem by using a standard form for all predicates and inference rules. This more constrained approach should produce a more manageable set of possible proofs, but it will still be necessary to check through a number of possible proofs. These are usually arranged in a proof tree; each link representing the application of a rule to a proposition represented by the node. The proof tree can be searched in much the same way as the rule tree in the previous section, i.e as various branches are explored we can either continue down these branches, or backtrack up through the tree to explore other branches. The control of searching and backtracking is central to much of automated theorem-proving. For the rest of this section, however, the discussion will concentrate on the methods used to represent the predicates and inference rules.

2.6.2 Conjunctive Normal Form

Logic provides a rich variety of methods of representing propositions and predicates. Indeed, it is possible to represent even the simplest proposition in an unlimited number of ways:

$$a, a \wedge a, a \vee a, a \wedge (a \vee a), \ldots$$

Fortunately the rules of logic also provide a number of standard forms for representing propositions. One of these, that is widely used in AI, is conjunctive normal form.

A (compound) proposition is in conjunctive normal form if it is in the form of a series of logical conjunctions (*and*) each of which is a disjunction (*or*) of simple propositions or the negation of simple propositions. The following are examples of propositions in conjunctive normal form:

$$a \wedge (b \vee c)$$
$$(a \vee b) \wedge (a \vee \tilde{d}) \wedge c$$
$$a \wedge \tilde{b}$$

It can be shown that any expression in propositional logic can be converted into conjunctive normal form. Though this will not be proved formally, an idea of the proof can be obtained by considering the truth table of a proposition; each cell of the table represents a disjunction of simple propositions or the negation of simple propositions. Since the table as a whole can be represented as the conjunction of its cells, the result follows.

Arbitrary predicates can also be represented in conjunctive normal form. The use of the quantifiers ($\forall$ and $\exists$), however, introduces some problems. One technique to overcome these problems is to regard predicates such as:

$$\forall x \exists y : p(x,y)$$

as defining a function S_p, called the 'Skolem function' of p, which given a value x produces (by some unknown method) a value y such that:

$$\forall x : p(x, S_p(x))$$

Any expression in predicate logic can then be converted to a form which does not include any existential quantifiers, and in which all free variables, such as x in the above example, are universally quantified. By regarding the basic functions, including any Skolem functions, as atomic propositions, the original predicate can be converted to conjunctive normal form in just the same way as propositions were. This standard form of predicates, often referred to as clause form, is the basis for automated theorem-proving using the resolution principle.

2.6.3 Resolution

Having shown how any expression can be represented in predicate logic, in a standard form, the next step is to provide a standard set of inference rules for manipulating these standard clauses. The pieces of evidence that are used to prove a goal, form a series of logical conjunctions, and since the individual pieces of evidence (the axioms to be used in the proof) can be converted into similar conjunctions, the whole can be regarded as a single series of logical conjunctions, the individual elements of which are simple disjunctions. We can convert the goal to be proved into a similar series of conjunctions, and the proof process would then consist of applying inference rules to the set of clauses in the axioms, to try and extend it to include all the clauses of the goal. In practice, however, it is easier to adopt a proof by 'refutation'; that is, we prove a proposition by showing that its converse leads to a contradiction. Thus we add the negation of the goal clause to the axiom set and apply the inference rules to reach a contradiction. (It is easier to direct the inference rules towards a simple contradiction, rather than towards an arbitrary goal clause, which may be quite complicated.) The problem, then, is how do we automate the application of inference rules to these standard clauses?

If a standard inference rule is to be applied to a standard set of clauses, then the rule must take one or more standard disjunctions and produce a new disjunction, also in standard form. The simplest example of this type of rule is the resolution principle:

$$a \lor b$$
$$\sim a \lor c$$
$$\overline{}$$
$$b \lor c$$

In fact, this one simple rule can be used to make all valid inferences from a set of standard clauses. Thus, by converting our axioms and goals to conjunctive normal form and applying the resolution principle, we can automate theorem-proving. Needless to say, there are numerous modifications that can be added. The most important of these are the methods of unification used to match the components of the standard clauses; however, a discussion of these is outside the scope of this chapter.

2.7 PROBABILISTIC REASONING

2.7.1 Introduction to Probabilistic Reasoning

All the knowledge representation methods that we have considered so far have been based on the assumption that the knowledge being represented is exact and certain. This is often not the case; we frequently need to reason with inexact or incomplete information. Even areas such as engineering, which are traditionally regarded as exact can give rise to uncertain reasoning. Examples of this would include such areas as estimating safety factors, interpreting client specifications, etc. Some of the research methods that might be able to deal with these problems are dealt with in Chapter 8. One method, probabilistic reasoning, has been developed to the stage where it is now regarded as a standard knowledge representation technique. The method is based on the theory of probability and in particular on Bayes' Theorem (see Chapter 10). In the remainder of this section the concept of inference nets will be introduced, it will be demonstrated how Bayes' Theorem can be used to calculate probabilities in an inference net, and discussion will proceed to how the use of certainty factors has replaced the use of probability in some applications.

2.7.2 Inference Nets

We have seen how it is possible use a series of simple logical *if–then* rules to represent knowledge. The need for dealing with uncertain information has also been discussed. Combining these gives the concept of a simple probability rule. These rules can be thought of as similar to the logical rule, but with a degree of uncertainty introduced:

> Rule:
>> *if E then H* probably

This can be interpreted as saying that if we have an initial probability of the hypothesis *H*, and a probability that the evidence *E* is true then the rule will allow us to calculate the new probability of *H*. The rule will provide us with a measure of the correlation between the evidence and the hypothesis. Later, it will be seen how Bayes' Theorem can allow us to quantify this correlation. For the moment, the point will be illustrated with a simple example. Suppose that the following hypothesis is to be investigated:

> *H* 'It will snow tonight.'

and that one of the pieces of evidence supporting the hypothesis is:

> *E* 'The temperature is below freezing.'

There will be some prior probability that it will snow on any given night, but the chances of snow are greater when the temperature is below freezing. This can be represented by a probability rule. It is interesting to note that in this case the converse of the rule:

> *If* the temperature is above freezing
> *then* it will not snow tonight

has a much greater certainty than the original rule. We will see later that it is possible to quantify, not only the original rule, but also its converse. Once the concept of probability rules has been understood, then inference nets can be defined. Just as a series of logical rules can be combined to form a rule tree, so a set of probability rules can be linked, with the hypothesis of one rule acting as the evidence for the next. This is called an inference net. However, unlike a rule tree where the propagation of logical values through the rule tree is fairly simple (based on the logical connections at each node) the propagation of probability values is more complex. Various methods have been proposed for calculating the probability of a hypothesis based on various pieces of evidence. Most of these are based on Bayes' theorem and conditional probabilities.

2.7.3 Conditional Probabilities

With reference to the more formal treatment of the subject, to be found in Chapter 10, $p(a/b)$ is defined as the probability of a given b. So in an inference net, we may wish to calculate:

$$p \left(H \mid E \right)$$

i.e. the probability of the hypothesis given that the evidence is true. In many cases it will be known that:

$$p \left(E \mid H \right)$$

i.e. the probability of the evidence being true given that the hypothesis is true, as well as:

$$p(H) \text{ and } p(E)$$

Bayes' theorem then gives us a method of calculating the first of these probabilities, given the other three. To see this relationship consider the contingency table in Figure 2.11:

	E	$\sim E$	
H	a	b	$a{+}b$
$\sim H$	c	d	$c{+}d$
	$a{+}c$	$b{+}d$	$a{+}b{+}c{+}d =1$

Figure 2.11 Contingency table for evidence and Hypothesis

In this table, a is the proportion of times H and E occur together, b is the proportion of times H occurs but E does not and so on. We can then express the various probabilities as:

$$p(E) = a{+}c$$
$$p(H) = a{+}b$$
$$p(E \mid H) = a/(a{+}b)$$
$$p(H \mid E) = a/(a{+}c)$$

From this, the standard form of Bayes theorem can be deduced:

$$p(H \mid E) = p(H) \cdot p(E \mid H) / p(E)$$

The simplest method of applying this to probability rules is by means of logical sufficiency and logical necessity factors.

2.7.4 Logical Sufficiency and Logical Necessity

If a probability rule 'If E then H' exists, then Bayes' theorem says that if the evidence

E is true, then we modify the probability of H by a factor $p(E|H)/p(E)$. This is called the logical sufficiency (LS) factor, and measures the sufficiency of E in the proof of H. Similarly if E is false, we modify the probability of H by a factor $p(\bar{E}|H)/p(\bar{E})$. This is called the logical necessity (LN) factor, and measures the necessity of E in the proof of H. Thus we have:

$$LS = p(E|H)/p(E)$$
$$LN = p(\sim E|H)/p(\sim E)$$

We would normally expect LS to be greater than unity, indicating that E does indeed support H, and similarly, it would be expected that LN would be less than unity, indicating that H requires E to be true.

It is now possible to give a formal definition of a probability rule:

Rule:
If E then H
 with logical sufficiency *LS*
 and logical necessity *LN*

2.7.5 Certainty Factors

So far, it has been assumed that, when building an inference net, it is possible to calculate the probabilities accurately. In some cases there may be sufficient historical data, but in others it may be possible only to estimate values for LS and LN. But as well as the need for estimation in building the inference net, there may also be problems in using it. It has been assumed that the client will be able to say whether or not a piece of evidence is true. In some cases the client may not know of certain events but may have some degree of certainty that an event has, or has not, occurred. This means that rather than expect the client to answer a question such as:

What is the probability of E?

the system should in fact ask:

How certain are you that E is true?

At this point, it is appropriate to introduce the concept of certainty factors. The idea of certainty factors is to allow the user to answer such questions, typically with a value from the range -5 to $+5$. As one would expect, -5 indicates that we are certain that E is false, $+5$ indicates that we are certain E is true. Superficially certainty factors are very similar to probabilities, the important point is that probabilities represent an objective measure while certainty factors are subjective. For this reason there is no one correct method to convert certainty factors to probabilities, or to combine certainty factors in the same way one might process probabilities. Individual AI tools and expert system shells adopt their own variants of Bayes' theorem to solve this problem.

2.8 AN EXAMPLE OF A KNOWLEDGE REPRESENTATION ENVIRONMENT

The main methods that are used to represent and process knowledge in AI systems have now been described. This section serves to demonstrate how these techniques can be combined into a working system. The system that will be used as an example is the IDEAS system that has been developed within the Information Technology Research Institute at Brighton. IDEAS is a framework for developing and evaluating tools for building knowledge-based systems, with particular reference to practical application in the fields of engineering. The system has been implemented in POP 11 (under the Poplog environment), and is now being evaluated at a number of industrial sites. The novel feature of the system is the way in which existing techniques have been combined into a set of coherent and easy-to-use tools (Winstanley *et al.*, 1989).

The first tool in the IDEAS environment, was a knowledge representation language, next came tools to help build knowledge bases using this language, and most recently we have been developing tools to improve the quality of the knowledge base and the consultation of the knowledge base. The Poplog system has provided a powerful environment in which to develop and test these prototypes quickly and easily. The design has been highly modular; each tool has been developed as an independent unit within the overall framework, and can be used as and when required. These tools demonstrate a range of new and interesting ideas for building expert systems; it must be emphasised, however, that although they have been tested in a number of practical applications, they are not, as yet, a marketable product. For a comparative treatment of commercial knowledge representation environments, reference should be made to Chapter 3.

2.8.1 Knowledge Representation

The knowledge representation language used in IDEAS is a frame-based system in which the links between frames may be established by rules; the pieces of evidence for these rules are predicates, in the slot values of the frame base. For this reason, the system is known as the 'Rule-Based Frame System' (RBFS), and further references to its structure and application can be found in Chapters 4, 5 and 6 of this text. The frame base has been implemented with features along the lines of the Goldstein and Roberts model (1979), including a comprehensive set of retrieval and manipulation functions. The rule inference mechanism, used to establish the frame links, is based on the Mycin model (Buchanan and Shortliffe, 1984). The following sections provide a brief description of the frame and rule base of RBFS, a definitive description can be found in Barber (1987).

Within the epistemology of the RBFS, the frame base represents a collection of statements about the objects within the world, and the relationships which do, or can, exist between them. This is represented by named nodes, or frames, and links or slots. The frame-slot notation in RBFS maps to the node-concept ideas of associative networks (Levesque and Brachman, 1985). Like KL-ONE (Brachman and Schmolze, 1985), no special restrictions are enforced in the naming of primitives. However, the names of nodes and links should be chosen to reflect the language used to describe the domain. In general, nouns are used to represent nodes, and verbs to represent the links. In the information model, we draw the traditional distinction between 'generic' entities and 'specific' entities (Brachman, 1983). General entities represent generalised types of objects, whereas specific entities represent instantiations of generic entities, and so represent what is factual, or what has been concluded.

In RBFS, the distinction is made between different types of links. A factual link represents a statement of a factual relation between two objects, e.g. 'Jim loves Joan' . There are two special links of this type, the '**is_a**' and the '**inst**' links. The '**inst**' link is used to denote that a specific object is a member of a generic object, e.g. 'Jim **inst** Teacher', whilst the '**is_a**' is used to build class hierarchies of similar objects, e.g. 'Teacher **is_a** Person' . These class hierarchies are a mechanism by which elements of a class can inherit common properties from a more general class.

The second type of link is the inferred link. This is used to represent the possibility that a relationship *may* exist between two generic objects. Rules must be provided to evaluate the certainty that this link exists, as a factual link, between any two specific instances of the generic objects. For example, a teacher may teach a pupil, but we need specific rules to say whether or not a particular teacher teaches a particular pupil. Obviously these inferred links may exist only in the general plane, and are evaluated only for entities which have already been *instantiated* in the specific plane. An RBFS rule consists of premises, similar to MYCIN premises, which test the equality (inequality, etc.) of slot values on specified paths through the associative network (beginning from a context variant which is instantiated by a calling enquiry). Specified relational slots can be concluded with a certainty value reflecting the strength of these premises.

A third type of link, which RBFS provides, is the instantiation link. This is similar to the inferred link, but with the difference that the specific instances in the target of the relation do not exist initially, but are created by the link. The process of creating a specific instance of a generic concept is termed 'instantiation'. At the simplest level, instantiation is the generation of a unique node linked to the generic node via an '**inst**' link. However, RBFS allows a certain amount of user control over the instantiation of concepts via *instantiation directives* in the conclusion of rules. For example, the relation '**has_parent**' could be defined as an instantiation link so that, if the parent of a particular child were needed, but was not represented in the knowledge base, the system could create an instance of type 'parent' and link it to the child. The instantiation directives may specify that one or more instances of the destination frame is created, and that the frame is named either automatically by the system or manually by the user.

2.8.2 Knowledge Acquisition

The full automation of knowledge elicitation is, in our view, an unattainable task with the present state of the art. However, systems do exist, which have been designed to cater to some degree with the problems of elicitation (see Chapter 4). In recognition of this, the focus has been on developing tools to assist the domain experts, in collaboration with knowledge engineers, to build their own expert systems and knowledge bases. The synergetic relationship between rule- and frame-based systems, that have been developed for knowledge representation, has been extended to knowledge acquisition by providing a common framework in which domain experts and knowledge engineers can work *together* to build expert systems (Eklund *et al.*, 1987).

The methodology of building expert systems with IDEAS, is to first map out the information as a frame base of possible concepts and relations, and then to determine the rules by which relationships and concepts will be confirmed in specific instances. In practice, the frame base and rule base are intimately related as information development provokes knowledge consideration which identifies additional information requirements. We have found that many domains can be modelled in this way, and it appeared to be an appealing idea to try to automate as much of this process as possible. The objective has been to produce visual and easy-to-use tools for effective interaction between the domain experts, knowledge engineers and the knowledge base. Two tools have been developed: a Visual Editor for the Generation of Associative Networks (VEGAN), and the Knowledge Engineering Tool (KET), which make the frame and rule structure of the knowledge base readily accessible through graphical interfaces.

VEGAN is a tool developed to allow the generation and modification of a frame base in the form of a visual representation of the underlying associative network. It is, in simple terms, a digraph editor which has been tuned to the particular application of RBFS. It facilitates all interaction with the frame base to be performed via a graphical interface, thus precluding the need for the user, be he domain expert or knowledge engineer, to have any contact with, or in-depth knowledge of, the underlying syntax of RBFS. A full description of VEGAN is given in Kellett *et al*.,(1986); the following is a brief summary of the main facilities of the interface.

VEGAN is designed to make the process of constructing a frame base to represent an information model as simple a task as possible. As nodes and links are created within VEGAN, the system automatically generates the appropriate frame operations. The user interacts with VEGAN via a WIMPs type environment (Windows, Icons, Mouse, Pull-down menus). The environment contains a view surface, on which the associative net is displayed, and a menu button box, to activate the various functions. The VEGAN menu features such functions as 'add node', 'delete node', 'add link' and 'delete link'. These are the simple primitive operations necessary for network generation; facilities for view control ('zoom' and 'pan' type operations), network editing and tidying ('move' type operations), and network enquiry (node/link enquiry, search facilities) are also provided. VEGAN has been deliberately designed with a very open architecture, with all its control and display routines available from RBFS, as Pop procedures attached to rules. This allows the animation of the displayed network during consultations. VEGAN caters for instantiation, by providing facilities for automatic placement of the instantiated nodes. It is possible to animate the networks, such that the user can 'watch' the creation of the specific model as the system runs.

KET was originally designed in order to facilitate the writing of production rules in RBFS. KET relieves the domain experts or the knowledge engineers from building systems in an AI language, by providing an easily used graphical interface. KET not only caters for visualisation of the rule aspect of RBFS, but also provides a way for the expert to think about how rules and their contents are related. In most rule-based systems the rules form a network or tree structure, with the conclusion of one rule being used as a premise for another. Just as VEGAN made the underlying semantic network visible, so KET makes this rule tree visible. A full description of KET is given in Esfahani and Teskey (1987). The following is a brief description of the system.

KET provides a method of creating individual rules and placing them in a rule tree. In addition it provides the standard graphical editor functions (similar to those in VEGAN) for this rule tree. There are three methods which RBFS uses to derive values for slots in a frame, namely inference by rules, calculations by procedures and requesting the values (via 'ask') from the user directly. The main modules in KET—Rules, Procedures and Ask—are based on these three methods.

These three methods of filling slots, are represented as boxes in KET. The knowledge engineer can indicate to KET that a module is required by selecting the 'add_box' option from the menu. KET then prompts for the type of the module to be opened, and depending on the result, creates the appropriate layout. When the modules are completed, the contents of the windows are saved; the window can then be closed, leaving only the rule name box or the 'Proc' heading. The user can expand these at a later stage, to see their contents by an appropriate menu option. On leaving KET, the contents of the graph are written in a file in RBFS format. This, together with the corresponding file from VEGAN, forms the knowledge base used by the user, in consultation mode.

2.8.3 Applications

Much of the impetus for the development of IDEAS came from the requirements of specific applications. The first, and most influential of these, was project management (see Chapter 6). Most existing computerised project management systems have little or no knowledge of the processes they are scheduling, and so are unable to make any intelligent decisions about the plans they are developing. By building models of the objects and actions in a project, and their relationships, it is possible to incorporate design knowledge in the planning system. Using these concepts we have developed a project management system in IDEAS called PIPPA (Professional Intelligent Project Planning Assistant) (Marshall *et al.*, 1987). The system can be regarded as a general framework for building expert planning systems for specific domains. At the centre of the system is a hierarchy of objects and actions with a set of predefined relations, over and above the **is_a** and **inst** relations. These additional relations include:

```
object_1 is_a_component_of object_2
object_1 is_manufactured_by action_1
action_1 is_preceded_by action_2
```

PIPPA contains appropriate rules for manipulating these relations. For example, if the system has not been told how to manufacture a specific object, then a first estimate would be to obtain all the actions required to manufacture all the components of the object, and then order these using the precedes relation.

2.8.4 Conclusions

The IDEAS system has provided a powerful test bed for developing methods for building expert systems. The Poplog environment, that we have used, has facilitated the rapid development and integration of a number of independent tools. The most mature of these tools are the knowledge representation language, RBFS, and the knowledge engineer's interface VEGAN/KET.

In summary the advantages of RBFS are that it:

- provides a powerful suite of manipulation and retrieval procedures to exploit the frame-base allowing default values, procedural attachment, inheritance from class hierarchies and associative network search paths;
- provides a distributed inference system by attaching rules to individual frame slots;
- provides an extremely good environment for rapid prototyping, by allowing objects to be developed individually and incrementally;
- supports on-line explanation facilities and post-consultation justification;
- is largely unconstrained, compared to expert system shell environments (see Chapter 3), and it is possible to define complex procedural definitions and incorporate these with a traditional production rule approach;
- facilitates some degree of reasoning with uncertainty in soft domains.

The main advantages of VEGAN/KET are that it:

- makes the frame and rule syntax simple and easy to use;
- provides a facility for fast prototyping, which the domain expert can understand;
- provides significant facilities for debugging and verification of the knowledge base; and
- facilitates working with large and complex knowledge bases.

The integration of these tools has opened the way for further research in the development of knowledge representation techniques in the field of engineering.

2.9 THE PROBLEM OF COMMON SENSE

In this chapter, we have shown how standard mathematical methods can be used to represent knowledge (see Chapter 10). These methods of representation have proved useful in building AI systems in a number of domains, particularly in 'hard' domains, such as medicine and engineering, where there is a well defined body of knowledge. These methods, however, have not proved successful in abstract or 'soft' domains where there is no clear set of rules and regulations (see Chapter 1 for a discussion on hard and soft problem definition). The main reason for this appears to be that the current methods of knowledge representation do not provide any effective method of representing common sense. They deal with fixed, quantifiable rules, whereas in many cases, we make use of a much more qualitative method of reasoning. In addition, we have a basic understanding of *how the world works* ; the so called 'naive physics' (see Section 8.6.3 for a discussion on naive physics). Attempts have been made, and are still being made, to develop methods for representing this type of knowledge.

REFERENCES

Barber, T.J. and Boardman, J.T. (1986) A Pragmatic Approach to Project Control Using a Heuristic Technique, *IEE Colloq. Digest*, no 1986/85.

Brachman, R.J. (1983) What IS-A is and Isn't: An Analysis of Taxonomic Links in Semantic Networks, *IEEE Computer*, **16**(10), 30–36.

Brachman, R.J. and Schmolze, J.G. (1985) An Overview of the KL-ONE Knowledge Representation System, *Cognitive Science*, **9**, 272–16.

Bratko, I. (1986) *Prolog Programming for Artificial Intelligence*, Addison-Wesley, Wokingham, England.

Buchanan, B.G. and Shortliffe, E.H. (1984) *Rule-Based Expert Systems: The MYCIN Experiments of the Stanford Heuristic Programming Project*, Addison-Wesley, Reading, MA.

Clocksin, W.F. and Mellish, C.S. (1981) *Programming in Prolog*, Springer-Verlag, Berlin.

Codd, E.F. (1970) A Relational Model of Data for Large Shared Data Banks, *Communications of the ACM*, **13**(6), 177–192.

Date, C.J. (1975) *An Introduction to Database Systems*, Addison-Wesley, London.

Eklund, P.W., Barber, T.J. and Teskey, F.N. (1987) An AI Environment For Knowledge Based Systems Development, *Proc. KBS 87 Con.*, London, 23–25 June 1987, On-line Publications, London, pp 127–35.

Esfahani, L. and Teskey, F.N. (1987) KET, A Knowledge Encoding Tool, *Proceedings of the first European Workshop on Knowledge Acquisition for Knowledge Based Systems*, September, University of Reading.

Goldstein, I.P. and Bobrow, D.G. (1980) Extending Object Oriented Programming in Smalltalk, *Proc. 1980 Lisp Conference*, Xerox Corporation, Palo-Alto.

Goldstein, I.P. and Roberts, B. (1979) Using Frames In Scheduling, *Artificial Intelligence: An MIT Perspective*, Ed. P.H. Winston, Vol. 1, MIT Press, Cambridge, MA, p 253.

Kellett, J.M., Winstanley, G. and Boardman, J.T. (1989) A Methodology for Knowledge Engineering Using an Interactive Graphical Tool for Knowledge Modelling, *International Journal of AI in Engineering*, **4**, 92–102.

Lacroix, M. and Pirotte, A. (1977) Domain Oriented Relational Languages, *Proc. 3rd International Conference on Very Large Databases, 6–8 October, Tokyo*, IEEE, New York, pp 370–378.

Lavington, S.H. and Azmoodeh, M. (1982) IFS—a Proposal for a Database Machine, *Proc. 2nd British National Conference on Databases*, British Computer Society, London.

Lenat, D.B. (1982) AM: Discovery in Mathematics as Heuristic Search, In: *Knowledge Based Systems in Artificial Intelligence*, Eds. R.Davis and D.B. Lenat, McGraw Hill, New York.

Levesque, H.L. and Brachman, R.J. (1987) Expressiveness and Tractability in Knowledge Representation and Reasoning, *Comput. Intell. (Canada)*, **3**, 78–93.

Marshall, G., Kellett, J.M., Lim, B.S. and Boardman, J.T. (1987) PIPPA : An Expert Project Planning System in Manufacturing Engineering, In: *Proceedings of the KBS 87 Conference*, London, 23–25 June 1987, On-line Publications, London.

Mellish, T. (1985) Generalised Alpha–Beta Pruning as a Guide to Expert System Question Selection, *Proceedings of the Fifth Technical Conference on Expert Systems*, Cambridge University Press, Cambridge, pp 31–42.

Minsky, M. (1975) A Framework for Representing Knowledge, In: *The Psychology of Computer Vision*, Ed. P.H. Winston, McGraw-Hill, New York.

Winstanley, G., Kellett, J.M., Best, J.T. and Teskey, F.N. (1989) IDEAS—For Expert Systems, In: *POP-11 Comes of Age: The Advancement of a Language*, Ed. James Anderson, Ellis Horwood, Chichester.

F. N. Teskey
IT Planning Division
National Westminster Bank plc
PO Box 114
Goodmans Fields
74 Alie Street
London E1 8HL

3 Knowledge-based System Shells and Languages

G. Winstanley and H.A. Heathfield

Brighton Polytechnic, UK

3.1 INTRODUCTION

Knowledge-based systems (KBS) are, as their name suggests, systems which use knowledge and reasoning to arrive at conclusions. They differ from traditional data-processing computer programs in their expressive power and their method of operation. In traditional programs, a predetermined sequence of actions must be followed, i.e. they are deterministic. Conventional programs are also based largely on linear relationships, and are optimised for numeric-processing, whereas the knowledge-based system concentrates on the representation and manipulation of information as symbols. Another noteworthy feature of KBS is their suitability for large and complex problem solution characterised by inexact, incomplete and uncertain information. Their structure includes an explicit body of embedded knowledge and a separate, identifiable inference mechanism. Using these facilities, the KBS builder is able to construct a mechanism capable of 'mimicking' human reasoning (the inference mechanism or inference engine), and the knowledge engineer is able to elicit and code expert knowledge which the inferencing mechanism may use to provide solutions to problems in a similar fashion to a comparable human expert. In spite of all this, however, knowledge-based systems are merely computer programs which have been written in a different way, in a deliberate attempt to isolate the various components of human (expert) problem-solving. The isolation of the program flow directives which represent components of knowledge in the form of rules permits an explicit body of knowledge to be created and enlarged/modified in a way which would be very difficult indeed in conventional data-processing programs. Knowledge-based systems have been shown to be particularly successful in situations dependent on expert knowledge: the range of problems, of which engineering is a prime example, which requires a high level of expertise in a very narrow (rather than broad) subject area. Suitable examples might be the generation of test vectors in a piece of automatic test equipment, a diagnostic system for diesel locomotives or a design configuration facility. In cases where expert knowledge is used in such circumstances, the system is commonly referred to as an expert system. The first part of this chapter deals with expert system

Artificial Intelligence in Engineering. Edited by Graham Winstanley
©1991 by John Wiley & Sons Ltd.

development toolkits and shells. The latter part concentrates on issues relating to the choice and use of computer programming languages in the development of dedicated expert systems.

3.2 EXPERT SYSTEM DEVELOPMENT TOOLKITS: ENVIRONMENTS AND SHELLS

The basic structure of an expert system has three basic components: a knowledge base, which contains facts plus information on how to reason with these facts; an inference mechanism capable of transforming user requests into reasoned information using data from the knowledge base; and the user interface itself.

An alternative diagrammatic representation may make the distinction between information, which represents facts, and knowledge, which is necessary to make specific expert decisions on the basis of a potentially large number of interrelated facts. This distinction is important in rule-based expert systems because the rules can be regarded as the vehicle for the representation of knowledge. In most cases, the rules require specific information 'outside the rule base' in order to satisfy their goals. In the Goldworks System, discussed later in this chapter, a distinction is made between passive knowledge, which represents facts, or assertions (information), and active knowledge in the form of rules, etc.

Fully developed expert systems, in regular use in industry, are still quite rare, partly by virtue of the specificity of applications. In most cases, an application has to be developed, almost on-site using an expert system which has the basic structure, but the domain-specific information and rules have not been encoded, only the general problem-solving knowledge embedded within the inference mechanism is included. Using such a tool, known as an 'expert system shell', it is possible to carry out the process of knowledge acquisition, knowledge base building and system evaluation. The process is known as knowledge engineering and is crucial to the success of the implementation and eventual routine use of the system.

One of the first published accounts of a successful expert system was Mycin (Shortliffe, 1976). It was designed to capture the decision-making model of clinicians, engaged in the diagnosis and treatment of bacterial infections. When it was recognised that the general principles of problem-solving embodied in Mycin could be potentially applied to alternative domains, an abstracted version of Mycin was produced, namely Emycin ('empty' or 'essential' Mycin). This consists of the Mycin system minus its domain-specific medical knowledge base, and provides an application-independent framework for constructing and running consultation programs. Thus the concept of the expert system shell came about.

Expert system shells are a common choice for the development of domain-specific applications. Their structure includes a structured, but initially empty knowledge base, an inference engine and control structure which operates in a predefined way, and a user interface which can often be 'tailored' to the specific needs of the application. Such systems quite often have built-in justification facilities which make use of the knowledge base content and structure, and the larger systems may be equipped with 'browsing' tools. Some even have the capability to access external databases. This is a powerful facility, especially when the eventual system, developed as an application of the expert system shell, is to be used as a shared resource within an industrial organisation, often as a distributed system.

For the prospective user of a shell, the checklist for a suitable tool would include the following questions:

- How can information be represented in such a way that the domain can be effectively modelled? Are the representational facilities powerful enough to satisfy the expectations of the user? Does the system use objects, frames, schemas, etc.? What types of data are allowed? Is inheritance supported, and if so, what type, i.e. single or multiple parents?

- In a rule-based system, what types of rules can be used: are they simple rules based on propositional logic? How does the system deal with the management of uncertainty (incorporation of certainty factors, probability measures etc.)? Is it possible for procedures to be used alongside rules? Is pattern-matching catered for?

- Does the inference mechanism use backward chaining in which a given 'goal' is satisfied by a process of subgoal satisfaction by 'working backward' from the initial goal, or is inference carried out by a process of reasoning forward from the available information, deducing new facts, and eventually leading to the final deduction of the goal? This latter process is known as forward chaining. An excellent intuitive analogy, based on the route planning problem can be found in Parsay and Chignell (1988).

- For the inevitable development and maintenance phases, what kind of development facilities are there? How is knowledge acquisition and implementation facilitated? What kind of implementation facilities (knowledge base building, rule generation, etc.) are there and how is debugging catered for?

- The user interface is perhaps the most important concern of the expert system designer, the application developer and the eventual user. Important factors here include the use of windows, on-line validity checks, use of graphics, explanation presentation, use of natural language to interrogate and control, and facilities for correcting, updating, etc.

- Very closely associated with development and maintenance, what language has been used in the development of the shell? Is it possible to extend/modify the system by writing extra functions/procedures? Is it possible to 'hook' the system to others, for example databases, spreadsheets, word processors, communications packages, other knowledge-based systems, etc.? What computers, and possibly more important, what operating systems will the system operate on? Is the shell interpreted or compiled (factors relating to speed of operation and memory usage)? How large is the system and is its performance impaired by the capacity (limitations) of the hardware platform?

- Is the documentation adequate? Is training provided? At what cost can the system be developed and eventually used effectively within an organisation? Is support available from the vendor, and is this support likely to continue during the lifetime of the product?

For the person charged with developing the application—the knowledge engineer—a fuller understanding of the technical factors concerned with representation and control is necessary, but experience gained with a particular shell often provides sufficient familiarity with the capabilities and constraints of the system.

At this time, there exists a fairly large range of systems, from large and complex systems which require a 'large system' hardware base to quite small personal-computer-based shells. Attempts have been made in the past to compare and comparatively evaluate the current range of popular (and somewhat less popular) shells (Chung and Kingston, 1988; Merrit, 1986). A brief description of a small, but representative cross-section of these systems follows, but it must be stressed that almost all of these systems are under constant development, and some of the criteria used in the comparison may be subject to quite significant change in the near future.

There are numerous expert system shells available on the market, designed to operate on various types of hardware and utilising different inferencing techniques. No attempt is made here to give either a comprehensive consumer guide or a detailed discussion of all the deep theoretical issues involved in designing expert system architectures. Rather the intention is to give the reader some technical insights into the functionality of typical tools. To this end, the treatment examines a set of expert systems development environments and shells. They range from large and sophisticated environments with all the necessary development tools built into them to small expert system shells. Specifically, these are:

- KEE: A large hybrid development environment, originally developed for large computer systems and specialist 'AI hardware'.
- ART: A very similar system to KEE, with some important variations. Both KEE and ART can be looked on as a single comparative section.
- Goldworks: A large hybrid development system, originally developed as a system to run on small(ish) personal computers. As such, it has a very different background to the two previous systems, but has 'matured' into a full development toolkit equal to the others in functionality. This, however, can be regarded (arguably) as a lower-level development system to KEE and ART.
- KES: An expert system shell rather than a full development system, even though it has development functionality. KES differs from the others in its choice and range of inference methodologies. One of its inferencing methodologies is ideally suited to the diagnostic application, and therefore examples from the biomedical domain are used to demonstrate its power (Heathfield *et al*., 1988).
- Crystal: A small expert system shell which lacks the development system support of the other systems, but finds much use, and indeed popularity, in situations which demand small and powerful expert systems and interfaces. Crystal is included to complete the comparison of systems from large to small.

The comparative survey concentrates on several key issues as outlined earlier. They are:

- knowledge representation,
- methods of inference,
- interfaces, development and user.

3.2.1 KEE (Intellicorp Inc.)

The Knowledge Engineering Environment (KEE) is one of the more mature of the 'large system shells', it having been in existence since 1983. Since that time, it has gone through several revisions and is now capable of operation on a wide range of hardware system platforms, including Symbolics and Texas Instruments Lisp machines, VAX workstations and mainframe, IBM 6150 and mainframe, and PCs which use at least the Intel 80386 microprocessor. The system itself is considered to be one of the 'top range' of shells, among such well known products as Inference ART and Knowledge Craft. It is known as a hybrid expert system development toolkit, i.e. it combines several well-proven AI methodologies, including frame-based representation, rule-based programming and object-oriented functionality. Various inference mechanisms are supported.

KEE provides a frame-based representation scheme in which frames are termed 'units'. The unit scheme includes certain extensions which give KEE developers the ability to exploit the object-oriented paradigm. As far as inheritance is concerned, any class unit may have any number of immediate superclasses. The unit in question is then able to inherit the collective slots of these parents. In fact, a number of class inheritance strategies are available, describing for example that a unit should not inherit, inherit only if it does not have local values, inherit all of the values of all the parents into a composite group, as well as others.

Each unit may be described as having two different types of slots: member slots (which are inheritable to all subclasses and members of the class) and own slots (which are confined to the unit). Slots in a unit may be attributed a certain amount of higher-level (meta) information, i.e. information which describes the nature of the slot, such as the range of values it may hold, comments, how inheritance may occur, how many values can be held and whether procedural attachment is made. Procedural attachment is achieved using 'Active Values', which are units containing procedures describing what should happen when a slot is accessed or modified.

The rule system in KEE describes rules themselves as units. This has the advantage that rules can be collected into classes, effectively factoring the knowledge base into a number of discrete problem-solving blocks. Each rule is written using KEE's TellAndAsk syntax, which is an English-like representation. The basic format of the TellAndAsk syntax is as below:

```
(the ?slot of ?unit is ?value)
```

where a question mark preceding a word indicates variable assignment. The following are valid statements in the TellAndAsk syntax:

```
(the voltage.level of testpoint1 is ?v1)
(a component of ?device2 is power.supply)
```

In addition to this simple syntax, other TellAndAsk expressions may also be embedded. For example, the following may form part of a rule:

```
(the voltage.level of (the testpoint of ?TX) is ?v1)
```

which corresponds to the general Lisp format. In fact, it is also possible to incorporate Lisp functionality directly in the syntax. For example:

```
(the voltage.level of ?TX is (* (the voltage.level of ?TX5) 3.142))
```

A simple KEE rule may look like:

```
(if (the skill of ?person is Instrumentation.engineer)
     (the pending.task of ?person is design.instrument)
     (the economic.climate of today is good)
 then (the current.task of ?person is design.the.instrument))
```

The generic TellAndAsk expression therefore, is loosely of the form:

(the attribute of unit is value)

Symbols preceded by a question mark denote variables, which may be used to bind references to units or to slots (attributes). Rules may themselves contain Lisp code to evaluate data which is not accommodated by the TellAndAsk syntax. Rules may be used in forward or backward-chaining mode, the mechanism of use being determined by the way in which inference is triggered—by making assertions or by posing a query. Mixed mode chaining is also possible.

A relatively powerful feature of KEE is its 'Worlds' facility. This allows many possible solution worlds to be considered in parallel. As worlds are generated, deductive rules consider the consistency of a solution and may 'poison' the world if it is found to be invalid. Once a world is poisoned in this way, it can no longer be considered for expansion. These deductive rules constantly monitor the status of the antecedents to check whether the consequents of the rule in question are still valid after it has fired. Whenever a fact is asserted or removed (which negates a previous conclusion) the conclusion status changes, i.e. non-monotonicity is built in.

Perhaps the biggest selling feature of KEE is its user interface which includes a good developer interface and excellent graphics support. KEE provides both KEEpictures (a library of object-oriented graphics primitives) and ActiveImages (a library of graphical dials, switches, graphs, etc.) which can be attached to slots of units. The ActiveImage can be used either to monitor the value of a slot, or to provide user input of values to a slot by 'mouse-activating' the image. For the developer, KEE supplies a heavily window-based interface comprising of unit and slot displays, inheritance hierarchy graphics, rule editor, etc. Every unit or slot reference shown on the screen can be mouse-activated to manipulate the representation of the entity. As debugging aids, a variety of graphical rule traces are provided for forward and backward chaining, method code traces, single stepper for rule evaluation in addition to those aids provided by the use of ActiveImages. At the time of writing, KEE does not provide, as standard, 'bridges' to external databases or to other languages, other than those supported by the Lisp environment. However, there are additional add-on packages that are available to cater for a variety of needs. Examples include KEEconnection (for attachment to SQL databases), KEE/C Integration Toolkit (for integration with the C language) and SimKit (for simulation).

The KEE system is a large, mature and powerful suite of programs which makes it ideal for the implementation of large and complex expert systems, and it has add-ons to support database management interfaces and communications functionality. It is good for rapid prototyping of applications by virtue of its intuitive rule format and user interface. The system is under constant development by the designers, Intellicorp, and is therefore likely to maintain the kind of support necessary for this kind of toolkit.

3.2.2 Inference ART (Inference Corporation)

ART is generally considered to be of the same 'class' as KEE. It is a relatively large system, originally written in Lisp. It has much in common with KEE by virtue of its rich variety of representation paradigms, object-oriented programming and use of rules. It supports, as does KEE, a powerful and 'friendly' user interface, and therefore would be used in the same range of situations as KEE, i.e. in large and complex problem-solving domains. ART-Lisp is the original ART, in which the knowledge representation techniques of facts, schema, forward and backward-chaining rules and hypothetical reasoning are incorporated. The system has been available since March 1985 and has since undergone continual development, becoming one of the most widely-used expert system development packages worldwide. As the name suggests, ART-Lisp is written in Lisp, and therefore benefits from the symbolic manipulation capabilities inherent in that language. Somewhat later, the company brought out the ART-IM system, which was designed to address the market for a 'mainstream' information technology applications platform with a significantly faster response to the earlier versions. In ART-IM, the inference mechanism is written in C and is deliberately designed to be small, fast and therefore eminently 'embeddable'. Evidence suggests that the system almost evolved from the changing needs of previous users, who were committed to the use of AI in the solution to their domain-specific problems, but were severely constrained by the performance of the system in high-volume information-technology applications.

ART is a system in which all data structures and knowledge are developed in a text editor (of the EMACS style), which is then compiled. This is actually not so tiresome as it may appear, since the development editor and the ART language support incremental compilation.

The base representation in ART is the 'fact'. A fact is a simple database entry of the form:

```
(power-dissipation component-12 500)
```

For certain models, it is more appropriate to collect a series of facts into 'schema' (or frame, unit or object) structures. A schema is a collection of related facts, which might have the appearance:

```
(schema output-module-12
        (instance-of output-module)
        (number-of-integrated-circuits 5)
        (stock-number PA315)
        :
        )
```

Schema organisation is not only easier to comprehend and compare, but it also permits exploitation of inheritance structures and generic representations. Thus, a generic output module might be described, which is segmented into subclasses, and eventually described in terms of specific individual output modules. Schema may have simple 'slots' which hold data, e.g. the number-of-integrated-circuits and stock-number slots as indicated above. Such slots may be attributed information which describes natural language assertion (input and output), templates and cardinality (how many values are in the slot). Ultimately, however, all schema details are compiled into facts within the database, by using an algorithm which tags individual patterns within schema/facts and associates them (with a pointer mechanism) with the rules which may assess or reference them.

Alternatively, schema may be described as having relations to other objects. Relationships may be attributed an 'inverse identifier'. If this is performed, then whenever the forward relation is asserted, the inverse is also asserted (or *vice versa*). The domain and range of relations may also be specified. Thus, an output module and an instrument might have a relationship called 'assigned-configuration', which would indicate the instrument type destined to incorporate that specific type of output module:

```
(schema assigned-configuration
        (instance-of relation)
        (inverse allocated-output-module)
        (is-a (?domain) (output-module))
        (is-a (?range) (instrument)))
(schema output-module-12
        (instance-of output-module)
        (number-of-integrated-circuits 5)
        (stock-number PA315)
        (assigned-configuration PPX15))
(schema PPX15
        (instance-of instrument)
        (allocated-output-module output-module-12))
```

Relations may be described as being inheritance relations, allowing information to be inherited between entities through relations other than merely `is_a` and `instance_of`. In this instance, the relation may be given a 'transitivity slot', which describes the route through which inheritance should be considered.

Basic ART rules are of the form:

```
(defrule<name>
        <if conditions>
        =>
        <then actions>)
```

but pattern-matching with schema is also possible by either using triplet equivalents or using the following:

```
(defrule<name>
        (schema ? sch-name
        (slot1 slot-val1)
        (slot2 ?slot-val12))
        => <actions>)
```

In addition to these capabilities, facilities exist to cater for fact assertion/retraction and modification, rule salience declaration and manipulation (used to stratify the rules into classes and permit priority setting), powerful pattern-matching and inclusion of logical constructs ('and'/'or'/'not' allowed in rule premises).

ART includes a feature known as 'viewpoints'. Viewpoints are used for hypothetical situations and includes a method of dealing with time variants. In fact, viewpoints are perhaps the most powerful feature of ART being similar in some ways to De Kleer's assumption-based truth maintenance system (De Kleer, 1986), which can support very complex hypothetical reasoning.

3.2.3 Goldworks (Gold Hill Computers Inc.)

Goldworks is marketed as a high-end expert system development tool. It certainly qualifies for that title by virtue of its hybrid nature, its mixture of paradigms (frames, rules and object-oriented programming) and its Lisp software platform. It is possible to run the system on a range of high-quality computing platforms, but the system was originally designed as a good PC-based expert system toolkit. Its main market was seen to be the high-end of the PC applications, and its reputation owes much to its 'accessibility' through that avenue. In some ways, however, it has 'outgrown' its very credible background to the extent that, although it is rightly marketed as a PC-based system, details such as the amount of memory the system has, the type of processor on which it is based and the hard disc capacity are crucial issues in the choice of a suitable hardware/software configuration.

Central to the system is a knowledge base comprised of frames, rules and objects. The inference mechanism, or engine, is an obvious central component, but perhaps more significant are the system components. Goldworks is equipped with three interfaces: a menu interface for high-level system building, interrogation and use; a developer's interface for lower-level and much more detailed application development; and finally an interface to external software. This latter facility has found great use in modern expert system applications where information is often required from existing computer-based sources, such as databases and spreadsheets. Two components of the structure are intimately associated with them: the windows and the graphics toolkit component. Goldworks operates within an environment which uses concepts such as windows, icons, mouse and pull-down menus (WIMP). For this, it needs a graphics-based interface capable of supporting such things and, perhaps more importantly, the windows interface needs to be portable across the range of computing equipment on which the software is designed to operate. On the IBM PC, the interface platform is Microsoft Windows, but the standard is quite different for the other two manufacturers and operating systems shown on the system architecture diagram. Apart from the difference in windows interface standards, each hardware platform is associated (as far as Goldworks is concerned) with its own Common Lisp system. On the IBM machines, the system is Gold Common Lisp, on the Apple Macintosh it is Allegro Common Lisp, and the Sun range of computers also has a Common Lisp. Each Lisp system has its own interface to the graphics routines specific to its hardware platform, and this in effect reduces the problems faced by the Goldworks designers in their quest for a portable system, i.e. Goldworks 'runs on top of' the Common Lisp software platform, and due to the syntax commonality imposed by that standard, a program written in one Common Lisp will run on another (in theory).

Goldworks is itself written in Common Lisp, and the inclusion of a Lisp interface adds enormously to the power and applicability of the system, especially in situations which require modifications to the functionality of the inference mechanism, etc. An interface to the C language is also included via the External Program Interface (EPI). This allows C applications to be linked to Goldworks and also allows the two software systems to share functions and data. In fact, it is possible to arrange for C programs to call Goldworks code, thereby facilitating 'embedded' applications of the expert system. Lisp, as a symbolic language, is discussed in more detail later in this chapter.

Knowledge structures in Goldworks are grouped into nine entities called objects. These include frames, instances of frames, relations, rules and rule sets, and assertions. Special objects known as attempts, sponsors and agenda items are used in the inference process.

The difference between frames and objects (in the object-oriented programming sense) in the system is mainly down to the use of Lisp functions embedded within frame structures to accommodate local program control. These functions, which are termed 'methods' in conventional object-oriented systems, are called 'handlers' in Goldworks (see the section later in this chapter on object-oriented programming). The handlers are used via message passing and individual handlers may be inherited down whatever frame hierarchy exists within a particular application. Although handlers provide the functionality necessary for a true object-oriented system, the basic frame structure is common to the frame and object constructs, i.e. an object is a frame with an attached function, and any frame can be equipped with such an attachment.

The basic structure, then, of a frame incorporates the following concepts:

Frame   A data structure which may or may not hold values. In Goldworks, frames do not store values—only instances of frames can do that. In effect, frames are used as templates for classes of objects in a hierarchy, and instances represent specific instantiations.

Top Frame An important concept which ensures that whatever hierarchy is built, there is always a common 'ancestor' (a root).

Slot   Slots are descriptions of attributes within a named frame. Frame instances contain specific values for those attributes.

Lattice   The Goldworks term for the inherent modelling structure. It incorporates a hierarchy which supports multiple inheritance. The way in which frames are positioned in the lattice is by assigning parent and child frames. The hierarchy ensures that inheritance is possible from parent frames , but in Goldworks, multiple parents can be assigned. It is not possible, however, to inherit 'up the hierarchy' (if this would ever be required).

As an example, consider a possible domain model for an electronic instrument distributor, who markets a range of high-technology devices, of third party origin, to a range of customers. In addition to acquiring knowledge of the market potential for a particular product and the subsequent cost of investment, he must also be aware of the possible manufacturers of the product and their ability to provide a good service. A top level frame (below top-frame) could be Instrument. Its structure might be:

Frame name: Instrument
Parent frame: Top-frame
Child frame: Oscilloscope
               Temperature-gauge
Slots: Weight
               Power-supply
               Input-impedance
               Cost-per-unit

The child frames would then inherit all the slots of the parent 'instrument' frame, but each would require their own 'local slots'. In the case of the oscilloscope, additionally relevant slots might be bandwidth and storage, and for the temperature-gauge they might include maximum-temp, auto-scaling, mode-of-display, etc. In fact, for electronic instrumentation, there exists a whole range of characteristics and figures of merit which can be associated with the 'generic' device, and there are a lot more characteristics which can easily be associated with individuals within the class of instrument. It is also possible to provide child frames from each of the instrument classes into individual subtypes having their own local attributes. In terms of the manufacturing company frame (which would also be a child of top-frame), relevant slots may be:

> Product-range
> Quality-control
> Manufacturing-capacity
> Manufacturing-site-location

Using the frame lattice structure and the knowledge-base building capabilities of Goldworks, it is quite a simple task to build a representative model of the company's activities. However, there are additional features of the frame representation used in Goldworks which have an important bearing on its functionality. The most important of these is the slot facet.

Facets are used to control the range or type of values that can be placed into a slot, and can be inherited in the normal way. They can be used to make an association between a slot value and a Lisp function. There are many possible facets in the range provided, including ones to set constraints on slot values, ones to set default values to slots, ones to set default certainty values, and a range of facets associated with on-line documentation and explanation. However, there are two particularly important facets which can significantly extend the functionality of any application built with Goldworks. These are:

- when-accessed
- when-modified

The 'when-accessed' facet causes a Lisp function (or functions) to be called when a slot is accessed. This can be used to cause a slot value to be computed when required, i.e. the attached function(s) must return a value for the slot. The 'when-modified' facet gives the system a kind of 'spreadsheet functionality', i.e. a selected value, whenever modified, will always cause some other actions to be performed (automatically). This could be the automatic readjustment of slot values which are either dependent on or directly linked in some way to another slot value. A simple example of this may be a person's salary which has a direct effect on that person's credit worthiness. Therefore, in a credit assessment system, any change in that kind of detail has an immediate and predicable effect on the credit worthiness value.

Facts that are believed by the system to be true are termed 'assertions' in Goldworks. In the most simple case, one type of assertion is a value placed within a slot of a frame instance. In this situation, a fact has been asserted as a 'slot-value assertion' and is known as a 'structured assertion', i.e. in this case it is structured into the frame lattice. Unstructured assertions are also catered for. These are assertions which are placed directly into the 'assertion base' without the use of instances and can be regarded as global facts. For example, within a digital electronics design system, the power supply voltage for conventional transistor–transistor logic (TTL) could be fixed at 5 V. Therefore, an unstructured assertion may be entered of the form:

```
(TTL-power-supply-voltage 5)
```

Object-oriented programming is incorporated into the Goldworks system by virtue of named procedures which can be attached to frames and used to operate on instances within the application. Once defined, these procedures (or methods) can be used to perform calculations or other data processing tasks associated with a particular type of object which a frame represents. In this way, the processing specified in the method can be invoked for that object via a 'message' passed to the object. A syntax of a message is:

```
(send-msg instance :handler {arg})
```

which will send a message to invoke the handler procedure attached to the frame instance. For example, if an instance of digital IC (say IC-7447) may require a mechanism for dissipating heat, a handler attached to the instance may be capable of determining whether a heat-sink is required for a particular application, and therefore what type should be used. The handler could be defined as 'heat-sink-if-needed'. The appropriate message to that instance would be:

```
(send-msg 'IC-7447 :heat-sink-if-needed)
```

For this simple case, no arguments are necessary. Handlers can also be invoked by rules.

The Goldworks system is equipped with a knowledge-base browser for viewing frames and frame lattices. The tool provides a familiar family tree hierarchy. In the context of frames and objects, the browser is used to display the gross structure of the lattice, whilst the 'Inspector' tool is used to examine a specific object's details.

Goldworks uses first order predicate logic in its rules. The basic syntax of rules is the familiar 'if–then' form, but complex relationships can be represented and manipulated using the logical constructs of 'and' and 'or'. As an example, consider a simple frame describing certain possible attributes of a company employee, and an instance of the frame, called 'Bill-Bloggs'. The structure can be represented as:

```
employee:              Bill-Bloggs
position:              Chief-Engineer
qualification:         PhD
progress:              good
promotion-potential:
```

The following rule might well be used in determining Bill's promotion prospects:

If
```
INSTANCE ?staff IS employee
        with qualification PhD
        with progress ?progress
```

Then
```
        INSTANCE ?staff IS employee
              with promotion-prospects ?progress
```

where the question mark prefix denotes a variable. For example, in the above rule, the variable '**?progress**' will be bound to the symbol 'good'.

This rule basically states that if a particular employee has the right qualifications, his (or her) promotion prospects are directly related to the known progress within the company.

This simple rule can be used to demonstrate some additional important concepts. The first is the notion of pattern-matching. Goldworks incorporates a sophisticated facility which matches patterns in rules to patterns representing facts in the knowledge base, to patterns representing goal states and to other objects (such as instances) in the system. Within a rule, it is possible to match against the kind of pattern shown in the antecedents of the simple rule above, which has three clauses, but it is also possible to match against:

```
        and        <pattern>
        or         <pattern>
        unknown    <pattern>
        retract    <pattern>
        print-out  <pattern>
```

And more sophisticated ones which can be used to invoke Lisp function evaluations, etc. As far as **and** and **or** is concerned, the simple company employee rule shows an implicit **and** relationship in the rule antecedent, and the same would apply to the consequent in a more sophisticated example. However, in cases where **or** relationships are required, these must be used explicitly. Both **and** and **or** can be used together to configure complex logical entities, and the use of parentheses allows for processing precedence.

Rules in Goldworks are capable of dealing with dependency. They can either be logically dependent or they can be dependent on a state change. In fact, the system employs a method of truth maintenance based on logical dependency, rather than the more sophisticated assumption-based truth maintenance systems as used in KEE. If the antecedents of a rule no longer match, then the assertions which result from the pattern match are either retracted or retained. The actual process depends on whether the rule is logically dependent or not, as determined in the rule's 'dependency status'.

The certainty, or belief of assertions is handled in Goldworks with a numerical range from 0.0 to 1.0 inclusive, where a certainty factor of 1.0 represents a known true value. The certainty factors associated with the various rules invoked during the inference process are combined by a certainty function resident in the system. The combining function operates by taking the lowest value (minimum) of all the antecedent assertion certainties, and multiplying it by the certainty of the rule. The algorithm is:

$$\text{(min } (\text{<cert-asser1><cert-asser2>}....)) * (\text{<cert-rule>})$$

So this means that each assertion, which is matched in the antecedent part of a rule, is compared. The minimum, or least certain value is chosen and multiplied by the certainty of the rule itself. For example, in the simple employee rule, there may be some doubt as to his progress; it may be assessed as good with a certainty of 0.8. His qualifications may be in doubt, say in the more subjective slot description of PhD-standard. This may be assessed at 0.9. If the overall promotion-prospect rule has its own certainty of 0.7, then the promotion prospects of our candidate are now good with a certainty of: ((0.8)*0.7))= 0.56.

A rule editor exists within the system to facilitate the building of complex rules, but it is also possible to define them from the developer's interface. A more useful and complete employee prospects rule would perhaps be:

```
(DEFINE-RULE EMPLOYEE-PROSPECTS
     (       :print-name "EMPLOYEE_PROSPECTS"
             :doc-string "very subjective and inaccurate
                          assessment of worth"
             :dependency T
             :direction :FORWARD
             :certainty 0.7
             :explanation-string
                   "Academic achievement and progress are used to
                   assess promotion prospects"
             :priority 0)
     (INSTANCE ?STAFF IS EMPLOYEE
                          WITH POSITION CHIEF-ENGINEER
                          WITH QUALIFICATION PHD
                          WITH PROGRESS ?PROGRESS)
THEN
     (INSTANCE ?STAFF IS EMPLOYEE
                          WITH PROMOTION-PROSPECTS ?PROGRESS))
```

which states that the rule has a logical dependency, it is a forward-chaining rule, and has a rule certainty of 0.7. A text string has been added to aid in documentation and some text has been included to cater for explanation. Priority is used in Goldworks to assist the application-builder in assigning rule-firing priorities during inference. Goldworks also maintains what it calls an agenda. All rules whose antecedents match patterns in the assertion base are placed as items on this agenda. Rules may be grouped into rule sets, and the inference process can be further partitioned and controlled in a hierarchy of 'sponsors'. A sponsor can be designated for a particular subtask in a large and complex application, in order to maximise efficiency.

Goldworks supports three inferencing techniques:

- forward chaining,
- backward chaining, and
- goal-directed forward chaining.

It also allows for a mixture of forward and backward chaining in the bidirectional mode. Rules and rule sets can be defined in terms of their inference direction. In forward chaining, a line of inference proceeds from assertions already existing in the assertion base to a possibly large range of final conclusions. In Goldworks, a successful match of antecedent patterns with patterns existing within the assertion base will cause an agenda item to be placed on the agenda. When this fires, the rule consequents assert new facts, and these in turn can cause more agenda items to be entered through the matching of other rule's antecedents.

Backward chaining is seen to be more efficient, in that a goal is asserted, i.e. the system knows that it has to reach that goal or prove its truth through the inference process. The inference engine attempts to fire the rule which provides the stated goal satisfaction, but may find that the antecedents cannot be matched. In this case it will attempt to match these antecedent patterns by looking in the consequents of other rules which would provide the necessary patterns. This process continues (in a backwards fashion) until one of the rules in the chaining path can fire. After this occurs, the system runs in a forward direction, successively supplying new assertions to all the relevant rules until the final rule fires to satisfy the 'top level' goal. Goal-directed forward chaining facilitates rule sets to be defined as forward-chaining rules, and therefore can be used as rule sets in a backward-chaining inference process.

Goldworks is classified as an expert system development environment because of its integrated suite of development tools, its widespread use of graphics and its ability to support the full process of expert system development from initial prototyping through to applications delivery (via the run-time package). It is not the only high-quality system to boast such facilities. KEE has a very similar 'feel' and in some ways sells to the same market. Goldworks uses windows and graphics, but it also uses hypertext techniques to smooth the transition from one system window to another and provide what appears to be a well-integrated environment. Facilities existing within the Goldworks system to support the development process include:

- A menu interface, which facilitates the definition and testing of frames, frame instances, assertions and rules. Applications can be built from conception to delivery (via the run-time package) from this interface without recourse to the underlying Lisp platform.

- A development interface, which permits a more detailed and complex development through access to Lisp functionality.

- The Lisp system itself. All Common Lisp systems, by virtue of their size, are equipped with their own development interfaces. They include sophisticated editors, error monitors and debugging aids. Some even have their own interfaces to external software and systems.

- The External Program Interface (EPI), facilitated mainly via the Lisp platform.

Goldworks' main disadvantages are related to its sheer size and operating speed. It does not have the multiple worlds of KEE (although it is possible to emulate it via programming), and its truth maintenance system is not quite as powerful as some comparative systems. In spite of this, Goldworks represents an ideal choice of system for serious expert system application development.

3.2.4 KES (Software A&E)

KES (Knowledge Engineering System) is a well-established commercial expert system shell written in C. It runs on a variety of computers including PCs, Sun microcomputers, Vax 11s, Hewlett Packard 9000s and Apollo workstations. The KES software provides multiple inference engines, as its designers see no single approach being well suited to all expert system applications. The inference engines are:

- Production Systems (PS),
- Hypothesise and Test (HT),
- Statistical reasoning (BAYES).

Whilst each of these subsystems is marketed as a discrete self-contained entity, inference model aside, they present the system builder with a number of common facilities.

We will first look at each subsystem in terms of knowledge representation schema and inference model, and then discuss additional features which are important to the development of a working system.

3.2.4.1 KES PS (Production Systems)

The KES PS subsystem uses rules to represent knowledge. Production rules are a popular method of knowledge representation and have a place in the majority of shells currently available. The advantages of a rule-based formalism include a modularised method of expressing chunks of knowledge and ease of incremental development. Rules are particularly well suited to expert system applications where the domain knowledge is already in the form of, or is readily translatable to, branching logic, or if–then rules. The KES PS rule format is:

```
if
        antecedent
then
        consequent
endif.
```

An antecedent is a conditional expression in the form of a logical comparison, which may be true or false. Multiple antecedents can be connected together using the operators 'or' and 'and' to create compound conditions. The concepts contained within rules are termed 'attributes'. These attain values during a consultation with the system, and must be declared, along with their permitted number and type of values, before they are used in rules. In the example shown below, the first attribute—size—is allowed a single value from the set: small, medium, large; whilst the second attribute nuclear features is allowed multiple values from the set: mitosis, bipolar cells, nucleoli. A rule which can determine cell type using these attributes is then given as:

```
attributes:
        nuclear size : sgl
        (small,
         medium,
         large).

        nuclear features: mlt
        (mitosis,
         bipolar cells,
         nucleoli).

        cell type :sgl
        (benign,
         malignant).
```

```
rules:
            Rule1:
            if
            nuclear size = large and
            nuclear features = mitosis or nucleoli
            then
            cell type = malignant.
            endif.

    %
```

The inference mechanism of KES PS attempts to find a value for a goal attribute by backward chaining through the rule base. It does this by first checking for all those rules which can assert the goal attribute's value in their consequent. Rules are consulted in the order which they appear in the rule base. The first rule is selected, and the inference engine begins by attempting to evaluate the rule's antecedent. If this is true, then the rule fires and the goal attribute is asserted the appropriate value. If the antecedent is false, the rule does not fire and the inference engine proceeds to the next rule.

It may happen that the values of the attributes in the rule antecedent are not known at the time of checking. In this case they cannot be immediately evaluated, and the inference engine must seek to determine the unknown values before it can finish evaluating the rule. Seeking a value in this way becomes a subgoal of the inference process. Subgoals are treated in the same manner as goal attributes, and can reveal further subgoals. When an attribute is found to have no rule which can be used in determining its value, the end-user is asked to provide a value. This phenomenon of chaining backwards, from goal to subgoal to further subgoal etc., is termed 'backward chaining'. For a more detailed explanation of backward chaining see *An Introduction to Expert Systems* (Jackson,1986).

The reverse of backward chaining is 'forward chaining' or data-driven inferencing. When any attribute becomes asserted with a value, the inference mechanism checks the antecedents of all rules. If any of these evaluate to true, given the new assertion, the rule is fired. If in turn this results in new assertions being made which match other rule antecedents, new rules may be fired and the process will chain forward. KES PS provides for forward chaining via the use of 'demons'. These identify rules which are specifically designed to act in a forward-chaining manner:

```
Demon1:
when
        patient age = over 80 years and
        cancer type = low grade
then
        treatment = none.
endwhen.
```

This type of rule will fire only when its antecedent becomes true, and no attempt is made to actively seek the values of the attributes contained within its antecedents. These rules are not included in the process of backward chaining.

Uncertainty is handled in KES PS by 'certainty factors' (*CF*s). These are based upon a numerical scale ranging from 1.0 to −1.0, and are intended to express the strength of belief in the presence of a specific attribute value. A *CF* of 1.0 for a value, indicates an absolute belief in the truth of that value. Conversely, a *CF* of −1.0 indicates an absolute disbelief in its truth. *CF*s are used in two ways. Firstly, to determine whether there is sufficient evidence to fire a rule, and secondly to express varying degrees of confidence in the consequent of a rule. In complex assignments involving multiple attributes and values, KES employs an arbitrary mathematical function to calculate their resulting *CF*s.

KES PS allows the grouping of objects that are related in some way using classes. Objects belonging to the same class, i.e. those sharing a common set of characteristics can be expressed using 'class inheritance'. This permits the organisation of classes into hierarchical relationships, with the most general classes being placed at the top level of the hierarchy and specialised suclasses at the lower levels. A subclass is a more specific version of its parent class, in that it contains all the characteristics of its parent, but may possess additional features that describe it more precisely. The format of the inherit clause is:

```
class name: [inherits: parent class name]
             attribute declarations ......
endclass.
```

An example use of this clause is shown below:

```
classes:

    infiltrating carcinoma :

            attributes :
            grade :              sgl (grade1, grade2, grade3).
            mitosis :            sgl (low, moderate, high).
            lymph node spread :  sgl (present, absent).

    endclass.

    infiltrating ductal carcinoma : (inherits : infiltrating
                                               carcinoma ).

            attributes :
            type :               sgl (papillary, mucoid,
                                       medullary, tubular).

    endclass.

%.
```

Here infiltrating ductal carcinoma is a specialisation of infiltrating carcinoma. It possesses inherited attributes, grade, mitosis and lymph node spread and newly defined attribute type (i.e. type of carcinoma).

In addition to string expressions, KES also supports numeric expressions. These can be composed of numeric constants, literal numbers , numeric functions and arithmetic operators. For example, the following expression computes the area of a circle.

```
3.14159 * circle radius².
```

Numeric expressions can appear in assertions of values belonging to integer or real type attributes or calculate clauses of integer or real type attributes, either in rule antecedents or consequents.

Pattern-matching in strings is useful in interpreting input values, and also to format output values. KES provides a number of clauses for use within pattern templates, to give greater control over the matching of substrings to variable components. These include 'alternatives', which allow the definition of a set of strings from which matches must be made; 'spans', which allows the matching of one or more strings listed in the clause; and the 'atleast' clause, which restricts the length of a substring that can match a variable component. Patterns are defined:

```
pattern name:
        [ component,..., component ].
```

where the components consist of variables, literals and clauses. For example patient status input may be matched to one of three alternatives:

```
patient status :
        [ alternatives : "improving", "stable", "deteriorating" ].
```

The system builder is provided with facilities to direct the order of rule evaluation and control dialogue between the end-user and the expert system, via commands in the 'actions' section. These commands are analogous to instructions in a computer language, although they present in an English-like formalism. Commands are of three types:

- Those which perform a task or cause information to be displayed. For example the 'askfor attribute' command which directs the inference engine to ask the end-user to provide a value for the specified attribute; 'assert attribute = value' which assigns a value to the named attribute, and 'obtain attribute' which causes the inference engine to immediately attempt to determine the value of the specified attribute.
- Control commands which determine the state of the actions section, e.g. 'break' causing the actions section to stop executing and take direction from the end-user, 'nextcase' erasing all attribute values and restarting the action section at the beginning, and 'stop' which terminates the session.
- Conditional and looping commands for controlling the order of command execution in the actions section. These include commands found in ordinary procedural languages such as 'if', 'forall' and 'while'. Conditional and looping commands can be nested to any level.

To aid the system developer in presenting an effective end-user interface, 'textual attachments' can be utilised. They allow the developer to associate supplementary textual information with a rule, an attribute or its value, and the knowledge base as a whole. Textual attachments can be displayed in a number of ways. The display command can be used in the actions section or rules section to display the text of attachments defined in the following way.

```
{attachment name: "text", "...", "text"}.
```

For example, we can specify some information concerning clinical stage, that can be displayed to the end-user at the appropriate time.

```
{clinical stage : "There are two aspects to clinical stage, they are",
                          " ...          ",
                          " ...          "}.
```

Alternatively, question text can be associated with an attribute, to be presented to the end-user when the system queries for that attribute value. For example:

```
pain: sgl
(yes,
no)
{question: "Does the patient feel pain in the area of the lesion?"}.
```

A query will then be displayed :

```
        Does the patient feel pain in the area of the lesion?
            1. yes
            2. no
            =?
```

Note that the numbering of the permitted values is automatically produced by KES when prompting for a value. This text-based end-user interface will be upgraded into a window-based interface in the next version of KES PS.

The explain and why attachments can also be tagged onto an attribute declaration in the same way as above. They will be displayed whenever the end-user issues the response why or explain in answer to a query.

It is possible to exchange information with other KES expert systems or user defined applications using the facility 'externals'. This allows one expert system to request information from another, the calling of a database to provide pertinent data, and the passing of control to the operating system. The 'read' and 'write' commands serve to read attributes from, or write attribute to, an external file in the format:

```
        read file name
                attribute or class
                attribute or class
                etc.
```

For example :

```
        read patientfile :
                age,
                previous therapy.
```

The 'message' command permits formatted messages to be exchanged between files. A KES expert system can be called from within another application through use of the 'embedding' command. This provides the user with the possibility of creating graphical interfaces, embedding an expert system as a module under control of another application, or adding an intelligent component on an otherwise traditional program. Embedding must be carried out using the C language and C routines provided within the KES package. In practice, this requires a good understanding of C, and an up-to-date version of the Microsoft C compiler.

KES provides no environment for system development. The application developer must first create a knowledge base using his/her own text editor, parse the file using the KES parsing program (this carries out syntactical plus some semantic checking), and finally run the parsed file using the KES run-time program. This process must be repeated each time any alterations or additions are made to the knowledge base.

A number of options can be specified during the parsing and running of the knowledge base. A useful one being the '-m' option. In situations where the default memory allocated to run the system is insufficient, it can be increased using this option. Despite this facility, the run-time program is unable to cope with large knowledge bases, exceeding several thousand lines, especially if heavy use is made of textual attachments. Additionally KES does not support modularisation of the knowledge base into separate text files. This can be a problem in developing, debugging and maintaining large applications. Execution of the run-time package is acceptable given small- to medium-sized knowledge bases, but again may become a problem with larger systems.

In conclusion, the KES PS system is a robust, easy to use, well documented and supported expert system shell, suitable for fast prototyping. It also provides a number of important control facilities necessary to the serious applications developer.

3.2.4.2 KES HT (Hypothesise and Test)

There are a number of problems associated with the use of production rules in some situations. Briefly these include:

- rules do not separate expert knowledge from the reasoning process;
- rules cannot conveniently describe the combinations of possible antecedents that may result in a consequent; and
- multiple simultaneous disorders are difficult to recognise.

In an effort to alleviate some of these difficulties the developers of KES PS produced the HT system. Hypothesise and test or abductive reasoning has been shown by cognitive psychologists to be very close to the way humans naturally solve a particular class of problems. The HT system is well suited to diagnostic, classification and selection problems.

KES HT represents knowledge in the form of frame-like formalisms, which describe a syndrome or disorder using a number of statements, in the format:

```
[description:
        statement-1,
        ............,
        statement-n; ].
```

Each statement consists of an attribute and its possible values. These attributes and their permitted values must declared prior to their use, in the the attribute section as described for KES PS.

Unlike the PS subsystem which treats uncertainty as numerical CFs, the HT system provides 'symbolic certainty factors'. These are intended to provide rough estimates of the frequency with which something occurs. Five symbols are given:

```
<a>  =  always
<h>  =  high likelihood
<m>  =  moderate likelihood
<l>  =  low likelihood
<n>  =  never
```

Symbolic certainty factors have two uses in KES HT. Firstly, they can cause a disorder to be rejected and so removed from any further consideration by the inference engine. Secondly, they can be used to calculate the scores of those disorders considered as possible alternatives at the termination of a session.

An example of a disorder description and use of symbolic certainty factors is given below:

```
flu <m>
        [description:
                temperature = raised <h>
                symptoms = runny nose <h>, cough <l> ].
```

This example states that flu is a moderately common disorder, and that a raised temperature and runny nose are highly likely features, whilst the presence of a cough is less likely.

The designers of KES HT claim that its inference mechanism acts much like a human diagnostician. For example, if a patient presents to the doctor complaining of a fever, the doctor will form a hypothesis about the patient's underlying illness. They will then proceed to gather more information from the patient in order to confirm or disprove this hypothesis. This process will continue until the doctor is satisfied that all relevant information has been collected. KES HT operates in a similar manner. Given an initial set of manifestations, i.e. attributes and their values, hypotheses are created to explain the first attribute value (called a manifestation in KES HT). Each of the manifestations is then considered in turn, and the hypotheses updated so that they explain each additional manifestation. When all initial manifestations have been explained, HT looks for additional manifestations (usually by asking the end-user) that can be used to refine its hypotheses. A simple example will illustrate the behaviour of the HT inference mechanism.

Consider a small medical expert system containing knowledge of four common diseases: cold, flu, allergy and streptococcal throat infection. A patient presents with symptoms of runny nose, high fever and sneezing. The inference mechanism chooses to explain the symptom runny nose first. To do this it creates a set of hypotheses called a 'FOCUS' which is a preliminary solution given the initial manifestation of runny nose. Each hypotheses within the FOCUS will contain a disease which can explain the runny nose e.g.:

```
{ cold } | { allergy } | { flu }
```

The symbol '|' stands for **or**, indicating that either a cold or flu or allergy can explain the symptom runny nose. HT then may proceed to consider the symptom high fever. All hypotheses in the 'FOCUS' that cannot explain the new symptom are removed. Colds and allergy do not cause high fever therefore they are removed:

```
{ flu }
```

Finally sneezing is considered. The only remaining hypotheses left in the FOCUS, flu, cannot explain sneezing. Therefore it follows that more than one disease must be causing this combination of symptoms. The inference engine then looks for all sets of two diseases that collectively account for this combination. The FOCUS is then updated:

```
{ flu & allergy } | { flu & cold } | {strep throat & allergy }
| { strep throat & cold }
```

The symbols '&' stands for **and**, indicating for example, that flu and allergy is one hypothesis that explains all the symptoms.

Once KES HT has explained all the initial symptoms in this manner, it will look for additional ones to help confirm or reject one of the hypotheses in the FOCUS. This continues until the diseases contained within the FOCUS have values attached to all their attributes. When termination point is reached, the remaining diseases are scored using their symbolic certainty factors. This provides a measure of fit between the symptoms found and those listed in the knowledge base descriptions. This inferencing algorithm is based upon the mathematical technique of minimal set coverage (Reggia *et al*., 1983).

An additional feature found in KES HT is setting factors. These mark manifestations that do not need to be explained in the final hypotheses. Setting factors often represent normal or default conditions. The presence of a setting factor will not cause a disease to be included in the solution, but it may cause some diseases to be rejected. Setting factors are indicated in the knowledge base by means of an asterisk.

The KES HT system does not provide classes or inheritance as in the PS version, nor does it allow the use of backward or forward chaining. However, it utilises similar action commands, textual attachments, externals and embedding facilities.

The HT system is particularly useful for a certain class of problems, providing an alternative to production rules and backward chaining. The success of an expert system built using this shell is based closely upon the ability of the knowledge base author to exhaustively list all possible manifestations under a disease description. Failure to achieve this may result in 'unexplained' findings' which disrupt the inference process.

3.2.4.3 KES BAYES

The KES BAYES subsystem performs statistical pattern classification based upon Bayes theorem. This approach is applicable only to situations where an extensive body of information expressed as probabilities already exists or can be compiled by the system builder.

Knowledge is represented as tables containing 'prior' and 'conditional' probabilities. Prior probabilities give the probability that a hypothesis is true in the absence of any specific evidence. Conditional probabilities state the probability of a hypothesis being true if the symptom or manifestation is true.

In the example below, prior probabilities appear in angular brackets immediately after the name of each value of the inferred attribute. Conditional probabilities are placed in a list after the prior probabilities:

```
        attributes:

            nuclear shape : (sgl)
            (round,
             irregular).

            features : (mlt)
            (bipolar cells,
             mitoses,
             nucleoli,
             chromatin).

            cancer grade : sgl
            (high              <0.254>
             0.25   0.82
             0.51   0.96 0.48
            medium <0.562>
             0.62   0.42
             0.73   0.58 0.62
            low      <0.899>
             0.98   0.43
             0.82   0.43 0.21 ).
```

Here, the table specifies that high cancer grade has a prior probability of 0.254, and conditional probabilities of 0.25 (nuclear shape—round), 0.82 (nuclear shape—irregular), 0.51 (features—bipolar cells), 0.96 (features—mitoses), 0.48 (features—chromatin).

This system will give reliable results provided that the data used is accurate. If empirically determined data is not available, it may be possible to use an expert's subjective estimates. However, this may cause problems, as studies have shown that experts estimated probabilities may vary widely from those actually occurring. The method by which KES BAYES utilises Bayes' theorem to calculate the final hypotheses probabilities is complex and will not be described here. For further details see Chapter 9 on the mathematical foundations of AI.

The BAYES subsystem does not provide inheritance, textual attachments, an actions section or explanation facilities as these are not compatible with its method of inference. It has limited applicability to those areas where vast amounts of accurate probabilistic data is available.

3.2.5 Crystal (Intelligent Environments Ltd)

This product differs from the previous ones in its structure, implementation language and applicability. It is written in the C programming language and falls into the 'low-end' expert system shell class. In spite of this, it has an excellent user interface, using screen forms and menus. The system is entirely rule-based, i.e. there is no explicit use of relational data structures such as frames, although it is possible to interface to databases and spreadsheets, etc. The rules within Crystal contain knowledge and information, and although this may appear to be something of a limitation, it provides a simplified representational methodology and less of a burden to memory within the computer system. It represents an ideal inexpensive prototyping tool which is capable of standing alone in somewhat less complex problem situations than those approached via KEE, ART or Goldworks. The inference mechanism is essentially backward chaining, although it is possible to provide something resembling forward chaining through judicious design of the rule base. It is a popular, low cost, expert system shell supplied by Intelligent Environments of London. Written in C, it runs on IBM compatible PCs and PS/2s. It utilises a production rule knowledge representation scheme and a simple backward-chaining mechanism.

The system builder must specify all rules using Crystal's in-built rule editor. This is intended to speed development and free the knowledge base author from any preoccupations with syntax. The rule editor presents a series of forms and menus into which rule components must be written. The order of entry is fixed and must begin with the rule consequent, followed by the antecedent list. Logical operations 'and' and 'or' necessary to combine multiple antecedents are generated automatically by the rule editor at the appropriate time, although in a somewhat clumsy fashion. An 'and' is produced by moving the cursor down one line, whilst an 'or' results from inserting a blank text line. The general form of a Crystal rule is shown below. The term 'condition' is applied to tests or actions that have a Boolean outcome:

```
Consequent

    IF

            condition 1   and/or
            condition 2   and/or
            ........
            ........
            condition n.
```

An example rule is:

```
Grade 1 cancer

    IF

            tubule formation is little

    OR      tubule formation is none

    AND     nuclear pleomorphism is marked

    AND mitotic rate > 3
```

Rules must be entered in a tree structure, commencing at the topmost rule. As rules are written, their antecedents can be expanded into further rules or subgoals, to create a hierarchical structure. For example selecting the antecedent 'nuclear pleomorphism is marked' from the above rule (by positioning the cursor above it and pressing the F10 key) the following template is produced by the rule editor:

```
    nuclear pleomorphism is marked

        IF
                ............
                ............
```

The antecedent has now become a consequent of a subgoal. The user can now fill in this rule's antecedent list. This is the only permitted rule structure in Crystal and may be inconvenient to the system developer who wishes to model a more complex system.

Moving between rules is achieved using the 'rule dictionary', which is presented as an alphabetically ordered menu window. Rules or rule segments can be cut, copied and pasted using the rule editor.

Inferencing is carried out by simple backward chaining and no means of forward chaining is provided. Crystal also does not support reasoning with uncertainty, the use of classes or inheritance.

Crystal offers a wide variety of already defined library procedures which can assist in obtaining and testing the validity of rule antecedents and consequents. These include time and date information, type conversion, terminal input and output, array manipulation, pattern-matching, DOS system calls and a range of mathematical functions; all of which are accessed via the function menu and automatically placed using the rule editor.

Crystal provides no procedural language or method of controlling inferencing, although it does recognise two rule types. 'Special' type rules can be used within loop constructs and evaluated and erased numerous times during a consultation session. 'Ordinary' rules can be evaluated only once.

Textual attachments can be placed as either 'conclusion' messages to be displayed by the system when a rule consequent is asserted or 'display' messages which can present text and variable values to the end-user at any time. These are appended by selecting the appropriate command from the menu and filling in the provided text box. Crystal provides no facilities to add specifically tagged explanation or justification text.

During run-time the system presents the end-user with an attractive, full colour, window-based interface. When requesting data from the end-user, the system automatically configures question–answer windows to input single values or menu selections, as required. The developer can tailor this interface, by changing the format of windows, adding question text, creating forms with specified fields, and changing the colour combinations used.

Crystal can be linked to application specific C programs, and allows both direct function calls and the passing of parameters in either directions. Parameter passing is conveniently achieved through memory rather than intermediate files, and can even cope with arrays. However, Crystal is itself the C 'main' function, which allows it to call external C routines, but not be called itself. Therefore it is not an embeddable system. Additional C functions written and linked in by the system builder are automatically detected by Crystal and added to its function library, giving access via the menu interface of the rule editor.

In addition to reading ASCII files, Crystal can interface to dBase III, Lotus 1-2-3, a business graphics package supplied as a Crystal extension and the Phillips range of laser video disc players. Plans also exist for a 'dynamic graphics' add-on, which will allow the creation of animated picture sequences using standard graphics packages.

It has been estimated by its suppliers that Crystal can cope with 1500–2000 rules, which is well within the range of most small-to-medium applications. However, if a large knowledge base is necessary and memory becomes insufficient to run it, Crystal can take advantage of expanded memory. In practice this does not increase its capabilities as much as expected. Expanded memory can only be utilised for the end-user interface screens (which often comprise 50% of the total memory used) not the rules or variables. An alternative option is to break the knowledge base into separate modules and load them separately, using the 'load' mechanism from within a Crystal rule.

In conclusion Crystal is a well designed and documented, easy-to-use expert system shell, which accounts for its high popularity in the United Kingdom. It provides a built-in rule editor and function library, plus automatic production of attractive menu-driven end-user interface. Its impressive interface capabilities and speed of execution make it an ideal choice for business applications. However, for more complex projects, perhaps at the research stage, its lack of forward chaining, procedural control and forced hierarchical rule tree may be a disadvantage.

3.3 SYMBOLIC LANGUAGES

In theory expert system applications can be written in any programming language, although someone who attempted to develop a system in assembler could be described as foolhardy. Many expert system applications have been built using traditional low level languages such as C, Pascal, Basic and Fortran. This is usually done for reasons of speed, efficiency and portability, and often takes place after initial prototyping has been carried out using an expert system shell or one of the specialist AI programming languages such as Prolog or Lisp.

Under the heading of 'Symbolic Languages' (i.e. those which manipulate symbols) we will introduce Lisp (a functional language) and Prolog (a logic language), both of which are well established within the AI community as programming tools. There are numerous other languages which are classed as symbolic, many of which are derivatives or enhanced versions of Lisp or Prolog, however it is not intended to cover those here.

A second section will describe the relatively new paradigm of object-oriented programming. We will look at the concepts of object-oriented design and programming, and then introduce two languages which support these ideas, namely Smalltalk and C++. Again these two examples represent only a small proportion of the object-oriented programming languages available. There is still much debate from those who advocate the use of object-oriented design and programming, concerning what constitutes the essential features of a truly object-oriented language. Those issues will not be discussed here.

3.3.1 Lisp

One of the main events in the history of AI was the conference held in 1956 at Dartmouth College in Hanover, New Hampshire, USA. The purpose of the conference was to discuss the possibilities of writing computer programs that could accurately simulate human reasoning and human behaviour. The major organiser of that fateful conference was John McCarthy, who was then an assistant professor of mathematics at Dartmouth, but went on to found and direct the AI laboratories at the Massachusetts Institute of Technology (MIT) in 1957 and at Stanford University in 1963. McCarthy is thought to have originally coined the term 'Artificial Intelligence', but perhaps more importantly, he is credited with the most famous AI computer programming language in widespread use today, Lisp. His work on differentiation program development at that time was used to identify the requirements of the language, and partly due to the computing equipment he used, the language acquired and retains some very strange syntax. Examples are CAR and CDR, which are built-in functions operating on elements of a list having specific positions in that list. The functions are important and realistic, but the function names are somewhat meaningless in their own right.

One of the language's big problems has been the number and diversity of Lisp dialects. Its capacity for change and extension has lead to many variations, and it has been notoriously difficult to ensure that a program written in one Lisp dialect will run on another. Quite recently, a consortium of academic and industrial users of Lisp began an attempt to define a language standard. The most significant individual in the group of people involved was Guy Steele Jr, who wrote what was to become the Lisp standard text '*Common Lisp: The Language*' (Steele, 1984). All Common Lisp systems adhere to this standard, even though they are usually equipped with their own extensions.

Lisp stands for *LIS*(t)*P* (rocessing) and, as its name suggests, the language is all about representing and manipulating information in lists. Examples of Lisp lists are:

```
(1 2 3)
(Mary Peter Harry Bill)
(resistor capacitor (transistor (field-effect bipolar)) inductor)
```

A list is composed of elements, which are themselves made up of either 'atoms' or lists in their own right. An atom is basically any non-list element. For example, in the lists above, the first list is made up of three numerical atoms and the second contains four literal atoms. The third list is somewhat different from the others. It is a list with a similar structure to the other two, but it contains atoms and lists. Each list (sublist) within the list is made up of atoms though. Looking at the structure of this last list, some sense can be made of it. If we indent it, the meaning becomes clearer:

```
(resistor
capacitor
        (transistor
                (field-effect
                bipolar))
inductor)
```

Another example of this kind of list structure is appropriate to demonstrate its potential in AI systems which would require the modelling of a domain, i.e. well-structured information storage:

```
(company ((board_room design_office stores canteen)
        (office_staff 30)
        (business instrument_manufacturer)
        (main_office Sweden)
        (established 1990)))
```

It can easily be seen from this list, that although it is just a list of elements within other lists, as the other examples, it is also a meaningful and realistic information structure describing some industrial company. Consider the next example:

```
(+ z (- x y))
```

This is yet another list, but this time it has elements which we recognise as arithmetic operators. This is very important in Lisp; data and programs are represented and manipulated in the same way. In this example, Lisp recognises the first element in the list as the predefined function, or 'keyword' +. The plus function is seen to have two arguments; z and another list. In Lisp, the function arguments are usually evaluated first before the actual execution of the function itself. Assuming that x,y and z have values associated with them (bound to them), then z evaluates to its bound value, but the evaluation of the other argument is a list. Lisp looks at the first element of this list and again finds that it is a function. The two arguments are evaluated and then the subtraction function is executed, providing a returned value to the addition function. This type of structure is actually known as a 'form'. A form is defined as a single complete program unit, which may be one line or extend to many. Lisp contains what it calls 'special forms', which handle program control and bindings. They have a predefined structure, take a predefined number and type of argument and normally produce a value (or values). Examples of special forms are:

```
if, quote, setq, catch,
```

For example, the quote special form merely returns its argument unevaluated. If the form:

```
(quote (* (- x y) (+ a b)))
```

were to be input to the Lisp system, then the value returned by the quote special form would be:

```
(* (- x y) (+ a b))
```

This is obviously a very useful device. Common Lisp has a selection of special forms, but the user may not define their own. It is, however, possible to define 'macros', which would provide a similar operation.

Programs in Lisp are functions or collections of functions. Unlike some of the procedural languages such as Basic, Pascal or C, which demand programs to be written as sequential lines of code with a predefined structure, the Lisp programmer must write functions and build information structures almost as an extension to the language itself. For example, the function shown earlier:

```
?(+ z (- x y))
```

where the ? symbol represents the system prompt will produce an error message :

```
Error: Unbound variable: z
```

Of course, the system has attempted to evaluate z, and found no value for it. z is a variable, and we can give it a value in the simplest case by using the special form 'setq' as below:

```
?(setq z 5)
```

which has the effect of assigning the value 5 to the variable z. In order to execute the function with no error messages, the two other variables must be assigned values.

This is an extremely trivial example, but it may well be that $z + (x-y)$ is an important mathematical formula operating on three critical variables. The method used to define the function is to use 'defun'. We will call the function 'scale_it':

```
?(defun scale_it (z x y)
        (+ z (- x y)))
```

Defun is actually a macro which defines functions. Once this form has been evaluated, the number of functions existing within the Lisp system (almost 700 in Common Lisp) has been increased by one. It is now possible to use the new function as one might use any of the predefined ones. For example:

```
?(scale_it 3 6 4)
results in:
5
```

Large Lisp programs are made up of sometimes large numbers of of functions such as this. Here is a slightly more complex example:

```
?(defun hypotenuse_calc (side_a side_b)
        (sqrt (+(square side_a)
                (square side_b)))))
?(defun square (p)
        (* p p))
```

Two functions are used here to demonstrate some important concepts. The first is that variables and variable names are local to the functions in which they are defined, and it does not matter that the argument names are different in the calling and the called functions, as long as consistency is maintained in the individual functions and that the number and ordering of arguments is consistent. It is possible to use lists, arrays, functions, etc. as arguments in functions and external functions can be invoked liberally within functions. Since data and programs are treated identically in Lisp, then the possibility exists for functions to dynamically modify and even create and initiate new programs automatically. It is this flexibility which has resulted in the widespread adoption of the language in the AI field. A good quote by Milner (1988) says: 'You can stretch it, mould it, fold it, build it, tear it down, and start all over again.'

There are almost 700 predefined functions in Common Lisp. Their uses include:

- list manipulation,
- mathematical operations,
- type checking,
- input/output,
- file handling

and many more. In fact, the language has developed alongside others and has been influenced by their extensions and improvements. Lisp certainly has all of the conditional branching capabilities of other languages and it also supports iteration and recursion. In fact, it is possible to identify many features of languages such as Pascal in newer versions of the language (such as Common Lisp). Perhaps the most important functions, however, are left over from the early days of Lisp itself, i.e. the basic list manipulation functions CAR and CDR and their derivatives.

Consider the list:

```
(a b c (d e) f (g h))
```

The CAR of that list is the first element a, and the CDR is the remaining structure:

```
(b c (d e) f (g h))
```

To carry out some manipulations to the list, it is easier to bind the list to a symbol name, say 'first_list':

```
?(setq first_list '(a b c (d e) f (g h)))
```

Notice the quotation mark before the list itself. This is a short version of 'quote'. It prevents the evaluation of the list, but rather binds the list as it is to the symbol 'first_list'. Now the list manipulation functions can be used:

```
?(car first_list)
A

?(cdr first_list)
(B C (D E) F (G H))

?(cdr (cdr (cdr first_list)))
((G H))

?(car (cdr (cdr (cdr first_list))))
(G H)
```

Notice how it is possible to 'nest' the functions within the structure of the form. With CAR and CDR, it is also possible to nest them directly. For example:

```
?(cddddr first_list)
(F (G H))

?(caddddr first_list)
(F)
```

Fortunately, Common Lisp has more meaningful function names like 'first' and 'last'

```
?(first first_list)
A

?(last first_list)
((G H))
```

Without going too deeply into Lisp, which would be beyond the scope of this chapter (and book), it is possible to identify certain attributes of the language which have become useful in the continuing development of expert systems. Three of these attributes have already been discussed, i.e. the fundamental list structure, the method by which Lisp treats programs and data in an identical way, and the way that programs can be built as an extension to the Lisp 'environment'. For a more focussed discussion, consider again the basic components of an expert system; the knowledge base, the inference engine and the user interface.

Most expert systems are equipped with a highly structured database which holds information relating to domain entities as the current beliefs within a model of that domain. For example, it may be necessary to hold information on electronic transistors within a stock control system. An entity within the computer system might be known as 'transistor', and may (for the purposes of example) hold the transistor types:

```
JFET, PMOS, BIPOLAR and UNIJUNCTION
```

Using the Lisp special form 'setq', it is easy to define such a list:

```
?(setq transistor '(JFET PMOS BIPOLAR UNIJUNCTION))
(JFET PMOS BIPOLAR UNIJUNCTION)
```

Having done this, it is very easy to retrieve a list of transistor classes using:

```
?(cdr transistor)
(PMOS NMOS BIPOLAR UNIJUNCTION)
```

and, of course, any of the list elements can be accessed via the the nested function structure of Lisp:

```
?(car (cdr transistor))
      or
?(cadr transistor)
PMOS
```

But the story does not end there. It is possible to relate or associate further information to each list element in a number of ways. Association lists represent one way, but a far more elegant construct in Lisp is the 'property list'. A property list is a list of properties (or attributes) and values associated with a symbol (where a symbol is defined in Lisp as a function or variable name). In order to give properties to symbols, the functions 'get' and 'setf' are used. In the transistor example, in order to place a cost to the device, of say 1.6, we would use the following:

```
?(setf (get 'PMOS 'cost) 1.6)
1.6
```

and in order to retrieve it, the following would be appropriate:

```
?(get "PMOS 'cost)
1.6
```

It is possible to examine all the property-value pairs for a specified symbol with the form:

```
?(symbol-list `PMOS)
(cost 1.6)
```

where, at this stage, only one property-value pair has been defined. It is also possible to remove a property-value pair:

```
?(remprop `cost)
#<Symbol Plist Locative : PMOS
```

Using these simple constructs, it is also possible to attach lists and other data structures as values to properties. Facilities are available to 'map' a function to successive elements of a list, thereby providing the means to build information structures around nested lists.

In Common Lisp, there is an even more powerful information structuring facility than property lists, which permits the user-definition of specialised data structures. Such a user-defined structure has 'parts', known as 'slots', and the new structure can be used as 'templates' for others (instances). One structure can even inherit another. The function name is 'defstruct'. Using the transistor example again, we could produce a structure which had the slots:

```
semiconductor_material
power_dissipation
construction_method
package_type
```

The appropriate Lisp form is:

```
?(defstruct transistor
        semiconductor_material
        power_dissipation
        construction_method
        package_type)
transistor
```

Having produced a transistor 'structure', or template, it is possible to make objects of that type:

```
?(setq PMOS (make-transistor
        :semiconductor_material `silicon
        :power_dissipation 20
        :construction_method `mol_beam_epitaxy
        :package_type `A123))
#S(TRANSISTOR SEMICONDUCTOR_MATERIAL POWER_DISSIPATION 20
CONSTRUCTION_TYPE MOL_BEAM_EPITAXY PACKAGE_TYPE A123)
```

It is instructive at this point to compare this type of program structure with the structures shown previously in this chapter as examples of commercial Lisp-based expert system development environments.

Once the structure has been defined and specific objects have been created of that type, it is then possible to retrieve information in the following way:

```
? (transistor-power_dissipation PMOS)
20

? (type-of PMOS)
TRANSISTOR
```

and quite a few more interrogation-type functions are available. It is possible to manipulate these structures in many powerful ways. The following represents some of the possibilities:

- make objects (i.e. instances of the new structure);
- access individual slot values;
- establish the type name of the structure;
- include another structure.

In fact, it represents a complete and very powerful frame base building and manipulation facility, which even has the feature of default slot values, specified data types and slot value ranges.

Lisp was one of the first (if not the first) computer language to be used for AI purposes. It was certainly designed to cater for some of the programming needs peculiar to the discipline. Its advantages are obvious, but its disadvantages are not quite so. These include:

- Most Lisps are interpreted. This makes them relatively slow, especially when used to run complex systems which demand complex Lisp functionality (as in powerful expert system shells and development systems). Most Lisps do however have some capacity for incremental compilation.
- The system tends to be 'large', i.e. take up a large amount of computer read/write memory and disk storage. In Lisp-based development systems, the Lisp system itself consumes 'space', the development system, also written (usually) in Lisp takes up significant space, and the application itself can grow to an inordinate size.
- One of Lisp's advantages is its inherent list-processing capability. This however produces a practical problem in most systems, by virtue of its use of the linked list. Effectively, this results in deleted data not being explicitly removed from the system; only 'unlinked'. All Lisp systems are equipped with a facility to automatically remove such unlinked data, which can no longer be accessed by the user, on a timed or demand basis. These are called 'garbage collectors' and they vary in efficiency from one Lisp interpreter to another. Any expert system using Lisp as a software platform will incorporate the garbage collector, which results in the application regularly appearing to halt for a short time.

However, in spite of these disadvantages, Lisp remains a excellent choice of programming language for the development of expert systems at this level (i.e. rather than from the level of development environment or shell).

In the early days of Lisp, the development environment provided by some manufacturers was notoriously poor. It was often difficult to build large programs and to inspect them later, especially when function nesting reached a high level. Testing and debugging facilities were also poor, but this was generally accepted because most users were academics, who were used to having to 'experiment' with their tools and 'put up' with difficult-to-use software and 'user-hostile' interfaces. However, users today are commonly to be found in industrial and commercial establishments, and the need for reliable and easy-to-use AI software for expert system development has significantly influenced Lisp system development itself. Modern Lisps are window-based, using graphics and sophisticated tools dedicated to the tasks of building applications, interrogating applications, testing and debugging. File handling has also improved, and it is now common to be offered interfaces (built-in or within separate libraries) to other programming languages (commonly C) and software packages.

3.3.2 Prolog

Prolog, an acronym for 'programming in logic', was created over a decade ago at the Faculty of Sciences at Luminy in Marseilles, France. Since then, Prolog has gained acceptance as a powerful vehicle for non-numeric programming, particularly in the AI field. It has been recognised by Japan's Fifth-Generation Computer Research Project, as the foundation of a new generation of computers that use logical deduction, instead of arithmetic calculation, as their principle of operation.

Prolog is the first practical and efficient language to have been inspired by research in logic programming. It embodies the idea that programs can be built using formal logic statements (these are usually based upon a type of logic named predicate calculus, see Chapter 10). It differs from traditional languages in that its basic data structures are not restricted to numbers or strings, but can consist of arbitrary structures of symbols. It also supplements the basic operations of comparison and assignment with processes for matching and constructing symbol structures. Thus, the Prolog approach is not restricted to numerical calculation, but also permits 'inference' i.e. allows conclusions to be drawn from a given number of assumptions about a problem. This inference is built into Prolog. When a Prolog program is run, computation results mainly from new facts that can be inferred and only partly according to explicit control information supplied by the programmer.

Specifying a Prolog program requires that the programmer view the problem in a non-traditional manner. In procedural languages, there must be explicit expression of the sequence of steps necessary to solve a problem. In contrast, the Prolog approach involves the description of facts and relationships within the problem, and the application of Prolog's in-built logical inference mechanism. There are two aspects to programming in Prolog:

- the definition of a set of facts and rules, describing objects and their relationships; and
- the interrogation of the program, i.e. asking questions about objects.

The second of these points relates to the fact that Prolog is an interpreted language. This means that code segments can be typed in and then immediately queried. There is no write, compile, run process as in non-interpreted traditional languages such as C.

The essence of Prolog is best illustrated using examples. We will employ the standard 'Edinburgh' notation. This is the most common notation although several alternatives do exist. In this schema, the names of all relationships and objects begin with lower case letters. Relationships are written first and the objects involved are written separated by commas and enclosed in parentheses. A period is placed at the end of each statement.

The relationship 'mother', that exists between two objects, 'women' and 'child' is represented:

```
mother (betty , sally).
```

The formal name for a relationship in Prolog is a predicate. The objects it acts upon are variables. Therefore the above fact has a predicate named mother and variables, betty and sally. Notice that the relationship has a particular order; betty is sally's mother but the reverse is not true.

Once a collection of facts (named a database) has been typed into the Prolog system, queries about these facts can be made. An example database is shown below:

```
mother (betty, sally).
mother (betty, simon).
mother (sally, joe).
mother (sally , ella).
```

If we wish to ascertain if betty is sally's mother, we could type the following query in response to the Prolog prompt '?-'.

```
?- mother (betty, sally).
```

Prolog then replies:

```
yes.
```

To have proved the existence of this fact , Prolog has examined each of the database entries in turn, attempting to match the query. In this case only one fact matches and Prolog replies with yes. An answer of no is given if a query cannot be matched:

```
?- mother (betty, philip).

no.
```

Questions may contain unspecified variables. These indicate objects that are unknown at the time the query is made. To find out if betty has any children the following query could be made:

```
?- mother ( betty, X ).
```

Here **x** is the unknown variable (these always begin with a capital letter). The Prolog system will search for any facts in the database that can match the predicate mother to betty. The identity of the third object in the relationship is not important. Two facts match this query:

```
mother (betty, sally).
mother (betty, simon).
```

Therefore the variable **x** is matched to both sally and simon and the system responds:

```
?- mother (betty, X).

X = sally;
X = simon;
no.
```

Note that after the first match is made and **x = sally** is printed, Prolog waits for a command from the user. Typing a semicolon will inform the system to look for another match. Typing return will terminate the search. The final statement no, indicates that there are no further matches in the database.

More complex questions can be formulated using conjunctions, allowing more than one query to be linked via a comma. (where the comma represents the **and** operator). To make the query '**is betty the mother of sally and simon**' we type:

```
?- mother ( betty, sally ),
        mother ( betty, simon ).

yes.
```

We can check if betty has a grandchild joe with the following query:

```
?- mother ( betty, X )
        mother (X, joe ).

X = sally.
```

Here matching the first statement '**mother (betty, X)**' instantiates **x** to **sally**. The value of **x** is then substituted into the second part of the query, and Prolog searches for a match to '**mother (sally, joe)**'. If the Prolog interpreter had encountered the fact '**mother (betty, simon)**' first, the X would have been bound to the value '**simon**', and no match would be found for the second statement. In this case Prolog would 'backtrack' to the previous predicate and attempt to find another match for it. Having found the fact '**mother (betty, sally)**', it would then proceed forward to the second statement of the query with **x** instantiated to the variable **sally**.

Backtracking is the name given to the technique which Prolog uses to interpret complicated queries. The predicate expressions of a query are considered left to right. If a predicate is satisfied Prolog moves on to the next. If not, it automatically backtracks to the predicate immediately before the one that failed, and searches for a new match. If this predicate fails, Prolog backtracks once again. The process of backtracking can continue leftwards until the first predicate is reached. If an alternative match for this is not found, the whole composite query fails. The purpose of backtracking is to allow an incorrect predicate to be erased and a new attempt made to match it. This is important in AI as many of their applications are based upon intelligent guessing and heuristics. These guesses may be incorrect and backtracking provides a means of recovering from these incorrect decisions.

Often it may be more convenient to express complex relationships using rules. Prolog allows knowledge to be modularised using rules or definitions. Rules allow the user to query new predicates without specifying new facts, by defining the new predicate in terms of old predicate names. For instance, we can represent the rule that a person is a grandmother if they are the mother of someone, who is in turn a mother of someone else:

```
grandmother (X) :-
        mother (X, Y),
        mother (Y, Z).
```

The rule is comprised a head, which describes the rule consequent and a body which lists the antecedents. The symbol ':-' implies **if** and the comma **and.** Therefore this example translates to 'X is a grandmother if X is the mother of Y and Y is the mother of Z'. Rules are added to the database alongside facts. When searching the database, if the system finds the head of a rule that matches the current goal, it will try to satisfy the subgoals in the body of the rule using backtracking. In satisfying all subgoals it can infer the rule consequent. Having defined the above grandmother rule, if we now make the query:

```
?- grandmother (betty).

yes.
```

the answer is affirmative. The Prolog interpreter has first matched the original goal to the grandmother rule. It then proceeds to match the predicate '**mother (X, Y)**' to '**mother (betty, sally)**'. Having instantiated the variable Y to the object '**sally**', it matches the predicate '**mother (sally, Z)**' to '**mother (sally, joe)**'. It should be noted that when matching occurs, rules and facts are considered in the order they appear in the database.

Facts in Prolog can describe simple relationships as shown above, but for more complex ones data structures are used. A structure is an individual object that is named by listing its component parts. They are written with a 'type' of individual and the components that identify the details of the individual. A simple description of a car can be constructed using the structure:

```
car ( Make, Colour, Year ).
```

Note that the objects beginning with a capital letter are unknown variables. We can now use this structure within another, to construct a list of cars for sale, including their price and sellers name:

```
sale ( car ( ford, blue, 1955 ), smith, 200 ).
sale ( car ( volvo, red, 1982 ), jones, 2500 ).
sale ( car (fiat, yellow 1976 ), brown, 750 ).
```

It is now possible to retrieve information without specifying all the components of these objects. Instead, we can just indicate the structure of objects that we are interested in, and leave the particular components in the structure unspecified or only partially specified. For example we can make the query:

```
?- sale ( C, S, P), P < 500.

C = car ( ford, blue, 1955 ),
S = smith,
P = 200.
```

Here the search is restricted to those cars which are priced less than 500 units. Structures in Prolog are a convenient method of representing very complex relationships.

Another important aspect of Prolog is its list-processing capabilities, which have been endowed with a simple syntax, enabling lists to be written very easily. Lists can contain objects, variables and other lists e.g.

[A, B, (C, D, E), F, G].

This is a list with five elements, the third one of which is another list. Lists are defined by their head, the first element of the list, and their tail, the remaining elements minus the head. This is described using the notation:

[X | Y]

Here **X** indicates the head of the list and **Y** the tail. This is useful in pattern-matching. For example, we can define a predicate 'member' which succeeds if its first argument is a member of its the list given in its second argument. This would allow queries of the form:

?- member (B, [A, B, C, D]).

yes

The predicate member is defined:

member (X, [X | Y].

member (X, [H | L] : - member (X, L).

The first of the definition states 'X is a member of the list if X is the head of the list'. The second states 'X is a member of the list if X is a member of the tail of the list'.

The second statement is unusual in that member is used within its own definition. This phenomenon is termed 'recursion' and is used extensively in Prolog. Consider the query:

?- member (b, [a, b c]).

The first statement of the member definition will fail because B is not the head of the list. The second definition is then applied. This splits the list **[a, b, c]** into the head **[a]** and tail **[b, c]**, and makes a recursive call to the first member definition as:

member (b, [b, c]).

This succeeds as **b** does now match the head of the list.

Given the facilities provided by Prolog, i.e. built-in deductive, chronological backtracking and pattern-matching via unification, it has been advocated as an ideal language for expert system development and AI applications. Problems can often be represented in a very clear, concise Prolog program, resulting in decreased development time. Prolog was designed to bridge the the gap between human reasoning and the machine's processing mechanism, allowing the programmer to express logical facts and rules without having the detailed programming knowledge necessary with other languages.

The promotion of Prolog in recent years has built up expectations in potential users, and it is often claimed that writing in a logic language such as Prolog implies that a program is logically correct. This is not the case. It is possible to make mistakes in Prolog as in any other programming language. Often the difficulty arises in translating the logical reasoning of the application to that of a Prolog program. Others point out that Prolog still shares the effect of the influence of von Neumann computer architecture on conventional programming languages. Thus, it is impossible to write a Prolog program just according to logical reasoning, and that there are unexpected pitfalls in the use of the language.

Another problematic area in Prolog is the implementation of tasks which are implicitly procedural in nature, such as specification of user-input. In these situations its declarative style may be clumsy. The absence of high-level block structure, type checking of variables or a documentation methodology in Prolog has acted as a deterrent to potential builders of very large applications. Nevertheless, Prolog gives an interesting insight into logic programming and is rapidly gaining popularity as a tool for prototyping.

3.3.3 The Object-oriented Paradigm

The concept of object-oriented design and programming has become extremely popular within the software engineering community. This approach can be seen to complement, or perhaps even replace traditional methods of software construction, such as structured design and data flow. Its development grew out of the dual concern to improve the quality of software produced and to encourage code reuse rather than reinvention. Object-oriented design and programming addresses the five criteria (as determined by Meyer, 1988) which are important in achieving the goal of software quality, i.e. correctness, robustness, extendibility, reliability and compatibility.

Object-oriented design can be viewed as a software decomposition technique which, unlike classical functional design, bases the modular decomposition of a software system on the classes of objects the system manipulates not on the functions it performs. Whilst the top–down functional approach can be adequately utilised if the application is static and will not require further development or adaption, it becomes problematic if consistent updating is necessary. In contrast, categories of objects designed using the object oriented approach, are likely to remain stable throughout the lifecycle. For example a text-processing system will always work on documents, chapters and paragraphs.

There are six facets of the object-oriented approach:

1. object-based modular structure;
2. data abstraction;
3. classes;
4. inheritance;
5. polymorphism; and
6. automatic memory management.

3.3.3.1 Object-based Modular Structure

This advocates that a system should be viewed as a set of modules based upon the objects it manipulates, not on the functions it performs. It is built on the idea that we naturally perceive the world around us as a variety of objects, and that it is possible to represent these as modular software concepts. Using this approach should make software more tangible to non-implementors who have application knowledge, and allow mass production of comprehensive libraries of autonomous modules with which to build software systems.

3.3.3.2 Data Abstraction

Software decomposition based upon the physical structure of objects does not provide protection against changes in data format. Studies have shown that 17.5% of the cost of software maintenance stems from changes in a program that reflect alterations in data formats (Lientz and Swanson, 1979). This emphasises the need to isolate programs from the physical nature of the objects they handle. Abstract data types achieve this by describing an object through its external properties rather than its data representation. It is the definition of the interface to a data structure that is the user's only access route to data that is otherwise private and hidden. This is often termed encapsulation or data hiding. An example of data hiding is the object 'time' which has operations defined for setting, reading and printing its contents. Any manipulation of the object's private data would have to take place via these operations.

Data abstraction is a powerful technique. In addition to supplying the value of data directly, the interface may compute new values from it, e.g. provide different representations or data conversions. It also allows the system builder to revise the internal structure of a data type, transparently to the user, provided the interface remains unaltered.

3.3.3.3 Data Representation

Abstract data types are represented in two sections. The first section 'private', containing the data internal to the type, i.e. accessible only within the type's defined operations (known as methods). The second section 'public', containing the interface, i.e. the methods available to 'clients' of the type. An abstract data type definition is often referred to as a class. For example a class may be defined as:

```
Class : TIME
        Private
                integer : day, month, year
                integer : hours, minutes, seconds
        Public
                method SetDate(integer day, month, year)
                method ReadDate(integer day, month, year)
                method SetTime(integer hours, minutes, seconds)
                method Print
    End Class TIME
```

The class definition is a template (its implementation is given elsewhere) and does not create an instance of the time object. This is usually done by an operation 'new' or 'create' e.g.

```
        new TIME birthday
```

Here an instance of the class 'Time' has been created, i.e. memory space for its private data has been allocated and attached to the object 'birthday'. The class object provides a mechanism for sharing operations or methods, i.e. the same code is used for each instance of the class.

An important benefit of this approach is that each object is self-identified, since it carries its type (i.e. the class type of which it is an instance) with it. This is useful especially during debugging, as all objects know their type and can be directed to display themselves.

3.3.3.4 Inheritance

Inheritance is the major feature distinguishing object-oriented programming from other programming systems. It can be seen as the extension of a previously defined class. When one class (the derived class) inherits from another (the base class), the derived class inherits the data structure and the methods of the base class, for example:

```
Class: APPOINTMENT
       inherit TIME
       private
              string location
              string purpose
       public
              method make(DATE when, string where,why)
              method print
End class APPOINTMENT
```

Appointment will inherit the data and methods from the class 'time', and add the new data 'location' and 'purpose', and the methods 'make' and 'print' (note 'print' is a redefinition of an inherited method).

Inheritance is an important organisational principle for concepts capturing natural mechanisms such as specialisation, abstraction, approximation and evolution. It encourages the production of general purpose classes which can be specialised for a specific application. In this way it provides for code reusability.

Single inheritance systems require that classes be organised into a tree structure. This is sometimes awkward, and results in deep inheritance structures that are difficult to use. To counter this, multiple inheritance is proposed. This allows a class to inherit from more than one base class, producing complex structures but with a tendency toward shallowness.

3.3.3.5 Polymorphism

It is first necessary to explain the concept of 'messages'. In object-oriented programming objects are communicated with via 'messages'. These messages invoke a specified method belonging to the object. The distinction between a method and a message is subtle but important. Since a method is part of an object if the objects 'line' and 'circle' both have a method 'draw', sending the message 'draw' to the line object will invoke its particular draw method, whereas sending the same message to the circle object will call a different method i.e. that belonging to the circle object. This ability of different objects to respond to the same message is known as polymorphism. It eliminates the need for large case statements, identifying a particular type before sending a common message to it.

3.3.3.6 Automatic Memory Management

This aspect of object-oriented programming is more of an implementation detail than an underlying principle. It seeks to free the programmer from the task of memory management i.e. claiming memory for objects and freeing that memory when the objects go out of context.

3.3.3.7 Object-oriented Languages

Whilst it is possible to follow the spirit, if not the detail of object-oriented concepts in classical non-object-oriented languages, such as C, it is better facilitated using languages that support its methodology more fully. These include languages such as Ada and Modula2 which offer encapsulation techniques, and the truly object-oriented languages such as Smalltalk, Eiffel and C++. In addition to these there are languages that have been developed within the AI community for the purpose of knowledge engineering, e.g. Hewitt's Actor, MIT's Flavours and Xerox's Loops. Indeed many of the knowledge engineering environments such as KEE and ART also provide facilities for object-oriented design.

Of those languages that which can be viewed as truly objected-oriented, there are two distinct types, the Smalltalk group of languages and the C group. The Smalltalk group are generally interactive and interpreted, tending to be used in applications that rely heavily upon the user interface. The C group generally consist of a preprocessor, generating C source code for compilation by a standard C compiler. These are used in a variety of applications. We will now look in some detail at an example from both groups. First, from the C group we will consider C++; secondly, from the Smalltalk group we will look at Smalltalk itself, concentrating particularly on the support provided for the six concepts of object-oriented methodology described above. Finally, we will present a brief comparison of the languages Smalltalk, C++, Ada and Modula2 in Table 3.1, to give an idea of their overall differences.

Table 3.1
A comparison of languages embodying some of the concepts of object-oriented programming

attribute	C++	Smalltalk	Ada	Modula2	Common Lisp
Objects only	no	yes	no	no	no
Inheritance	yes	yes	no	no	yes
Dynamic binding	yes	yes	no	no	yes
Overloading	yes	yes	yes	no	yes
Initialisation and deletion	yes	yes	init	init	yes
Compiled	yes	no	yes	yes	yes
Interpreted	no	yes	no	no	yes
Garbage collection	no	yes	no	no	yes
Environment	no	yes	yes	no	yes

3.3.4 C++

C++ was developed by Bjarne Stroustrup of AT&T in 1983 (Stroustrup, 1986). The aim was to make the benefits of object-oriented programming available to experienced C programmers who are committed to using C, and have vast libraries of reliable C functions. C is a versatile succinct and relatively low-level language that is suitable for most system applications. It is also highly portable and forms part of the popular Unix programming environment. For these reasons AT&T chose to use C as the language on which to base C++.

We will now look at how C++ facilitates object-oriented programming by examining its support for the six facets described in the previous section. We assume that the reader has a basic knowledge of C.

3.3.4.1 Objects

To elaborate upon the concept of objects in object-oriented languages we must consider the notion of 'type'. It is this that distinguishes true object-oriented languages from others such as Ada and Modula2. In these an object is purely a syntactic construct, used to logically relate program data and procedures. In contrast, object-oriented programming languages view an object as a semantic denotation or type.

Standard C is a strongly typed language, in that there are a number of predesigned types such as 'int' and 'char', plus those explicitly defined by the user. C++ is also typed and its objects have a type (this allows the compiler to check for inconsistencies in a type's use, therefore leading to fewer run-time errors). However, unlike Smalltalk which classes everything as an object, C++ does not consider basic constructs such as integers or characters as objects. This makes C++ a hybrid system combining mechanisms of both object-oriented and procedural programming.

3.3.4.2 Data Abstraction and Classes

A class is a user defined type in C++. It is the definition of a data structure and its interface, for example:

```
class Window
{
        int column,row;                 // Window co-ordinates
        int length,width;               // Window size
        public:
        void setposition(int r, int c);  // Set window coordinates
        void setsize(int l,int w);       // Set window size
        void readposition(int* r, int* c); // Read window coordinates
        void readsize(int* l, int* w);   // Read window size
        void display();                  // Show window
        };
```

The variables—column, row, length and width—are the private data of the class, they cannot be accessed except by the member functions (or methods) belonging to the class. The methods defined below the keyword 'public' form the interface to the class. The implementation of methods can be within the definition, however for ease of maintenance a preferable scheme is to hold the implementation separately, e.g.

```
        void Window::setposition(int r, int c)
        {
                row = r;
                column = c;
        }
```

The method is associated with the appropriate class through the double colon operator.
An object of type '**Window**' may be created by declaring an instance:

```
Window window1;
```

Messages are sent to the object through use of the dot operator:

```
window1.setposition(0,0);
```

This then invokes the setposition method associated with '**window1**'.
Objects may also be dynamically created using the '**new**' operator:

```
Window *windowptr;
windowptr = new Window;
```

The first line declares a pointer to the type '**Window**', the second line creates an instance
of type '**Window**' referenced through the window pointer. Messages are sent to the instance
by use of the point operator '->':

```
windowptr->setposition(0,0);
```

3.3.4.3 Inheritance

Assuming the definition of the class '**Window**' as given above, a specialised text window
class may be defined:

```
class TextWindow : Window
{
        char* text;

        public :
        void settext(char* text);
        void display();
};
```

Two points are important here. Firstly, the methods of the derived class '**TextWindow**',
cannot access the private data of the base class '**Window**', despite having inherited them as
class data. Therefore the following would be incorrect:

```
void TextWindow::display()
{
        int lineno = row;      // Incorrect
        ....
}
```

Instead, private data of the base class must be accessed by use of the public methods of the
class. Messages to inherited classes are sent via the double colon operator:

```
void TextWindow::display()
{
```

```
            int r,c;
            Window::readposition(& r,& c);
            int lineno = r;
            ....
    }
```

Secondly, although member functions of the derived class can access the public methods of the base class (as above), instances of the derived class cannot:

```
    TextWindow twindow;
    twindow.setposition(0,0);      // Incorrect
```

However C++ provides flexible control of such access. In the example above, access by instances of the derived class 'TextWindow' to the basic sizing and positioning methods of the base class 'Window' would be useful. This is achieved by use of the keyword 'public':

```
    class TextWindow : public Window
    {
            ....
    };
```

This allows instances of the class TextWindow access to the public methods of the class Window.

In addition the 'friend' mechanism allows access by a specific method or function to a class's private data:

```
    class Window
    {
            friend void TextWindow::display();
            ....
    };
```

Alternatively all the methods associated with one class can be given access to the private data of a second class:

```
    class Window
    {
            friend class TextWindow;
            ....
    };
```

The use of the 'friend' mechanism can invalidate the concept of data hiding, and so should be used with extreme caution.

In C++ a derived class can be a base class, thus class hierarchies can be constructed. However, since a derived class can inherit from a single base class only, such a hierarchy is a tree. To produce a more general graph structure would require multiple inheritance. This will be available in the next release of C++.

3.3.4.4 Messages and Polymorphism

As described earlier, messages are passed to objects using the dot or point (->) operator with the structure:

```
object.message or & object->message
```

An additional facility of C++ is 'function overloading', in which methods can share names. The compiler will decide which method is appropriate by examination of the type and number of arguments passed, e.g.

```
box.draw();
box.draw(colour);
```

In order to use derived classes in a more useful manner than purely as convenient shorthand denotations, C++ provides dynamic binding. Consider the following problem:

```
Window* windowptr;
windowptr = & window1;
windowptr = ( TextWindow*) & textwindow1;
```

In C++ a pointer of type base may refer to an object of the (public) base class or an object of a derived class. Thus it is allowable to declare a pointer of **Window**, and reassign it to the derived class **TextWindow** as above. The problem arises when we call a member function using the base pointer:

```
windowptr->display();
```

The display method has been redefined in the derived class **TextWindow**, therefore we would want to call this specific method in the above statement. However this will call the display method belonging to the base class **Window**, which is incorrect.

C++ provides virtual functions to deal with this problem. A method declared as a virtual in the base class can then be recognised in a derived class:

```
class Window
{
    ....
    public :
    virtual void display();
    ....
};
```

Having declared display as a virtual method in the base class and redefined it in the derived class, the following statement will now call the correct display method:

```
windowptr->display();
```

The concept is termed dynamic binding. The correct method to display is decided at run time rather than compilation time.

3.3.4.5 Instance Initialisation and Memory Management

C++ provides mechanisms by which methods can be invoked at both the time an object is first created and when it goes out of scope (i.e. no longer accessible). These methods are known as constructors and destructors. A constructor contains initialisation code, and a destructor 'clears up' after an object (e.g. frees any memory allocated during the object's lifetime) . Their syntax is shown below:

```
class Window
{
        ....
        public :
        ....
        Window();          // Constructor
        ~Window();          // Destructor
};
```

3.3.4.6 Conclusion

Many of those that advocate the use of object-oriented programming languages, do not view C++ as a suitable language in which to learn the concepts. They recommend instead Smalltalk, due to its strict adherence to all the concepts of the object-oriented paradigm (Smalltalk is described in the next section). Despite this C++ is perhaps the most likely to be widely used (especially in the short term) for the following reasons:

1. availability on traditional machines;
2. availability under traditional operating systems;
3. it can compete with traditional programming languages in run-time efficiency;
4. it can cope with most applications; and
5. its close association and compatibility with C.

Particularly in such areas as real-time programming, C++ can provide the necessary speed and efficiency. It also allows the incorporation of assembly code sections, which are still occasionally required for especially time critical applications. Thus it enables the programmer to use object-oriented techniques, whilst working within the constraints often forced upon her by practical applications.

C++ is a strongly typed language (as opposed to Smalltalk). While strong typing may force the programmer to complicate the inheritance hierarchy and require a longer period to implement the program, it has the advantage of detecting problems earlier in the development cycle. The result is that programs produced with strongly typed languages are often more robust.

C++ does not provide automatic memory management. There are two views on memory management: one, memory management is so important that it cannot be left to the user; two, it is so important that it must be left to the user. Automatic memory management is often less than optimum, but may be permissable for systems where the overhead is acceptable, and the user wishes to be freed from such implementation details. Conversely, some applications require hand-crafted memory management schemes to achieve the required speed and memory usage efficiency. C++ falls into the latter category.

One of the main disadvantages of C++ when compared to other advanced languages is its lack of a coherent support environment. This situation is now changing with the introduction of C++ development systems containing tools such as intelligent editors, interpreters and browsers, (e.g. Designer C++, Oasy Inc.).

In conclusion, C++ can be seen as a realistic application of object-orientated techniques, providing the programmer with support for the new object-oriented methodology, yet retaining all the flexibility (and some of the problems), of its base language C. However, any programmer attempting to choose a language for a practical application should consider the issues involved very carefully.

3.3.5 Smalltalk

The ideas behind Smalltalk were first conceived by Alan King at the University of Utah in 1970, and were influenced by both the Simula67 programming language and Lisp. Simula67 provided the concepts of classes and inheritance, whilst Lisp contributed the concepts of dynamic binding and an interactive environment. Smalltalk was consequently developed into a commercially available product and released in 1980 (Smalltalk-80, Robson and Goldberg, 1983). The vision of its designers concentrated on two principal areas of research: a language description (the programming language) which serves as an interface between the models in the human mind and those in computing hardware; and a language of interaction (the user interface) which matches the human communication system to that of the computer. Their goal was to create a powerful information system, within which the user can store, access and manipulate information, whilst allowing the system to expand in unison with the user's ideas.

We now examine a number of the facilities provided by Smalltalk to support object oriented programming.

3.3.5.1 Data Abstraction and Classes

A class in Smalltalk describes a set of objects that express commonality. The individual objects of that class are named 'instance variables'. A class defines the form of an instance's private data structures and the method in its interface. Classes in Smalltalk are presented in two ways. Firstly as a 'protocol', which lists the messages available in the instances interface. Each message should be accompanied by a comment, describing the operation that will be performed when the message is received. A protocol is intended to provide a programmer with sufficient information to manipulate instances of that class e.g:

```
window

        initialisation
                xposition : x yposition : y   set window coordinates
                length : l width : w          set size

        read
                readx : x ready : y           read window coordinates
                readlen : l width : w         read window size

        display window
                display                       display window
```

The messages of the class window are grouped into three user-specified categories: initialisation, read and display. This is intended to make the protocol more readable to a programmer and does not affect the operation of the class.

Secondly, each class needs an 'implementation' description. This contains the class name, a declaration of the instance variables, and implementation code of the interface methods defined in the protocol e.g.

```
class name                  window

instance variable names     column
                            row
                            length
                            width

instance methods

initialisation
        xposition : x yposition : y
                xposition : x yposition : y
        row <- x.
        column <- y.

        . . . . . . . . . . .
        etc.
```

The operator '<–' causes variable assignment from right to left. Note here, that unlike C++ which declares its instance variables in the class definition, Smalltalk declares them in its implementation.

3.3.5.2 Objects

In Smalltalk everything is an object, including all constants and the contents of all variables. The only entities which do not denote objects are the message selectors (method or procedure names), comments and a few punctuation characters. However, Smalltalk is a weakly typed language; objects are a type, but variables are not. Therefore, any variable may refer to an instance of any object. Smalltalk emphasises dynamic binding and no type checking is performed. The determination of whether a method can be legally applied to an object occurs only at run-time.

Given a definition of a class window, an object window is declared:

```
window1 <- window new.
```

3.3.5.3 Messages

Messages allow interaction between the objects of the Smalltalk system. A message requests an operation on the part of the receiving object. For example:

```
window1 display.
```

sends the message display to window1.

Arguments are passed to methods by stating the variable name, a colon and the value, at the end of the message statement:

```
window1 xposition : 0 yposition : 0.
```

It should be noted that in Smalltalk messages are global symbols which do not need prior declaration. Therefore a message may be syntactically correct but will give a run-time error if it is not understood by the object it is sent to.

3.3.5.4 Inheritance

Smalltalk creates new classes as extensions to existing ones. This extension is named a subclass. It specifies that its instances will be identical to instances of the class it has inherited from (its superclass), except for the differences that are explicitly stated. Every class must have a superclass, thereby forming a tree called the 'object hierarchy'. At the top of this hierarchy is the class named 'object'. This is a system class which describes the similarities of all the objects in the system. Thus, every class will be at least a subclass of object.

A subclass will inherit all variables and methods from its superclass. This superclass will in turn inherit all variables and methods from its own superclass, and continue on upwards in this fashion until the class 'object' is reached. A subclass will inherit from its superclass in the following manner:

```
class name              textwindow
superclass              window
instance variable names text

instance methods

    display
```

Here the subclass textwindow inherits all the variables and methods from the class window. It also declares the additional instance variable text, and redefines the method display.

It is important to note here that any subclass has direct access to the variables of its superclass (and that superclass's predecessors in the hierarchy). This contrasts with C++ which enforces stricter data hiding and does not allow a subclass to access variables of its superclass except by its defined interface methods.

If a subclass redefines a method with the same name as that of a method defined in the superclass, its instances will respond to messages of that name by automatically executing the new class method. This is termed 'overriding' a method (i.e. polymorphism).

When a message is sent to a class instance, the methods of that class are searched for a matching method. If none is found, the methods of the superclass are next searched. This continues up the hierarchy until a match is found. If the topmost class object is reached and a match cannot be made, an error message is generated.

Smalltalk provides two mechanisms for controlling this method search more explicitly. The term 'self' directs the method search to begin in the instance's own class, whilst '**super**' instructs the method search to commence in the superclass of the class containing the method.

Multiple inheritance is not generally available in Smalltalk, although an experimental version is available for Smalltalk-80. However it is not being promoted by the suppliers.

3.3.5.5 Memory Management

Smalltalk provides automatic memory management. Any object which is no longer referred to may be deleted, and its storage returned to the common pool in a manner that is transparent to the user.

3.3.5.6 The Smalltalk Environment

Smalltalk is a graphical, interactive programming environment, providing tools for locating, viewing, writing and running smalltalk methods. Interaction is mouse-driven and takes place through various types of windows and pop-up menus. The system may be viewed via the system browser. This displays the class hierarchy and allows the user to select any method for display and editing. Classes and methods may be added and deleted interactively. Running Smalltalk code may be interrupted and the debugger used to investigate the the state of the computation and alter it before recommencing.

Building an application in Smalltalk cannot be viewed in the traditional sense of writing a program. Rather it should be seen as adding the required extra functionality into the system. Any user-defined classes are incorporated into the existing hierarchy. Whilst this approach can be extremely productive for an individual, it poses two problems. Firstly, the code developed is distributed throughout the system, and so requires sophisticated tools to track it. Secondly, all all code, including source code, is available to any system user and can be changed or augmented. Large projects involving a team of programmers need strict control to guard against this.

3.3.5.7 Conclusion

Programming with Smalltalk is incremental by nature, and therefore lends itself well to the progressive development of modules by stepwise refinement. It also assists application development by providing a user friendly environment and a dictionary of built-in base classes available to the user. Smalltalk views all system components as objects, including classes. The advantages claimed for this approach are as follows:

- it yields conceptual consistency, i.e. a single concept, that of the object, is applicable to all Smalltalk notions;
- it strongly encourages the programmer to model their application in terms of objects;
- by making classes part of the run-time operation, it thereby supports the development of interactive tools, such as symbolic debuggers, class browsers and other tools that require access to class text at run-time.

However, Smalltalk is a weakly typed language, which eases the task of creating an initial working program, but may allow the inadvertent creation of errors which are extremely difficult to locate (the Smalltalk environment provides some assistance in finding such errors). Programs are often less robust when produced in weakly typed languages and so are less suitable for applications that must be extremely reliable. To overcome this problem of weakly *versus* strongly typed languages, it has been advocated that one should prototype using a weakly typed language thereby gaining speed and ease of development, then create a more robust production version using a strongly typed language. There are two arguments that direct against this approach, firstly software is not often rewritten from scratch. Even when this is done, it usually occurs late in the development cycle so that architectural deficiencies have become well established, and must be supported in the reimplementation. Secondly, programs often need complete redesign for translation into strongly typed languages, thereby producing problems of speed and cost.

Because any Smalltalk application is automatically incorporated as part of the system, any potential end user of the application requires access to the complete Smalltalk system in order to run it. This has acted as a barrier to the commercial development of applications using Smalltalk, but will be alleviated with the recent introduction of a run-time package.

To conclude, Smalltalk provides for truly object-oriented programming, within an elegant programming environment. It is an excellent tool for fast prototyping, but is not yet sufficiently well equipped to produce the robust, efficient and portable programs required in many application areas.

BIBLIOGRAPHY

Chung, P.W.H. and Kingston, J. (1988) State of the Art: Knowledge-Based toolkits, *Proc. Fifth Annual Meeting of the British Medical Informatics Society 'Expert Systems in Medicine'*, Royal Free Hospital School of Medicine, London, 29 March 1988.

De Kleer, J. (1986) An Assumption-based Truth Maintenance System, *Artificial Intelligence* **28**, 127–162.

Heathfield, H. A., Winstanley, G. and Kirkham, N. (1988) Computer-assisted Breast Cancer Grading, *Biomedical Engineering Journal*, **10**(Oct), 379–386.

Jackson, P. (1986) *Introduction to Expert Systems*, Addison-Wesley, Reading, MA.

Lientz, B. P. and Swanson, E. B. (1979) Software Maintenance: A User/Management Tug of War, *Data Management*, (April), 26–30.

Merrit, B. (1986) Trends of Expert System Building Tools: Personal Consultant Plus, *Texas Instruments Engineering Journal*, **3**(1).

Meyer, B. (1988) *Object-Oriented Software Construction'*, Prentice Hall, New York, pp 160–163.

Milner, W.L. (1988) *Common Lisp: A Tutorial*, Prentice-Hall, New York.

Parsay, K. and Chignell, M. (1988) *Expert Systems for Experts*, Wiley, Chichester.

Reggia, R.A., Naut, D.S. and Wang, P.Y. (1983) Diagnostic Expert Systems Based on a Set Covering Model, *Int. J. Man-Machine Studies*, **19**, 437.

Robson, D. and Golgberg, A. (1983) *SMALLTALK-80: The Language and its Implementations*, Addison-Wesley, Reading, MA.

Rowe, N.C. (1988) *Artificial Intelligence Through Prolog*, Prentice Hall, New York.

Shortliffe, E H. (1976) *Computer-Based Medical Consultations: MYCIN*, Elsevier, Amsterdam.

Steele, G. L. Jr (1984) *Common Lisp: The Language*, Digital Press, New York.

Stroustrup, B. (1986) *The C++ Programming Language*, Addison-Wesley, Reading, MA.

Winstanley, G. (1987) *Program Design for Knowledge-Based Systems*, Sigma Press.

G. Winstanley
Information Technology Research
  Institute
Brighton Polytechnic
Lewes Road
Brighton
East Sussex BN2 4AT

H.A. Heathfield
Intelligent Medical Systems Laboratory
Information Technology Research
  Institute
Brighton Polytechnic
Lewes Road
Brighton
East Sussex BN2 4AT

4   Knowledge Engineering: Tools and Techniques

J.M. Kellett and G. Marshall

Brighton Polytechnic, UK and
AI Unit, Middx, UK

4.1 INTRODUCTION

4.1.1 What is Knowledge Engineering?

Knowledge Engineering is a term which was coined by Edward Feigenbaum (1977), which he defined as:

> the art of bringing the principles and tools of AI research to bear on difficult applications problems requiring experts' knowledge for their solution.

This, then, is a broad discipline covering the whole process of knowledge-based system development. In his book on machine learning, Richard Forsythe (1986) states:

> Knowledge Engineers are people who build expert systems, and their job is not easy.

Of course, the complexity of the knowledge engineering process is largely dependent upon the scale of the problem to be solved. Knowledge-based systems vary in many ways. There are applications developed on small shells by the end-user for his/her own personal use. For example, an expert may wish to capture a small part of his knowledge which he feels it would be useful to automate. There are larger, stand alone systems, which rely upon the expertise of a team of individuals (which have much more of a widespread impact). At another level of complexity still are applications which 'embed' knowledge-based systems into existing processes. Comparisons have been made between knowledge engineering and systems analysis. Indeed, there are many similarities between the two processes; an expert system is, after all, a computer program. However, Hart (1989) identifies the fact that one of the main differences between the two processes is that, in a knowledge-based application, the user is unlikely to be the expert from whom the knowledge has been elicited, and thus has different requirements.

Artificial Intelligence in Engineering. Edited by Graham Winstanley
©1991 by John Wiley & Sons Ltd.

The task of the knowledge engineer* is often more complex than that of the systems analyst. The most obvious example of this, is that the first major phase in standard software development methodologies is to specify the system accurately. However, the specification of expert systems is notoriously difficult. The seasoned systems analyst, however, will recognise that many of the techniques described in this chapter owe much to the work performed in their field.

It is possible to break down the tasks involved in knowledge engineering into four:

Identify the problem

It is vitally important to clearly define the scope of the problem to be solved. This is not in itself an easy task, many failed systems have resulted from a failure to identify and fix the system boundaries. Another important factor has been the desire to utilise knowledge-based technologies to build all-singing all-dancing systems which attempt to solve problems which are larger than those solvable by existing methods. It is also important to bear in mind that the use of an expert system to solve certain problems may be wholly inappropriate. Waterman (1986) identifies three levels of justification for the use of expert systems. First, one must identify that the use of an expert system is possible. This will clarify whether the task involved is one which may be solved via a knowledge-based approach. Features such as whether the task involves common-sense reasoning, whether experts exist, whether there is a general level of agreement amongst experts should be covered.

Justify the use of AI

Or to be more specific, an expert systems approach. The cost of developing an expert system can be very high, and so the expenditure must be justified. Common reasons for the use of expert systems include the loss, or rarity of a particular type of expertise, the need to utilise expertise over a broader area, and the tuition of novices. These may be justifications, but in a commercial application, there must always be a cost benefit. DEC had one of the first major commercial successes with expert systems with XCON (originally called R1) (McDermott, 1982). This system is used in the configuration of VAX computers, a task which is highly complex because of the large range of options offered by DEC. The task is one which no single expert could undertake, because of its complexity, however it is one which is vital to DEC's trading. The system is a major financial success, and is estimated to save $35 million per annum.

A final level of justification is to check the appropriateness of an expert systems approach. Firstly, is the task possible

---

* To avoid any confusion I will make the disclaimer here that the use of the male gender with reference to knowledge engineers, domain experts and other parties is purely to avoid the proliferation of terms he/she his/her etc.

using standard programming techniques (are we using a sledgehammer to crack a nut?). Is the task essentially heuristic or rule-based, does it involve high degrees of symbolic representation?

Acquire the knowledge   This is widely recognised as being the major bottleneck in the knowledge-engineering process. Knowledge acquisition (or elicitation*) can be described as the process of attempting to elucidate the structure and body of the knowledge which the expert uses, or which is required to be brought to bear upon the problems of the domain. It is thus a term normally associated with expert or knowledge-based systems, as opposed to being a term necessarily used throughout the entire AI discipline. Knowledge acquisition, then, entails interacting with the domain expert in such a way as to develop a complete (or as complete as is possible) model of the knowledge and information of the domain. Areas which must be investigated include the structure of the domain, the main features of the problem and solution, the rules which enable conclusions or inferences to be made, and the manner in which reasoning is carried out. At the heart of the problems associated with knowledge acquisition is the ability to communicate, or share, ideas effectively between the 'domain expert' and the knowledge engineer. Many factors will affect this necessary communication, ranging from simple personality conflicts, to the failure to understand important concepts, to the inability to express important concepts—it is not at all easy to tell someone how you solve a particular problem. The majority of this chapter will cover the techniques which can be applied in this process.

Design the representation
and inference models   This phase is based upon the results of the knowledge acquisition phase. The boundary between these two phases is not well defined, as the development of knowledge-based systems has traditionally been an iterative process, with some form of prototype system being developed.

4.1.2 What is Knowledge? Where is it Kept?

The term 'knowledge' is a highly emotive one (as indeed is 'artificial intelligence'!). How do we decide what exactly constitutes the knowledge which we hope to embody in our

---

* In this chapter I will use the terms 'knowledge acquisition' and 'knowledge elicitation' interchangeably. Philip Slatter (1987) gives one possible definition of how these two processes differ. He describes knowledge elicitation as the process of acquiring knowledge from a human expert, as opposed to inductive techniques based upon raw data.

knowledge-based systems. Is there a difference between knowledge and expertise? If so, what is it that we need to capture—knowledge, expertise, or both? Studies have been made, which would tend to suggest that, as a person develops from novice to expert, the structure of their stored knowledge undergoes significant changes. As a beginner, the problem-solver will tend to apply his textbook knowledge as his main strategy. However, the expert will more often than not apply his recollection of similar cases to facilitate his problem-solving (Dreyfus and Dreyfus, 1986). One example often quoted is of the grand-master chess player *versus* the novice. Both have the same (or comparable) ability to analyse the current position and to predict the future directions of the game. The expert, however, has experienced many situations, and is able to recognise features of these previous games in the current one, and react accordingly. The recognition of this fact is becoming more widespread, one example being the research effort being placed in the domain of neural networks which essentially attempts to train a program with scenarios which represent these experiential situations. Another series of psychological studies (McKeithen *et al.*, 1981) examined the structures held by expert, intermediate and novice level Algol programmers. The resulting structures differed greatly between levels of experience; however across the expert level programmers the structure showed great consistency.

In terms of the design of knowledge-based systems, we tend to classify the knowledge which we wish to acquire as consisting of:

- the objects, or features, of the domain;
- the relationships between the domain objects;
- the rules, or heuristics, used in problem-solving; and
- the strategy applied to the problem-solving.

The final remaining question is that of where this knowledge resides. In many engineering disciplines it will be argued that the knowledge of the domain exists in the many textbooks that fill the technical libraries. Whilst this type of knowledge is undoubtedly used by experts to solve problems, it is not the actual knowledge, but the application of that knowledge which is vital to efficient problem-solving. Most textbooks do not carry details on how the knowledge is applied, and so there is an onus upon the knowledge engineer to obtain that information from some identifiable source—the expert.

There are two main problems associated with this. Firstly, the expert may well recognise that he is the repository of this 'magical asset'. However, he may find that it is beyond his capabilities to express enough information about it to make it useful for anyone else. When posed with the question: 'How do you do what you do?', the domain expert is often actually at a loss to attempt to explain. Much of his knowledge cannot be immediately expressed in terms of facts. How does a test engineer with many years of experience formalise a 'gut feeling', such that it can be expressed in adequate terms?

Secondly, the expert may well be making use of knowledge of which he is unaware. As outlined above, the expert will build rich connections between concepts within his domain, which then enables him to act as an efficient problem-solving agent. However, he may well not keep any model of the causality of the connections he maintains. If this is the case, then we must accept that there are some sources of knowledge which we cannot tap through knowledge elicitation techniques. These sources are stores of what has been referred to as 'tacit knowledge'. Some techniques from cognitive psychology are, however, being applied in knowledge engineering to attempt to deduce tacit knowledge from an analysis of expert behaviour.

Putting the issues of tacit knowledge aside, a further complication of the recognition of the human expert as the source of the required knowledge is characterised by the 'multiple expert' situation. This is not a problem when the experts agree, but the nature of experience may lead us to believe that it is a 'personal facet'. There is great scope for disagreement between the experiential models of different people (refer to Chapter 1 of this text for a discussion on the implications of an individual's 'world view'; the *Weltanschauung*). In this situation, there are three approaches. One is to abort the attempt to utilise an expert system (as proposed by Waterman (1986)), the second is to choose to believe one of the experts, and the third is to apply some form of averaging technique.

Another, often overlooked source of the required knowledge, is that of historical databases. These records of the cases upon which experts themselves have built their expertise should, in theory, enable us to extract the knowledge we require. Of course, the expertise we require is in an encoded form, and the data must be processed in some way to extract it. This has led to the use of machine-learning techniques as a method for knowledge elicitation.

4.1.3 The Expert and the Knowledge Engineer

Is an expert someone who is intelligent? Perhaps it is someone who is very knowledgeable. Perhaps it is someone who has gained much experience over an extended period of time. Is it possible to have an expert who is not intelligent? In the realm of knowledge-based systems, the answer to the last question must be 'yes'. The commercial (reality-based) objective of expert systems is not to build computer programs which can think for themselves, but more to build computer programs which can accurately simulate the behaviour of someone who has been identified as an expert. The expert is then identified as the person who currently undertakes the task which has been identified for knowledge-based system application!

In terms of choosing the domain expert* to work on a particular project, there are some guidelines which will assist in the project, if they are addressed adequately. The expert should be someone who has been performing his task for several years, and who has gained true expertise, rather than text book knowledge. It is vital to choose an expert who is truly interested in the project. An expert who has been forced into cooperation will not be very much use. An additional characteristic of the expert to look for is experience in teaching about his domain. This experience of communication of concepts will prove valuable.

Of course, the main problem with any expert will be that they tend to be very busy people. This factor has a great impact upon the knowledge acquisition process, placing an emphasis upon the knowledge engineer to carefully plan the elicitation sessions, in order that they may 'fit in' with the expert's timetable. The knowledge engineer must also be prepared to be flexible, graciously accepting that interviews will be cancelled at the last minute.

In some small-scale applications, it may be the case that the expert and the knowledge engineer are the same person (although this is not generally regarded as a good solution). The reasons for this lie in the fact that it is very difficult to extract knowledge from oneself.

---

* The question of one expert *versus* many is not addressed by this chapter in detail. The criteria for selection of the expert, however, can be applied for multiple experts.

Although many of the interpersonal barriers are removed, the process will depend totally upon introspection, leading to an inadequate description of the knowledge.

The knowledge engineer, then, will normally be an external agent, unfamiliar with the domain. His task is highly dependent upon interacting with personnel in a manner which puts them at ease. Thus, the knowledge engineer must be an 'all-round-nice-guy'. Hart (1989) identifies some of the interpersonal skills required as:

- communication skills,
- intelligence,
- diplomacy,
- patience and persistence,
- logic,
- versatility.

4.1.4 Knowledge-based Systems and Prototyping

The development of knowledge-based systems is historically an ill defined 'art'. However, with the rapid growth of interest in such systems, and with the plethora of available programming paradigms, interest is being generated into exactly how to control the development of such systems. In this section, the intention is to give an impression of the nature of expert systems development, and how it differs from conventional software programs.

Figure 4.1 shows what can be viewed as the 'traditional' lifecycle for conventional software projects. In this, there are well defined phases of feasibility study, requirements specification, system design, system production, system acceptance, and system

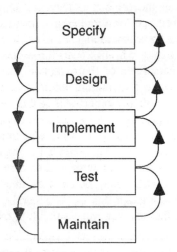

Figure 4.1 A typical software lifecycle

maintenance. There tends to be an iteration between adjacent levels in the process, although too much iteration can lead to the two levels becoming indistinct. It can be said, then, that the conventional software engineering cycle depends upon the ability to prespecify the problem.

Historically, however, the development of knowledge-based systems has been associated with an evolutionary development process, much removed from the traditional lifecycle model. The larger part of expert systems development has revolved around the development of prototype systems. Initially, it was hoped that the development of prototype systems was a short step from the development of deliverable, larger scale systems. This, however, has proved to be incorrect. The debate concerning what role, if any, the prototype should play in expert systems development 'rages on'. In this section, the intention is to present a view of the use of prototyping in various roles within the development process. Current expert-systems programming environments provide a broad range of facilities which enable the rapid development of software modules. These include features such as incremental compilation, which allow the code to be updated and tested in a very interactive cycle.

The prototyping of hardware products is a commonplace practice. It was soon recognised that their use to demonstrate quickly and cheaply the viability of a new product, as well as providing a good deal of assistance in defining that product's specification and initial problem-solving, was of great value. In software engineering, prototype systems are much less common. This situation has arisen from a lack of tools that really support software development in ways which are conducive to a prototyping approach.

Common prototyping approaches in knowledge-based system development can be partitioned into two broad categories (Figure 4.2):

1. prototyping for definition; and
2. prototyping for development.

The first category is based upon the use of 'throw-away' prototype systems. A prototype is developed to enable the system developers to investigate features of the problem. These can include features of the user–interface, the way in which knowledge is used, the identity of that knowledge (or a definitive subset of it). The prototype is a working subset of the final system which is used to gain the confidence of the experts (and the funding body of the project). The use of this approach is based on the belief that it is very rarely the case that the transition from prototype to full-scale system is simply a matter of magnification in scale. The second category has, in the past, been the most common. It would also seem that it has been the source of most of the problems. This process operates on the principle that the final system is an 'evolution' of the initial system. The prototype is simply extended to achieve the final implementation. The problems with this approach have been mentioned above, but there are also benefits to be gained. It is unlikely that an initial specification, even at the prototype system stage, will remain fixed. As cultures within organisations change, this will impact upon the software requirements. Thus, if an evolutionary approach is taken to software development, then these changes can be taken on board. Of course, this has its negative side in that too much evolution may lead to complete disaster (initial objectives being negated for example). Another problem associated with this approach is that it is very difficult to monitor, control and document. What are the criteria for stopping the evolutionary cycle?

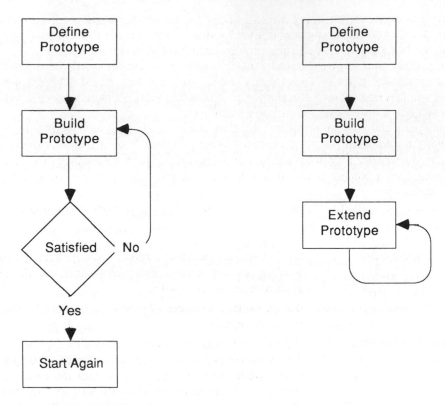

Figure 4.2 Two models of prototyping

The choice of which approach to take to systems development remains unclear, but there is a wide acceptance that prototyping is a useful tool in knowledge engineering, and specifically in the knowledge acquisition cycles. The ability to generate a working system quickly provides a great amount of impetus to a project. The domain expert is quickly able to receive feedback from the seemingly tedious knowledge acquisition process, and as such his interest is 'caught'.

4.2 PHASES IN KNOWLEDGE ACQUISITION

There is no definitive description of the process of knowledge acquisition, so in effect there are no real phases in the process. However, a general trend is appearing which tends to point to at least three distinct phases. Breuker and Wielinga (1987a) describe these three phases as Orientation, Problem Identification and Problem Analysis, as part of their knowledge acquisition and document structuring (KADS) methodology. Orientation is the acquiring of a basic understanding of the domain, its concepts, its terminology and other features. In the problem identification phase, the aim is to structure knowledge about the domain concepts, in order to identify the task to which the system will be applied. The final phase—problem analysis—performs an in-depth acquisition of the knowledge required, the user requirements, and forms a detailed specification of the final system. Despite these formative attempts at generating a methodology for knowledge acquisition, it would still appear that the knowledge engineer is armed with a variety of methods from which he must choose the best one for the particular problem, and level of detail.

This chapter aims to give the reader a 'feel' for some of the commonly used techniques and the particular types of knowledge to which they are best suited. The chapter then proceeds to describe some of the knowledge acquisition tools which have been developed, and how these utilise the aforementioned techniques.

In any particular problem under investigation, one must ask questions, such as: Is the task essentially a rule-based diagnosis problem? What part does 'rule-of-thumb' experience or generalised heuristics play in the solution of the problem? If it is established that the task is rule-based, then a process of identifying the series of rules which need to be applied must be embarked upon.

There are basically four broad approaches to obtaining knowledge from the chosen expert(s) :

Direct Approaches	the knowledge engineer interacts directly with the expert to obtain an explanation of the knowledge that the expert applies in his problem solving.
Observational Approaches	the knowledge engineer observes the expert in the performance of his task.
Indirect Approaches	the knowledge engineer applies methods through which it is hoped that the expert will reveal his expertise. This approach differs from the direct approach, in that the expert is not encouraged to try to verbalise his knowledge. The structure of his knowledge is derived from the results of the elicitation techniques.
Machine-Based Approaches	the knowledge engineer attempts to elicit the knowledge through the use of either knowledge-engineering languages or through induction from databases of domain examples.

Each of these approaches will be described in this chapter, and criteria for the application of each particular technique will be identified. However, before any of these particular techniques can be applied, an important phase must first take place; basic 'groundwork'.

4.2.1 Groundwork—Orientation

There is no substitute for basic domain knowledge. The first task of the knowledge engineer will be to identify suitable texts, perhaps with the help of the domain expert. These books should serve to provide the knowledge engineer with a basic understanding and a grounding in the vocabulary of the domain. The level to which the knowledge engineer must become educated is a matter of much debate. There is obviously a requirement for the domain expert and the knowledge engineer to be able to communicate about features and problems within the domain, and thus an understanding of the terms used is a prerequisite. Of course problems will arise in domains where the terms and jargon used are not consistent or standardised. In these cases, some common ground must be found to ensure that, in the very least, the expert and knowledge engineer are conversing in common terms.

It must be emphasised that the objective of the groundwork phase is not to become in any way expert in the domain. Problems with too much domain expertise can arise when the knowledge engineer attempts to utilise what has been learned to explain the expert's behaviour. This has two main effects. Firstly, it has already been mentioned that studies have indicated that the expert structures his knowledge in a different manner from the novice, thus attempting to apply the novice's structure to reveal the expert's can be counter-productive. At a more personal level, it will be very easy to alienate the expert, who may already be feeling defensive, if the knowledge engineer gives the impression that he believes himself to be able to predict the expert's behaviour.

By the end of the orientation phase, the knowledge engineer should have succeeded in answering questions relating to the possibility, justifiability and appropriateness of an expert system development (as described in the problem identification stage earlier). In addition, an appreciation of the personnel involved with the project should have been gained— this includes the prospective users and the experts. The knowledge engineer should have identified the relevant texts, and thus gained an appreciation of the domain vocabulary. Another important factor to research is that there are other examples of similar systems reported in the journals, as this will provide a powerful aid in the control of the project.

4.2.2 Direct Approaches

The objective of these elicitation approaches is to provide a framework in which the expert is able to verbalise his thoughts about his problems, and how he approaches their solution. However, as a result, the elicited information is not necessarily an accurate representation of what the expert actually does. All that is produced is verbal data, which must be interpreted by the knowledge engineer. It is widely accepted that the use of verbal data has certain associated problems. The expert is possibly not accustomed to explaining himself, and as a result may generate explanations which he thinks are correct (in a theoretical sense), rather than the ones he would actually use. The expert may forget facts, or may withhold information, or may not verbalise that which he considers common knowledge. Finally, language is often ambiguous. The expert will use his own knowledge to support his explanation, thus not appreciating where misinterpretations may arise through a lack of that knowledge. Some of the following techniques attempt to address some of these issues, but it must always be accepted that there is a limitation on the accuracy of the behaviour of any system which is based purely upon knowledge elicited through verbalisations.

4.2.2.1 Interview Techniques

Most knowledge elicitation strategies will revolve around interviewing the domain expert. Interviews serve many purposes, and can act at most levels of detail within the knowledge acquisition phases. However, the process can be slow, and sometimes unrewarding. Quinlan *et al*. (1987) state that one 'rule-of-thumb' puts the rate of knowledge acquisition at between two and five rules per man day!

In the early stages of knowledge acquisition, interviews with the expert can serve as useful sessions to increase the knowledge engineer's familiarity with the language and concepts within the domain. These interviews will tend to be very unstructured at first, but as time progresses, the purpose of the interview will change from general orientation to specific detail acquisition. Unstructured interviews will usually consist of sessions in which the expert is allowed to be anecdotal, and generally very relaxed. The analysis of transcripts of unstructured interviews is not guaranteed to reveal much of value to the final system, but unstructured interviews certainly form an important part of the education of the knowledge engineer and the domain expert. They also form an excellent means for the interviewer and the expert to become better acquainted, in a technical environment. This is an important point; the human issues of knowledge acquisition cannot be underestimated. To be effective, it is important for a relationship to be built up between the knowledge engineer and the expert.

It is vitally important that some form of record is maintained of each interview. The data which is generated in these sessions will form the basis of analysis, and is thus valuable. In addition to this, the knowledge engineer needs to keep the interest of the domain expert for extended periods, over several interviews, and this interest will be quickly lost if the expert is forced to repeat the same information several times. The methods which can be used for interview recording include taking notes, using audio-recording equipment and the use of video-recording equipment. The benefits of using recording equipment are the accuracy of the transcripts, and the use of video also captures any gesticulation or body language which may carry some meaning. It must be remembered that any form of formal recording must be employed with some care, as most experts will find the idea of being recorded oppressive, especially with video equipment. If the chosen medium is to take notes, then it may be a good idea to introduce either another knowledge engineer to take notes or perhaps a stenographer. The use of a third party allows the knowledge engineer to concentrate more on the direction of the interview. However, the expert may again find this uncomfortable. Another factor to take into consideration is the level of commitment which the expert is willing to give. This is often dependent upon the approach taken when defining the role of the proposed system. An expert who has been instructed by 'higher authorities' to attend interview sessions with some 'computer boffin', is likely to be less helpful than one who has been pre-educated as to how the proposed system will help him personally and how he can help the project (the latter consideration may perhaps be a more powerful psychological ally).

It is important to note some of the major hurdles to interviewing techniques. These can be broken down into two main groups, those which are avoidable and those which are unavoidable. The problems which are avoidable arise from a failure to truly enlist the help of the domain expert. For example, the expert may be worried that the expertise he may describe is too obvious, thus lowering his status as expert. Similarly the fear of 'replacement by a machine' is one which must be firmly laid to rest before successful knowledge acquisition can take place. The second class of problem exists because there may be a difference between the way in which an expert performs a task and the way in which he perceives that he performs that task. Indeed, it is common to find that the expert cannot describe how certain conclusions were reached, in effect they cannot state what it is that they know.

Bearing these points in mind, it is important to lay down some basic guidelines for interviewing. These are described in Reitman, Olsen and Reuter (1987) as:

Enlist expert's cooperation   As outlined above, this is very important to the success of the whole project. It is really a matter of marketing the product (the expert system) in such a way as to capture the imagination of the expert.

Ask free-form questions   As an initial step, ask free-form questions. Become more specific as the interview progresses. In this way the knowledge engineer becomes familiar with the vocabulary before attempting to identify the relationships and inferences the expert makes.

Do not impose understanding   Your own understanding should not be imposed on the expert. The knowledge engineer should only interrupt to ask about things that he does not understand.

Limit the sessions   Limit the sessions to coherent tasks, recognising fatigue and attentional limits. This is a careful balance. It is important not to make interviews too long, but it is also important not to interrupt the interview at an inopportune moment. As Olsen and Reuter point out, it is almost impossible to 'take up where we left off' if an interview finished with important features not finalised.

A large body of work exists, largely produced by cognitive psychologists, which describes different interviewing techniques for knowledge acquisition. The following sections describe the more notable of these.

4.2.2.2 Tutorial Interview

In this situation, the expert is asked to give the knowledge engineer an introduction to the domain through a tutorial. The tutorial session should be recorded, in order that a transcript can be generated, from which the knowledge engineer can draw information. In the tutorial, it is important for the expert to outline the main features and vocabulary of the domain. The benefit of this approach over textbook analysis is that a clearer impression of the way in which the expert relates domain concepts is given. It is also important to note that in many domains, there is no total agreement over techniques, and so the views of the domain expert may be 'at odds' with those of the expert who 'wrote the book'. The use of tutorials goes some way towards overcoming this by presenting the views of a particular expert.

4.2.2.3 Teachback Interview

This technique places a joint effort on the knowledge acquisition process between knowledge engineer and domain expert. The definitive description of this technique can be found in Johnson and Johnson (1987), which describes the use of the technique on two case studies. The first uses a child as the expert and attempts to capture knowledge about arithmetic. The second is an example from VLSI design.

The basic process is for the expert to 'drive' the interview by describing a feature of his domain. The task of the knowledge engineer is then to teach that concept back to the expert, in the language of the domain. This cycle is iterated until the expert is satisfied that the knowledge engineer has fully grasped the concept. At this stage it is said that the expert and the knowledge engineer share the concept. However, the process does not consider the sharing of the concept as being the same as understanding that concept. In order to achieve understanding, it is necessary for the expert to describe how the concept was reconstructed. Again, the knowledge engineer must teach this explanation back to the expert until he is satisfied that the knowledge engineer understands. These two levels of interviewing are based upon Conversation Theory (Pask, 1975), where Pask's Level 0 is the shared concept, and Level 1 is the knowledge about the concept. When this state has been reached, the expert introduces a new concept and describes the way that this relates to an established concept.

The benefits of this technique are gained from the fact that between them, the domain expert and the knowledge engineer are defining a formal, shared, definition of the domain. In addition, this technique will overcome any problems which may have been caused by the knowledge engineer attempting to impose some of his own personal understanding of the problem, as this will be highlighted during the teachback process.

4.2.2.4 Introspection

This is a general term for techniques which are used to produce a verbal report of the expert 'imagining' how he would solve a particular problem. Three techniques are used to facilitate this: retrospective case description, critical incident reports and forward scenario simulation. In retrospective case description, the expert is asked to recall how he dealt with a selection of typical cases; in critical incident reports, the expert is asked to describe how he dealt with extraordinary cases; and in forward scenario simulation the expert describes how he would deal with a hypothetical case. This hypothetical case can be chosen either by the expert or by the knowledge engineer. Of course, if the hypothetical case is chosen by the knowledge engineer, then much care must be taken to ensure that it is a realistic scenario. In the use of retrospective case description, it is important to ensure that the cases described are typical, whereas in critical incident cases the opposite is true. It has been noted that problems can more readily arise with retrospective accounts as human memory is fallible (Ericsson and Simon, 1984), and there may be a tendency to 'retrofit' the explanation with hindsight. The benefit of the critical scenario method over the retrospective method is that there is a tendency to remember critical incidents because of their interest factor. Hart (1989) also identifies that, in the description of typical cases, the expert is more likely to omit mundane details.

4.2.3 Observational Approaches

4.2.3.1 Verbal Protocol Analysis

The technique of verbal protocol analysis is designed to provide a detailed account of the way in which the expert solves his problem. The expert is required to give a description 'out-loud' of the tasks he performs as he performs them. The objective is to describe what it is that he is doing, why he is doing it, what other factors he is considering, and where he hopes it will take him. A recording of this is made, and from this a transcript is prepared. Various analysis techniques can be applied to this protocol, from which it should be possible to extract the rules that the expert uses. The text can be analysed to identify the dominant concepts used by the expert, and the steps that he takes to solve the problem. 'Maps' can be made of the problem-solving stages, highlighting where the expert backtracks, forward chains, applies heuristics, etc. The definitive work in this field is given in Ericsson and Simon (1984).

The benefits of this technique are based in the fact that the expert is not placed in an unnatural situation. He is observed actually performing his task, and as such some of the pressures of explaining about his knowledge are removed. Here the emphasis is much more on obtaining data to enable the knowledge engineer to derive the knowledge.

The problems with this technique, however, are widespread. Many experts will find it uncomfortable to explain in this manner, and may regularly need reminding of the objective of the technique. In some cases, the effort that the expert needs to apply in order to verbalise his thoughts will interfere with those thought processes. In addition to these (personality) problems, there are problems with verbal data. It is possible, for example, that the expert will make verbal explanations of processes that do not actually occur, in order to provide some foundation for a particular step. In addition, there are problems with the expert actually considering certain factors as *common-sense* and, as such, not worthy of articulation.

4.2.3.2 Dialogues

Often in the development of an expert system, the needs of the actual system users are overlooked. This is an important factor, as the user is not usually intended to be the expert. The recording of dialogues between the expert and prospective system users is thus an important phase of any system development. In addition to this, an analysis of these dialogues will often reveal much information relating to the concepts that the expert considers, and the strategies he employs in the problem-solving process.

There are several examples of the application of these techniques in the literature. Presented below are a selection, which may be used to elucidate the variety of methods which can be used, and the data that they identify.

One broad grouping are those techniques which record the dialogues between the expert and the user. An example of this is described by Fenn *et al.* (1986), in a medical application. The technique involved three parties: two experts and the system user. The roles of the experts are as patient (to give information about the illness and its symptoms accurately) and as expert system. The expert system was physically separated from the user and the patient, with a communication channel via an intercom. This enabled the user to undertake his task and interact with the system to obtain the consultation he required. This scenario provides a good model of the user interaction required. Similarly, the additional information required by the expert in order to answer the user's query, provides information relating to the strategies employed by the expert.

Another, similar approach, is that of the 'Wizard of Oz' described by Diaper (1987). This technique exists as part of an overall expert systems development methodology (POMESS— a People-Oriented Methodology for Expert Systems Specification). In this system, the user is led to believe that he has at his disposal an expert system (perhaps for trial, etc.). In reality, the system consists of a piece of terminal driver software which allows the user to communicate with the human expert via an electronic medium. This again highlights both user-interface design information and expert inferential processes. Diaper also mentions a tool, known as the 'Wizard's Apprentice', which will perform an analysis of the dialogues and encode these into rule structures which will assist in the generation of the final expert system.

An alternative emphasis was placed by Stefik (1989), when he described a scenario in which the knowledge elicitation is driven more by the expert. He described a methodology in which the expert is entirely divorced from his normal environment. For example, in the scenario of fault diagnosis, the expert engineer is placed away from any source of information about the fault. The expert then utilises some domain novice as a remote problem-solving operative. In this methodology, a clear model of the data required to solve the problem is obtained. The tests that the expert performs and the ordering of these is also detailed.

4.2.4 Indirect Approaches

4.2.4.1 Repertory Grid Analysis

This is arguably one of the best developed of knowledge acquisition methodologies. The technique is based upon the work on personal construct theory in clinical psychology, reported by Kelly (1955). The technique operates upon the principle that each of us holds our own personal model of the world about us (see Chapter 1 for a discussion of this). In theory, then, if we can capture this model we can capture the knowledge we use to solve problems. The ideas have been adapted for knowledge-based system by Boose (1986) and Shaw and Gaines (1986).

There are two stages in the use of the 'Repertory Grid'. The first stage is the development of the actual grid. This grid is a mapping of features of the domain/problem, which act as an example set for attributes which are common between these examples. The terminology commonly found refers to the examples as elements, and the attributes as constructs. It is important that these constructs are bipolar, i.e. heavy/light, fast/slow. They are normally attributed a scale of 1–3 or 1–5, although for acceptability these integer values are often substituted with textual equivalents, i.e. very-fast, fast, medium pace, slow, and very-slow. There must always be the same number of grades for each construct, as the final analysis relies on a scaling technique which measures the 'distance' of one concept from another on the chosen scale. This form of analysis is often referred to as Multi-Dimensional Scaling (MDS). The choice of construct is performed by presenting the expert with sets of three elements. The expert is asked to identify a construct which distinguishes two of the elements from the remaining one. This process is repeated until the expert is unable to identify new constructs or until all possible combinations of elements have been examined.

As an example, consider a domain which involves motor cars of types A, B, C, D, E and F. If we ask our expert to identify relevant constructs and then assign values between 1 and 3 to all the cars, we could generate a grid as in Figure 4.3. The layout of the grid

	A	B	C	D	E	F	
Prestigious	1	3	2	1	3	1	Not prestigious
Sports car	2	3	3	2	3	1	Saloon car
High passenger comfort	2	1	3	3	2	2	Low passenger comfort
Good fuel economy	2	1	1	2	2	3	Poor fuel economy
High service costs	2	3	2	2	3	1	Low service costs
Rare	1	3	2	1	2	1	Common

Figure 4.3 Repertory grid for car types

is such that the rows represent the constructs, with the left-hand element showing the high value end of the scale (i.e. in the first row the value 3 represents Prestigious, and 1 Not Prestigious).

This grid forms the basis of several forms of analysis. The objective of these analyses is to form an understanding of the similarities or relationships between the elements and constructs defined. The technique is useful for identifying classifications of domain objects, making access to the expert's personal model of how his domain is structured. Also, as the technique relies on simple grading tasks, and attempts to derive structure from these values, it can produce groupings of which the expert may not have been aware.

The remainder of this example presents an analysis method which can be applied to the repertory grid. The initial point for this analysis is a grid of measures of distance between concepts, and in this case we will take the absolute value of the difference in value for each particular element/construct pair. This gives the grid in Figure 4.4. To clarify, in order to calculate the value for A *versus* B, we take all the absolute differences of the values in the first two columns of the initial grid (Figure 4.3) and sum them.

	A	B	C	D	E	F
A	—	8	5	1	5	3
B		—	5	9	3	11
C			—	4	4	8
D				—	6	4
E					—	8
F						—

	A	B	C	D	E	F
A	—	33	58	92	58	75
B		—	58	25	75	8
C			—	67	67	33
D				—	50	67
E					—	33
F						—

For AB, differences are 2+1+1+1+1+2 = 8.
Converting to a percentage gives : (-100*8)/(2*6) + 100 = 33

Figure 4.4 Spacings for MDS analysis

Using this grid of spacings, it is possible to identify that elements A and D have a similarity rating of 92. Thus we decide to create a group (AD). From now on in the analysis, A and D will be considered as the same object. This means that the grid will need to be reconstructed with the spacings recalculated. This is performed in different ways, the most obvious being to take either the minimum value for the grouped pair, the maximum or the average. However, the method chosen must be used for the remaining analysis. For example, in this example we choose (arbitrarily) to take the minimal values. Thus for AD *versus* B, we take:

$$\min (A,B)(B,D) = \min(33,25) = 25$$

The reduction process is repeated until the grid is 2×2. This gives us the series of grids in Figure 4.5.

Once the grid has been reduced, it can be displayed as a hierarchy diagram, as in Figure 4.6. In this hierarchy, we see a structure of classes of objects, where the expert may perhaps give labels of 'economy cars' *versus* 'sports cars' at the top level.

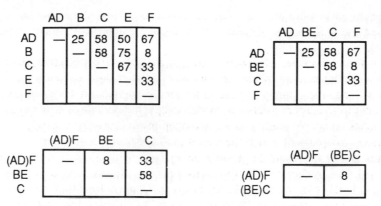

Figure 4.5 Reduction of repertory grid

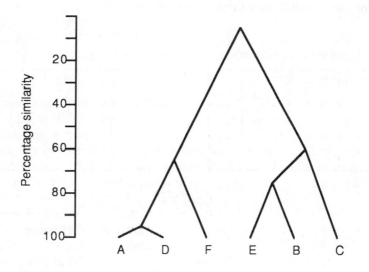

Figure 4.6 Resulting hierarchical clustering

This hierarchical clustering of the domain concepts reveals groupings which help to clarify the expert's model of the domain. Similarly, this type of analysis can be applied to the actual constructs used in order to evaluate similarities between them. In a later section, it will be seen how this technique has been automated in a tool called ETS, which can generate production rules directly from the rating grid.

4.2.4.2 Card Sorting

This technique, also referred to as conceptual sorting, is useful for eliciting the hierarchical structure of domain concepts or similarity judgements between concepts, as in the repertory grid technique. The common form for the application of this technique is for the knowledge engineer to identify a set of domain concepts which he then 'writes down' on individual cards. These cards are presented to the expert, who is then asked to sort the cards in various ways. One method of sorting involves asking the expert to divide the cards into piles, giving the criteria for membership of each pile. This process is repeated with each of the resulting piles, producing subgroupings.

One technique is to sort the cards into as many small groups as the expert can identify. These groups are then combined into higher level groups, forming a hierarchy of concepts in a bottom–up manner. Gammack and Young (1985) report the use of these sorting techniques in a project dealing with central heating systems. Here the expert was presented with 75 concepts, from which 25 groups were identified. Criteria for each group were identified, and then the groups were amalgamated. In this case the expert was asked to give a verbal account (see verbal protocols, discussed earlier) which was found to be of great use.

An alternative approach is to sort the cards in a top–down manner. Here the cards are sorted into large groups, followed by successively smaller groupings. This technique has been reported by Wright and Ayton (1987). However, Neale (1988) makes the point that, although this method may seem the more natural way to sort, a higher level of risk of duplication of classes within the resulting hierarchy exists.

4.2.5 Machine-Based Approaches

4.2.5.1 Machine Induction

A body of work exists which concerns itself with the issues of attempting to identify knowledge, or general principles, from a set of specific instances. This work has a direct impact upon the knowledge acquisition process as, in theory, its use can circumvent problems associated with the direct elicitation of knowledge from human experts. The induction process takes a set of examples and from that set attempts to define general rules about the examples. These rules, when applied to further examples, will reach (correct) conclusions.

An example of induction could be, on the basis of observing swans, the observer may induce a rule that all swans are white. However, this induced rule is made upon an observation of swans which is not necessarily representative of all swans. Thus the rule does not take into account the existence of black swans. Thus the results of an inductive system can never be guaranteed to be accurate, it can only be as good as the data it has been given (see Chapters 2 and 8 for a further discussion of this kind of anomaly).

Prior to using an inductive algorithm, a certain amount of preprocessing of the data set is required. Initially, the domain expert will be required to check that the data set (the examples) is representative of the domain/problem. This, of course, is a subjective process, but the examples can be easily updated on the basis of the results of the induction. In addition, to verify the completeness of the examples, the expert will identify features which will allow him to make statements about, or comparisons between, members of the set (the attributes). There is little constraint on the nature of these attributes, although early algorithms could only deal with finite classifications such as colour, rather than continuous scales such as numerical values. Finally the expert must identify a set of classes. These classes are, essentially, the set of possible outcomes (normally a classification) for any object in the example set. These must be chosen such that any object belongs to only one class. Once these phases have been undertaken, it is possible to use an algorithm to build rules (normally in the form of a decision tree) which correlate an object's attributes to its class.

There are several well documented experiments which used machine based rule induction to generate rule bases to solve particular problems. The most widely quoted example is that of Michalski and Chilausky (1980). In this example, an induction system was used to derive rules to aid in the diagnosis of soybean diseases. In this experiment, an example set of 630 cases was used. From this set, 290 were actually chosen as training examples for the induction system, by a system called ESEL, whose objective was to identify examples which adequately cover the domain (the remainder being used for testing). Within this, 35 attributes were identified which described features such as environmental conditions like temperature as well as identifiable features of the plants, such as seed size, root condition, leaf condition. From these attributes, the examples were classified as belonging to one of 15 possible classes, i.e. 15 soybean diseases. The system used for the actual induction is known as AQ11. This example is interesting, as both machine induction and direct knowledge elicitation were used and the resulting systems compared. The system, based on human knowledge actually did not perform as well as the induced one, although it must be pointed out that the knowledge elicitation techniques used are not well documented, and that the amount of time involved in actually acquiring this knowledge was around twenty hours.

Another very well documented programme of work is that which is based on the ID3 (Iterative Dichotomiser 3; (Quinlan, 1983) algorithm and its derivatives, which in turn is based upon the Concept Learning System (Hunt *et al*., 1966). The ID3 algorithm develops a decision tree based upon a set of examples, and was originally developed to classify King–Rook *versus* King–Knight chess end-game moves. The algorithm works in the following way. First the system selects a subset of the training data, known as the window. The CLS algorithm is applied to this window, which takes the form of the algorithm searching for a test which discriminates most successfully the data set. This process then repeats on each of the two newly formed classes, and then iteratively until each subset contains only one element. At this point, the system holds a postulated decision tree. The ID3 algorithm then takes control, with the system searching the database for exceptions to the current decision tree. If any are found, then they are inserted into the window and the process repeated. Finally, the system will come to a satisfactory tree.

The original ID3 algorithm was fairly limited, as the data upon which it was developed was very well defined and well structured. In addition, the choice of attribute was limited to those with finite, discrete membership. The results of the algorithm are highly dependent upon the quality of the initial training set of examples, and upon the choice of sufficiently adequate attributes. Nevertheless, a form of the ID3 algorithm forms the basis of a package called EXPERT-EASE, and has proved to be successful in a number of examples. Criticisms levelled at the ID3 algorithm were that it did not perform well on so-called 'noisy' data, that the selection of attributes was a very difficult task, and that the rules produced were too large and complex for human consumption. In addition, because the induction system holds no knowledge about the domain, the ordering of the questions which may be asked in the problem-solving process is not known. As a result, it is possible for the system to ask questions in a nonsensical order; a trait which must be captured and corrected by the knowledge engineer.

The use of machine induction as the sole method of knowledge acquisition is questionable. However, its use as an initial stage in the process is one which holds great promise. To begin to discuss the content and nature of an expert's knowledge from a 'blank page' situation is very difficult. However, to present the expert with a decision tree, based upon data from his domain, gives the process an initial boost, identifying specific subjects for discussion, and some form of structure around which to discuss/experiment.

4.2.5.2 Knowledge Engineering Tools

There are a variety of systems which can be used to support the knowledge acquisition process. These vary in the amount of support given, but can be roughly categorised as :

Programming Languages	With these, little support is offered for actual knowledge acquisition tasks, but many of the languages (Lisp, Prolog, Poplog, SmallTalk) offer advanced development facilities which greatly enhance the system development and test phases of knowledge engineering.
Shells	These are usually systems in which the scope of knowledge representations is limited. They tend to be based upon production rules, and some may offer both forward-chaining and backward-chaining inferencing. Of course, as with all areas of software, the technologies offered are regularly increasing, and it becomes more difficult to categorise a particular product as either a shell or a programming environment.
Expert Systems Programming Environments	These offer much support for a wide variety of well known knowledge representation and reasoning paradigms. The best known of these are KEE, ART and Knowledge Craft. They tend to be very flexible, in fact one of the major criticisms aimed at these products is that they offer no clear methodology to the developer. It is up to the knowledge engineer to choose from 'a bewildering array of possibilities' and utilise them in an appropriate manner. (Reichgelt and van Harmelen, 1986).
Knowledge Engineering Systems	These systems are environments designed around some of the knowledge acquisition methods described in earlier sections.

Descriptions of programming languages, shells and expert systems programming environments can be found in Chapter 2, with appropriate examples. Therefore, the remainder of this chapter will be restricted to the knowledge engineering systems in the above classification. It is important to note that these systems are mostly research systems, and, as such, are not (generally) commercially available.

4.2.5.3 TEIRESIAS

The TEIRESIAS system (Davis, 1979) was developed at Stanford University as an aid to the automation of expertise transfer from human expert to computer system. In this case, the computer system was MYCIN, one of the original expert systems which was also developed at Stanford. The aim of the TEIRESIAS system is not necessarily to assist in the development, or identification, of a knowledge representation scheme, but rather to introduce some level of 'expert' control over the correcting and updating of system knowledge bases. Thus, its use lies in the iterative refinement cycle of expert systems development, as at the outset it requires a prototype knowledge base with which to work. Davis clearly states that TEIRESIAS makes an initial assumption that the problem-solving approach of the expert system is correct, i.e. the choice of inference, knowledge representation scheme, etc., and that TEIRESIAS acts as a supervisor to the extension of what the system knows rather than how it works.

The task of updating and extending a knowledge base is certainly non-trivial. Due to the nature of the interrelatedness of the data structures used, one change can have wide-reaching effects. To attempt to control this, TEIRESIAS holds models of the knowledge structures of the expert system that it is working with, and how they may relate. Thus, it holds information about typical objects, attributes and rules within a particular application. In this way, it can trace the possible effects of changes, identifying where other conflicts will have to be resolved. One example may be that in the course of defining a new rule, the expert may introduce a new concept that TEIRESIAS recognises as being unknown to the expert system. From its understanding of the knowledge structures, TEIRESIAS will identify the type of the concept and hence a schema which defines its structure. From this TEIRESIAS can guide the user through the process of inputting information about that concept, which will be necessary for the expert system. In effect, TEIRESIAS is a canny knowledge-base editor, which uses an understanding of the behaviour of the shell (Mycin), in terms of its knowledge structures to control the updating or correction of the knowledge base.

The typical flow of interaction with TEIRESIAS would be as follows. The user, who has been identified as an expert user and thus has higher privileges than an 'everyday user', interacts with the expert system, giving a case study. When the expert system has reached its conclusion, TEIRESIAS will prompt the expert for his opinion as to the correctness of that decision. There are basically two types of error possible within the expert system :

(1) The conclusion is incorrect. An incorrect conclusion will be reached in one of two ways: the concluding rule has a condition which should not have been satisfied; or the concluding rule has a number of premises missing which would have prevented the rule firing in this situation.

(2) A different conclusion is missing. In this case: either rules which should have fired have not done so; or there are rules missing.

In the case of an incorrect solution, TEIRESIAS will step backwards through the inferencing process, presenting the expert with the train of information (conclusions) which led to the final rule firing. The expert can identify where an incorrect premise exists, or where a preventative premise is missing. In both cases, TEIRESIAS will assist the expert in updating the faulty rule. In the case of missing conclusions, TEIRESIAS will locate all the rules which could have lead to that conclusion, and highlight why they did not do so. Again, the expert is assisted in the task of correcting these rules, if that is appropriate. Alternatively, there may be a missing rule which TEIRESIAS will assist in generating.

TEIRESIAS can also have an impact upon the initial knowledge model development. In essence, TEIRESIAS holds 'knowledge about representations'. Using this, it can help the knowledge engineer to define the template knowledge structures for use in a particular application. These templates are the mechanisms through which TEIRESIAS is able to control the knowledge refinement process, and as such their definition is vital. Thus, in much the same way as for domain specific data, TEIRESIAS holds models of general-purpose knowledge structures, which it uses to guide the process of definition of domain specific structures.

4.2.5.4 ROGET

Another system, linked to the MYCIN project, is ROGET (Bennett, 1985). This is a tool which assits in the domain conceptualisation phase of system development. ROGET holds knowledge about how a knowledge engineer approaches the problem of identifying and structuring domain models. ROGET also holds a library of conceptual structures, which are models of knowledge structures which have been found to be useful for particular diagnostic applications. The system aims to help the knowledge engineer to define his particular knowledge structure by evaluating and classifying his particular problem. This is performed by ROGET holding a dialogue with the expert in which concepts, relationships and inference strategies are discussed. Through the use of a thesaurus system and a parser, ROGET attempts to classify the diagnostic task being proposed and hence identify the appropriate structure. One of the results of this phase is that ROGET is able to provide an evaluation of the feasibility of the particular project. Once the domain model has been identified, the user is assisted in filling out the structure, which can then be automatically translated into an EMYCIN* representation.

ROGET only currently works on domains which are similar to that of MYCIN, although Bennett claims that it could be applied to other problem types such as planning.

4.2.5.5 ETS—Expertise Transfer System

The Expertise Transfer System (ETS) system (Boose, 1986) is based upon the repertory grid technique, which was described earlier. One of the features of the repertory grid is that it is particularly well suited to automation, thus allowing the expert to undertake a transfer of expertise directly into a machine representation. In common with TEIRESIAS, ETS attempts to facilitate expert system development via a process of gradual refinement. However, ETS requires no initial rule base.

ETS undertakes an automated interview process with the expert, with the initial aim being the identification of a set of domain elements. These are usually the concepts into which the final system is to classify any problem. In the example detailed by Boose, this is a set of possible cities for a holiday destination, with the final system being a travel advisor. Once a set of elements has been identified, ETS presents the user with sets of three elements, and prompts for traits which differentiate two from the other one. This involves identifying both poles of the trait, for example Warmer/Colder. Once this point has been reached, the initial grid has been formed, and now must be filled in. ETS prompts the user to give ratings of 1–5 for each element against each trait, in addition it is possible for the user to rate an element as 'B' or 'N' (both or neither).

Once the grid has been completed, ETS undertakes an analysis phase, whose objectives

---

* EMYCIN is the expert system shell which was derived from the MYCIN system.

are to refine the grid to remove redundant traits and elements, and to identify the need for additional ones to complete a sufficient solution. The first stage is an analysis of the implications of the grid in terms of the relationships between the traits. Here, the equivalence of traits is highlighted, along with the location of reciprocal traits (i.e. trait A implies trait B, which in turn implies trait A).

At this stage, the user is asked to identify labels for each of the traits, and to rate their importance in this problem (again on a scale of 1–5). From this information and the rating grid, ETS is able to generate a set of rules which perform two tasks. One set of rules are termed 'intermediate rules' with the second set 'conclusion rules'. Conclusion rules are those which enable the final system to reach a conclusion about the classification of an object, and are generated from the grid ratings. Intermediate rules are those which relate traits together. The rules are generated with a certainty factor (a numerical representation of strength of belief) calculated from the information in the grid and from the relative strengths allocated to each of the traits. For example, in the motor car grid given in the section on repertory grid techniques, a conclusion rule can be generated which states that :

> *If*   prestige = High Prestige
> *Then*   car is B with certainty (0.8)

Similarly, a rule can be generated that links the trait of economy with that of maintenance costs :

> *If*   economy is Good Fuel Economy
> *Then*   maintenance costs is Low Service Costs (0.8)

At this point, ETS has built a simple expert system which will perform a classification task. However, it is unlikely that this system will be anywhere near complete. In recognition of this, ETS supports techniques for extension and refinement of the grid, and hence the rule base. In an interaction, the system will enable the expert to identify implications of traits that he does not feel are accurate, and give examples of exceptions.

Refinement of the knowledge base takes place through a variety of techniques supported by ETS. At all times, the expert can volunteer new elements and traits and is able to edit those that have already been given. In addition to this direct manipulation, there are automated techniques available. Firstly, the results of the implication analysis can be reviewed, with the expert correcting errors caused by the following faults:

- Incorrect ratings in the grid. Often the grids can become large, resulting in an increased probability of input error.

- Too coarse a granularity of the grid. This results in the system making implications between traits which are correct according to the grid, but not according to the expert. This is resolved by the expert presenting an exception to this trait and rating this new element in the grid.

- Inconsistent use of traits. A trait can be quite easily used in an inconsistent way. For example, a car may be classed as 'sporty'. However in some cases the expert may mean that the car *looks* 'sporty' and in others it may mean the car's *performance* is 'sporty'. ETS uses a laddering technique to resolve some of these ambiguities. Laddering uses a series of 'why' and 'how' questions to identify superordinate and subordinate relationships between the traits in the model. Superordinate traits tend to be the most important, but can only be expressed in terms of their subordinates. Through this technique, the inconsistent use of a trait can be recognised because the sub/superordinates will be different.

A second process is that of similarity analysis. Here ETS calculates a numerical value for the similarity of two elements from the grid values—in much the same way as shown in the motor car example. Those whose similarity is greater than a certain threshold value are reported. Again in the car example, cars A and D have a similarity of 92%. ETS will prompt the expert for a new trait which will enable the final system to distinguish between the two elements. This trait is then added to the grid, and values for all elements are given.

ETS can be used to generate prototype systems in very short periods (Boose talks about timescales of 2 hours). However, it is questionable whether ETS can truly be used as the sole method for large expert systems development. This has been acknowledged, and ETS now forms part of a larger system, AQUINAS (Boose *et al*., 1988), in which it performs an initial prototyping task.

4.2.5.6 KADS—Knowledge Acquisition and Document Structuring

The KADS toolkit (Breuker amd Wielinga, 1987b) is one of the most well developed to date. It has evolved as a set of tools which have been designed to support a methodology for knowledge-based system design and implementation, and now forms the heart of a European ESPRIT project.

KADS can be regarded as an attempt to move the development process of knowledge-based systems more towards that of conventional software. This is performed by providing a methodology which attempts to prespecify the problem accurately. The output of KADS is a detailed 'documentation handbook', which is used in the coding phase of the project. In this way, KADS decouples the analysis and design phases of the solution method. The main part of the KADS methodology deals with the analysis phase of system development through three cycles of knowledge acquisition. The cycles are referred to as 'Orientation', 'Problem Identification' and 'Problem Analysis'. The aim of these cycles is to allow but also control a form of iterative refinement. The KADS methodology identifies four main issues for investigation, these being 'Modality', 'Feasibility', 'Knowledge' and 'Task'. The various phases will identify different aspects of any of these issues. In the orientation phase, the goals are to identify the domain vocabulary, the characteristics of the domain and the tasks that are performed by the expert. In the problem identification phase, the knowledge engineer aims to identify the structure of the knowledge within the domain. A functional analysis is also performed to clarify the way in which the final system will be used, along with an analysis of the actual tasks that the system must perform. The final phase is the problem analysis phase. Once the problems have been identified, it becomes possible to analyse them in more detail. The KADS methodology also identifies pertinent acquisition techniques for each of these phases, mainly based around verbal protocols, dialogues, introspection and interviews.

In KADS, a four-level model of expert knowledge is proposed. At the base is domain knowledge, which is characterised by object/relationship knowledge. The second level is the inference level, which models the types of inference that can be made. The third level— the task level—holds knowledge of goals and tasks. Finally, the strategy level holds plans, meta-rules and repairs. The aim of the system is to obtain a conceptual model of the domain, which can be translated into a design for the final system. The acquisition philosophy is based around the idea of an interpretation model, which acts as a template knowledge structure designed to assist the knowledge engineer in obtaining the correct 'ingredients' for a particular application. The choice of appropriate interpretation model is based upon an analysis of the task and inference data, which is gained through interview techniques, and the analysis of verbal protocols. The product of the KADS knowledge acquisition cycle is an appropriate conceptual model, which forms a high-level functional specification of the expert system.

KADS includes two tools called PED (Protocol Editor) and CE (Concept Editor). The PED tool allows the user to analyse a piece of text, identifying fragments of the text, relating fragments, attaching notes to fragments and relating fragments to concepts. The results of PED can be viewed graphically, in terms of a fragment network. Similarly, CE allows the generation of concept networks, which are presented graphically.

4.2.5.7 KEATS—the Knowledge Engineer's Assistant

The KEATS project (Motta *et al*, 1989) is a product of the Human Cognition Research Laboratory of the Open University. The idea of KEATS, is that it is a software environment which, as opposed to the approach of KADS, aims to produce a viable expert system. This means that KEATS is the final 'delivery vehicle' for systems developed under its methodology. KEATS uses a variety of tools to support the iterative development and refinement of the final system; these tools being CREF, GIS, and Acquist. In its methodology, KEATS concentrates upon two phases, termed 'data analysis' and 'domain conceptualisation'. Data analysis is the phase in which the knowledge engineer reviews the results of interviews (referred to as 'raw data'). The objective is to identify important concepts and information about them, and remove redundant information. This can be thought of as preprocessing the data before attempting to derive any structure from it. The second phase—domain conceptualisation—is the process of creating a structure of the domain knowledge.

CREF is a Cross-Reference Editing Facility, which supports the knowledge engineer in the analysis of the transcripts of interviews, protocol analysis, etc. In order to support these analyses, CREF enables the knowledge engineer to identify blocks of text (segments) and block them together into groupings (collections) according to two criteria. These criteria are semantic (related because of meaning) and syntactic (similar to aliasing). The output of this tool is a useful 'knowledge encyclopaedia' or a verbal model of the domain.

GIS is the Graphical Interface System which provides the knowledge engineer with a graphical approach to building models of the domain. These models are essentially graphs of nodes and links, with the nodes representing domain concepts and the links representing relationships between them. As the user builds his models, these are translated to an underlying frame-based representation which is continuously checked for consistency. GIS was originally intended to support the domain conceptualisation phase.

More recently, the KEATS project has concentrated upon the development of Acquist, a tool which has arisen from the work on GIS and CREF. It was discovered that, although CREF and GIS could be related, they needed to be integrated in a much more powerful manner. The tasks of data analysis and domain conceptualisation cannot be totally divorced; more reasonably a level of iteration between the two is needed. Acquist is a highly flexible textual analysis tool which has a much higher level of functionality than CREF. It concentrates around the definition of 'units of knowledge' which are classified as 'fragments, concepts, and groups'. Fragments are pieces of text within the initial transcript (which is generated in any of the ways detailed in the techniques section) which the knowledge engineer identifies. They can be of any size (i.e. from character through to whole transcript) and may even overlap. The fragments are treated as basic units of knowledge, from which it is possible to build a domain model. This phase is referred to as domain conceptualisation. Concepts are higher level collections of fragments which are created to enable the knowledge engineer to classify fragments. The creation of concepts arises from an analysis of the text, either through an automated lexical analysis of texts, identifying all the words which are used within a transcript or through the knowledge engineer locating the need for concepts to attach fragments to. In addition, Acquist allows the definition of 'filters', which remove words from the lexicon, which the knowledge engineer has identified as uninteresting. Finally, groups are collections of concepts, allowing a hierarchical structure of fragments, concepts and groups to be built. Acquist also provides facilities for linking fragments, concepts and groups together. A set of domain independent link-types have been identified, which can be augmented by any domain specific relationships that the knowledge engineer feels are necessary. Examples of link-types are :

> SPECIAL-CASE : a fragment is a more specific example of another
> REINFORCES : one element strengthens the other

Acquist also provides support for a graphical presentation of the results of the domain conceptualisation process. In much the same way as GIS, Acquist facilitates 'maps' which are networks of nodes (concepts, fragments, groups) and links (the relationships between them). The maps provide a powerful mechanism for organising the large amounts of data held in the textual transcripts. As the name suggests, the maps act as a navigational aid through the data.

The authors of Acquist refer to these domain models as theories. Their power lies in the fact that, through the use of Acquist, a detailed record of the rationale for the creation of concepts and their relationships is maintained. Thus, new knowledge engineers can easily work on the project and inherit the 'work done' by previous staff. This is an important point, as one of the perceived values of expert systems is that they capture knowledge and represent it in such a way as it becomes a useful commodity. Acquist supports the maintenance of multiple theories, and provides methods for the merging of different theories.

4.2.5.8 RBFS—the Rule-based Frame System

This is a system which has been developed to support research work at the Information Technology Research Institute at Brighton Polytechnic (Barber et al., 1987). It consists of several modules which exist to provide graphical representations of the underlying knowledge structures. The philosophy behind this system is to provide a medium through which the knowledge engineer and the domain expert can communicate their ideas. As such, it shares its philosophy with that of the KEATS system, with an equivalent tool to GIS, called VEGAN (a Visual Editor for the Generation of Associative Networks), (Kellett et al., 1989). The system uses an intermediate representation format based upon semantic, or associative, networks. This representation is translated into an underlying frame database, which is augmented by a production rule system. This forms the 'runnable expert system', thus sharing with KEATS the idea of the system in effect being the output of the knowledge acquisition process.

VEGAN is a tool which was initially developed to display the underlying frame models held by expert systems developed using RBFS. However, it was quickly discovered that the use of graphical representations of the data structures of the system impacted greatly upon the level to which the domain expert could be involved with the development process. As a result, the work has concentrated on the development of intermediate representations of knowledge which can be used to drive knowledge elicitation sessions.

The representational philosophy behind RBFS is a two-level model of knowledge; the higher level model is a map of what is possible within the domain, and the second level model is a map of what is actually true in this particular instance of the problem. These are termed 'the generic model' and 'the specific model' respectively. Reference should be made to Chapter 2 for a discussion of the representational structure of RBFS in the wider context of the AI representation problem). The development of an expert system using RBFS concentrates upon the generation of a generic plane model of the problem. This is performed by using VEGAN to build a network of nodes and links. In these networks the nodes are domain concepts and the links represent the ways in which these concepts can relate. In this way, RBFS, KEATS and KADS share a common goal: to identify a domain structure. RBFS differs from KEATS and KADS in that the representation used is directly 'runnable'.

Representation of domain structure takes the form of a network, such as that shown in Figure 4.7. Here we see a simple model of the class hierarchy, represented by the 'is_a' link. This models type relations between general concepts. Thus we see in this example, that 'pcb' is a type of 'electronic_component'. Similarly, 'computer' is a type of 'office_equipment'. VEGAN also supports the use of the inst (instance) relationship to represent examples. Thus 'pcb#1' is an instance of a pcb. These two special case relationships (is_a and inst) are in-built to RBFS/VEGAN. The user is able to specify any kind of relationship between concepts, but must then also define a derivation method for that relationship. This derivation method can be rule-based, procedural or 'askable'. The existence of rule-based relationships is displayed graphically as a link with the suffix '.i'. In the figure, we show that it is possible for a computer to have a subcomponent (sub) pcb. The relationship shows the '.i' suffix, and so a rule must exist to evaluate whether in this case the specific computer has a pcb subcomponent. The concept of building these models in real-time, with the domain expert and knowledge engineer jointly defining concepts has been found to be a source of motivation for the experts.

Much in line with the ideas of interpretation models (KADS) and conceptual structures in KEATS, the work on RBFS has highlighted the existence of generic models of particular problem-solving paradigms, which can be used to define solutions to problems in similar domains. The most developed of these is in product planning (planning is considered in detail in Chapter 6).

The elicitation of production-rules is performed in a manual method, via any of the techniques described earlier. It has been found, though, that the use of the graphical representations assists greatly in the development of rule structures. Tools also exist to aid in the representation of rule structures (Knowledge Encoding Tool (KET), (Esfahani and Kellett, 1988), and in the display of the inference tree (Graphical Explanation System (GES). KET is used to present the hierarchies of rules to the system user. In addition to this it incorporates facilities for rule-base modularisation, highlighting the need for intermediate concepts in the VEGAN model. GES is a system which provides a graphical trace of the deduction process of the system. It portrays paths of dependence upon data; in effect how a conclusion was reached. Both of these systems serve to provide the domain expert with a picture of the behaviour of the expert system; a facet which the authors believe is important. If the expert can see a clear direction to the knowledge elicitation process, then he is better armed to help. This, of course, can be counter-productive, as the expert may attempt to rationalise his knowledge in terms of inappropriate structures.

The RBFS/VEGAN tools are used as a means of providing a prototyping approach to expert system design and implementation. It has been found that the use of their particular intermediate representation of knowledge structures enables the use of common language between the domain expert and the knowledge engineer.

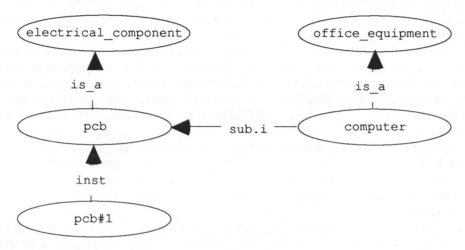

Figure 4.7   VEGAN knowledge representation

4.3 CONCLUDING REMARKS

This chapter has presented a few of the common techniques used in knowledge acquisition. In Table 4.1 we present a summary of those techniques, and the types of knowledge for which they are most suitable. It can be seen from the table, that the techniques vary in the types of knowledge that they identify, with interviews and protocols proving the most general. These techniques are particularly suitable for the early stages of knowledge acquisition. In the latter stages, when the knowledge engineer has gained an appreciation of the form and structure of the domain, then the more specific approaches can be applied to elicit more detailed information. No one technique is sufficient; successful knowledge acquisition will be achieved as a result of using a variety of techniques over the lifetime of the elicitation phase.

The chapter has also provided overviews of some of the developing knowledge acquisition toolkits and methodologies. These are roughly categorised as those which provide support for production rule elicitation (TEIRESIAS and ETS) and those which aim to elicit the overall structure of domain models, through the identification of concepts and their relationships (KADS, KEATS, ROGET, RBFS).

Knowledge acquisition remains, however, as a growing science. At this current time, there

Table 4.1
A summary of elicitation techniques

	Objects	Relationships	Rules	Strategy	User requirements	Tacit knowledge
Interviews	•	•	•		•	•
Protocols	•	•	•	•		•
Sorting	•	•				
Induction		•		•		
Dialogues			•	•		
Repertory grid	•	•				•

is no clear-cut methodology; no definitive solution. The knowledge engineer should still be regarded as a 'gifted individual', who brings his expertise to bear upon a problem through the choice of tools to acquire and represent domain knowledge. Future developments will no doubt provide toolkits which support the domain expert (to release his own knowledge into expert system shells), but these may well be some way off. It is hoped that this chapter has provided the reader with a 'feel' for the problems involved in expert systems development, and an appreciation of the ways in which these problems have been approached.

REFERENCES

Barber, T.J., Marshall, G. and Boardman, J.T. (1988) A Philosophy and an Architecture for a Rule-Based Frame System (RBFS), *Applications of Artificial Intelligence*, **1**, 67–860.
Bennett, J.S. (1985) ROGET: a Knowledge-Based System for Acquiring the Conceptual Structure of a Diagnostic Expert System, *Journal of Automated Reasoning*, **1**, 49–74.
Boose, J.H. (1986) *Expertise Transfer for Expert System Design*, Elsevier, Amsterdam.
Boose, J.H., Shema, D.B. and Bradshaw, J.M. (1988) Recent Progress in AQUINAS : A Knowledge Acquisition Workbench, *Proc. European Knowledge Acquisition Workshop (EKAW 88)*, Eds. J.H. Boose, B.R. Gaines and M. Linster, GMD, West Germany, pp 2.1–2.15.
Breuker, J. and Wielinga, B. (1977a) Use of Models in the Interpretation of Verbal Data, In: *Knowledge Acquisition for Expert Systems : a practical handbook* Ed. A.L. Kidd, Plenum Press, New York, pp 17–44.
Breuker, J. and Wielinga, B. (1977b) Knowledge Acquisition as Modelling Expertise: the KADS Methodology, *Proc. First European Workshop on Knowledge Acquisition for Knowledge-Based Systems*, Reading University, Reading, Section B1.
Davis, R. (1979) Interactive Transfer of Expertise : Acquisition of New Inference Rules, *Artificial Intelligence*. **12**, 121–158.
Diaper, D. (1987) POMESS : a People Oriented Methodology for Expert Systems Specification, *Proc. First European Workshop on Knowledge Acquisition for Knowledge-Based Systems*, Reading University, Reading, Section D4.
Dreyfus, H. and Dreyfus, S. (1986) *Mind Over Machine*, Free Press, New York.
Ericsson, K.A., and Simon, H.A. (1984) *Protocol Analysis: Verbal Reports as Data*, MIT Press, Cambridge, MA.

Esfahani, L., and Kellett, J.M. (1988) An Integrated Graphical Approach to Knowledge Representation and Acquisition, *Knowledge Based Systems*, **1**, 301–309.

Feigenbaum, E.A. (1977) The Art of Artificial Intelligence: Themes and Case Studies of Knowledge Engineering, Proc. Int. Joint Conf. on Artificial Intelligence—IJCAI 5, Morgan Kaufmann Publishers Inc., CA.

Fenn, J.A., Worden, R.P., Foote, M.H. and Wilson, R.G. (1986) An Expert Assistant for Electromyography, *Biomedical Measurement Information and Control*, **1**(4)

Forsythe, R. and Rada, R. (1986) *Machine Learning: Applications in Expert Systems and Information Retrieval*, Ellis Horwood, Chichester, p22.

Gammack, J.G., and Young, R.M. (1985) Psychological Techniques for Eliciting Expert Knowledge, In: *Research and development in Expert Systems*, (Ed. M.A. Bramer), Cambridge University Press, Cambridge, pp 105–112.

Hart, A. (1989) *Knowledge Acquisition for Expert Systems*, (2nd edn), Kogan Page, London.

Hunt, E.B., Marin, J. and Stone, P.(1966) *Experiments in Induction*, Academic Press, New York.

Johnson, L. and Johnson, N.E. (1987) Knowledge Elicitation Involving Teachback Interviewing', In: *Knowledge Acquisition for Expert Systems: A Practical Handbook*, (Ed. A.L. Kidd), Plenum Press, New York, pp 91–108.

Kellett, J.M., Winstanley, G. and Boardman, J.T. (1989) A Methodology for Knowledge Engineering Using an Interactive Graphical Tool for Knowledge Modelling, *International Journal of Artificial Intelligence in Engineering*, **4**, 92–102.

Kelly, G.A. (1955) *The Psychology of Personal Constructs*, W.W.Norton, New York.

McDermott, J. (1980) R1: The Formative Years, *AI Magazine*, **2**(2), 21–29.

McKeithen, K.B., Reitman, J.S., Reuter, H.H. and Hirtle, S.C. (1981) Knowledge Organization and Skill Differences in Computer Programmers, *Cognitive Psychology*, (13).

Michalski, R.S. and Chilausky, R.L. (1980) Knowledge Acquisition by Encoding Expert Rules Versus Computer Induction from Examples: A Case Study Involving Soybean Pathology, *International Journal of Man–Machine Studies*, **12**, 63–87.

Motta, E., Rajan, T. and Eisenstadt, M. (1989) Knowledge Acquisition as a Process of Model Refinement, Human Cognition Research Laboratory Technical Report No.40.

Neale, I.M. (1988) First generation Expert Systems: A Review of Knowledge Acquisition Methodologies, *Knowledge Engineering Review*, **3**, 105–145.

Pask, G. (1975) *Conversation, Cognition and Learning: A Cybernetic Theory and Methodology*, Elsevier, Amsterdam.

Quinlan, J.R. (1983) Learning Efficient Classification Procedures and Their Application to Chess End Games, In: *Machine Learning: An Artificial Intelligence Approach* Eds. Michalski, J.G. Carbonell and T.M. Mitchell, Tioga, Palo Alto, USA, pp 463–482.

Quinlan, J.R., Compton, P.J., Horn, K.A. and Lazarus, L.(1987) Inductive Knowledge Acquisition: A Case Study, In: *Applications of Expert Systems*, Ed. J.R. Quinlan, Addison-Wesley/Turing Institute Press, Reading, MA. pp 157–173.

Reichelt, H. and van Harmelen, F. (1986) Criteria for Choosing Representation Languages and Control Regimes for Expert Systems, *Knowledge Engineering Review*, **1**(4), 2–17.

Reitman Olsen, J., and Reuter, H.H.(1987) Extracting Expertise from Experts: Methods for Knowledge Acquisition, *Expert Systems* **4**, 152–168.

Shaw, M.L.G. and Gaines, B.R. (1986) Interactive Elicitation of Knowledge from Experts, *Future Computing Systems*, **1** 150–190.

Slatter, P.E.(1987) *Building Expert Systems: Cognitive Emulation*, Ellis Horwood, Chichester.

Stefik, M. (1989) Lecture given in the Knowledge Systems Laboratory Seminar Programme, Stanford University, CA.

Waterman, D.A. (1986) *A Guide to Expert Systems*, Addison-Wesley, Reading, MA.

Wright, G., and Ayton, P. (1987) Eliciting and Modelling Expert Knowledge, *Decision Support Systems*, **3**, 13–26.

J.M. Kellett
Brighton Polytechnic
Lewes Road
Brighton
East Sussex BN2 4AT

G. Marshall
AI Unit (S 5090)
PO Box 10
Halton Cross
Middlesex TW6 2JA

5   Engineering Design, Manufacture and Test

T. Katz

Brighton Polytechnic, UK

5.1 INTRODUCTION

Applications of AI in engineering cover a wide range of domains, some of which will be discussed in this chapter. However, of concern to the majority of engineers is the production of goods and services. The function of production is of great importance within a company (and nationally), in that these activities are the prime generators of wealth, through a process of adding value to raw materials. This 'added value' is due to capital investment and subsequent engineering effort. Resources require careful management, in order to maximise the returns that render the wealth creation process viable. The capital investment strategies and policies are the concern of the financial institutions and government, and therefore will not be discussed here. However, management and project planning aspects are certainly the concern of the professional engineer, and are discussed in some detail in Chapter 6. This chapter will concentrate on the technical aspects of product provision, and how AI is and can be exploited in this regime.

The quality, and therefore commercial viability of a company's product, is dictated to a large extent by the calibre and availability of an appropriately qualified and motivated workforce. The complexity of a company's product line and the success of the company in the market place is commonly mirrored in the consequent complexity apparent in the management structures. Competent decision-makers are therefore in very high demand and are increasingly required to be more and more technically oriented. AI, and Expert Systems in particular, provide an opportunity to distil some of the domain expertise and replicate the skills of a human expert (albeit in relatively narrow domains) in a formal and consistent manner.

This chapter is developed as a guide to a practical piece of AI as used in engineering. The engineering domain is rich in specialised disciplines, each having its own particular range of problems suitable for the application of AI. However, there is a common 'core' of issues in engineering, which can be partitioned into design, manufacture and test. There is a place for embedded expertise in all of these important (and interrelated) areas of

Artificial Intelligence in Engineering. Edited by Graham Winstanley
ⓒ1991 by John Wiley & Sons Ltd.

all engineering disciplines, and all have one thing in common: the product. Design is a complex process, partly cognitive, partly intuition and partly skill. However, there are well established guidelines or methodologies for design (see Chapter 1), and these can be usefully employed as embedded rules to assist the designer. Manufacture is largely dependent on a knowledge of the product and the processes necessary to manufacture it. Manufacturing planning is discussed in Chapter 6. The testing of complex products is becoming more and more important, especially in the electronics (and computing) industries. The traditional electronic test process, when performed on automatic test equipment (ATE), is performed sometime after the product is manufactured. The process usually involves 're-learning' the functionality of the system and partitioning what is likely to be a very complex product into a number of functional 'blocks', each having its own predictable input/output characteristics.

This chapter discusses some of the issues relevant to the application of AI to the Design, Manufacture and Test (DMT) regime, but concentrates to a great extent on the problems of automatic test. A large-scale example is discussed in detail, in order to fully acquaint the reader with the processes involved in the application of expert systems technology to one important aspect of the DMT regime. This example uses much of the material to be found elsewhere in this book. For example, the framework for the system (the expert system itself) is based on frames and rules. Reference should be made to Chapters 2 and 3 for a comprehensive coverage of these concepts. The use of this example gives the chapter a focus in reality, and includes generalised methodologies applicable to DMT. However, it is not representative of the whole spectrum of DMT. Therefore, reference should be made to the texts concerned exclusively with engineering design, manufacture and/or test.

5.1.1 Introduction to Design, Manufacture and Test

Design, manufacture and test (DMT) are the normal functions referred to when a product is generated. They are commonly modelled as serial activities, with design occurring first and test last. However, there is usually a link from test back to design, describing an iterative process, as shown in Figure 5.1(a). Obviously, this scenario is a naive over-simplification. Whenever these activities are carried out in isolation, which is often the default case due to increased technical complexity and specialisation, it is found that many iterations are required before a satisfactory product results. This leads to very long lead times and massive production capital waste, or the inclusion of costly post-production modifications.

It is much more sensible to model the overall process with a parallel architecture (Figure 5.1(b)), sometimes referred to as product engineering, and ensure communication throughout the product team from the start. However, this still ignores the commercial functions and assumes that a requirement and specification have already been identified and included in the 'concept'. Again, further refinement should include these (and other) elements, to provide a fully integrated approach. However, DMT, as an independent concept, has much momentum within industry and can be useful as a broad classification to define the scope of this chapter, so long as its limitations are appreciated.

5.1.1.1 Design

The design arena was probably one of the first to become computer aided, from its utilisation as a clever calculator, through its use in a statistical analysis role, to full functional

(a)   (b)

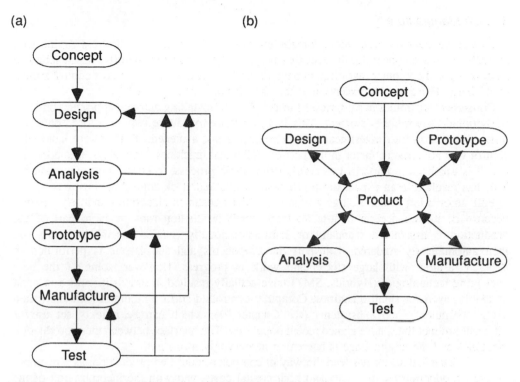

Figure 5.1 The DMT cycle: (a) serial and (b) parallel models

simulation. This purely numerical or algorithmic approach has been complemented by the continuing development of graphical capabilities, which have evolved from draughting through to three-dimensional symbolism and the visualisation of design information. Engineers' uptake of these tools had initially been a slow process, confined to fairly major companies with large budgets or visionary management.

Hardware developments in both silicon and display technologies, coupled with some major software developments, has brought about a sharp reduction in the cost of computing power, and this has resulted in the emergence of the engineering design workstation network. Computer-based tools are now readily available in various engineering disciplines, and some have been successfully integrated into larger systems having a wider utility. This approach has been assisted greatly by improvements in display technology, which have facilitated the efficient information dissemination to engineers at different 'levels'. Examples include:

- Schematic capture, simulation and layout of electronic components, as either a printed circuit board (PCB) or on silicon. Documentation (parts lists and user guides) and artwork are also produced at this stage.

- Structural design of buildings, including analyses of stresses and bending, and the provision of site drawings.

- Three-dimensional modelling of engineering components, leading to finite element stress and strain analyses, and the provision of suitable mechanical working drawings.

5.1.1.2 Manufacture

The manufacturing, or production, domain has varied somewhat in the different engineering sectors. In civil engineering, the fact that nearly all projects are 'one-offs' has resulted in automation and computer involvement being limited to estimation and project control almost exclusively. For a notable exception to this, see Tommelein *et al*. (1987).

Consumer electronics has progressed to the point of installing automated production lines or automatic assembly equipment. This large capital expenditure requires close control in order to maximise on investment, but flexibility is not a precedent. Therefore, computer control was not a major factor in the past on individual machines. Nowadays, it is accepted that it is much more important to monitor the whole process, to ensure optimum running. This has resulted in an expansion in process analysis and stock control applications.

Full automation remains largely elusive in the consumer electronics industry, mainly because of its reliance on high-technology, small production runs. A large part of the production is assembled manually or semi-automatically (guided assemblers, etc.) and then automatically soldered. Transport of components and subsystems is performed in a batch fashion, with large stocks and work in progress. However, some of the new packaging technologies (Hybrids, SMT) have actually resulted in more difficulties in hand assembly, even for small quantities. Computer-controlled surface mount assemblers (often using computer vision subsystems (see Chapter 7)), which can be quickly set up for different product lines, have gained recent popularity. The interface between these production machines and the design stage is becoming increasingly important.

The mechanical sector has lead the way in computer-controlled production, as a response to mass production requirements and high capital costs, and with the incorporation of the flexible machining centre (FMC) concept. The numerically controlled, multi-tool machining centres that were initially programmed by paper tape, were soon given direct computer control for added flexibility. It was later realised that small groups of these machines could efficiently share the workload as long as the 'feeding' of the machines could be adequately controlled. Automatic workpiece transfer (rovatrain, etc.) soon came under the overall control of the master computer. The interface between the design and machining processes is also now commonplace, and links have been developed to allow improved inspection (sometimes carried out using computer vision). For these reasons, this domain has come closest to a true flexible manufacturing system (FMS), and is most likely to develop much further towards computer integrated manufacture (CIM).

5.1.1.3 Test and Inspection

The electronics sector is an ideal candidate for the application of computers to product testing, where it has become prevalent in all production facilities. This domain is dealt with in great detail in Section 5.5, so little need be said here beyond the fact that process control of electronic production relies heavily on reliable fault diagnosis. Therefore, efficient testing and diagnosis are the prerequisites of process management, where the results of various testing procedures are statistically evaluated and analysed.

The functions of test and inspection in the mechanical domain have become more automated, employing both destructive and non-destructive testing. Non-destructive testing is usually performed by a programme controller specific to each test rig, rather than by any integrated testing system. Integration is difficult by virtue of the large variety of techniques, ranging throughout the spectrum from inspection to functional testing, and the operation of mechanical systems can be very diverse. Destructive testing, on the other hand, is usually performed for design verification or process optimisation, and requires significant sample preparation if diagnosis is to be effective. This, together with the fact that normally only a small minority of the production run will be tested in this way, makes it difficult to achieve full integration, although a computer-controlled rig with data logging is quite normal in the monitoring of a particular process.

In civil engineering, the requirement is more one of verification and inspection. Testing remains important on subsystems and processes, for safety reasons, but the individual and massive nature of most projects renders destructive testing irrelevant. Here the emphasis is on safety factors, and inspection and monitoring over many years. Therefore, the application of computer technology is often restricted to data logging and analysis.

5.1.2 The Interdisciplinary Nature of Engineering

The majority of manufactured products are constructed of many subsystems. The complexity and sophistication of some modern products results in the incorporation of a large electronic component (to improve functionality), together with precision mechanical engineering. Typical examples of this interdisciplinary nature can be seen in, say, automobiles or video recorders. These examples certainly rely on diverse engineering disciplines, whose integration in a particular application takes on a specialism of its own. Even for more mundane products, it is only possible to isolate the mechanical domain occasionally.

However, it is obvious that the engineering processes are often similar. Although few engineers will have identical technical backgrounds, they should all recognise good practice. Fundamental understanding of one domain can be easily transported into another; it is the intention here, that any engineering reader will be able to gain insight into the AI principles and applications from the examples that follow.

5.2 SYSTEM DESIGN APPROACHES

When it is considered appropriate to attempt an AI solution to an engineering problem, there is much to be taken into account before commitment to a particular course of action. This section summarises these prerequisite decisions, together with a brief explanation or further reference. Hopefully, their importance will become more apparent as the actual examples are discussed. However, it should be noted that practically all present-day commercial engineering AI applications are confined to expert systems (ES). A useful check-list of the basic steps to be taken when developing an ES application (Adeli, 1988) is now given:

(1) Selection of the program language, environment or shell. Refer to Chapter 3.

(2) Selection of the AI representation and inference mechanisms. Refer to Chapters 2 and 8.

(3) Analysis, acquisition and conceptualisation of the knowledge to be included in the knowledge base. Refer to Chapter 4.

(4) Formalisation and evolution of the knowledge base. Refer to Chapters 2, 3, 4 and 8.

(5) Development of a prototype system using the knowledge base and tools. Refer to Chapters 3 and 4.

(6) Evaluation and review of the system, then further expansion. Refer to Chapter 4.

(7) Refinement of the user interface. Refer to Chapter 3 and 4.

(8) Maintenance and updating of the system. Refer to Chapters 6 and 8.

Note that some of the later decisions and processes are constrained by the initial decisions, and cannot be altered without expensive or time consuming effort.

5.2.1 Choice of Expert System Type

The choice is between a shallow or a deep expert system (SE and DE respectively) (Abu-Hana and Gold, 1988). A SE is one that makes anecdotal or empirical connections between various 'pieces' of knowledge. Typically, an action will be recommended when a particular circumstance is recognised; this action having been successful before. In this situation, there is no understanding within the system of the underlying causal relationships. However, the DE does possess an internal model of the fundamental process leading to a particular conclusion. The 'triggering information' initiates a search along a causal network until it arrives at a reasonable solution. Obviously this is much more difficult to implement (for example, which model is the most appropriate?) than the SE and can be far less efficient in actual operation in most cases, but it has several advantages. Further reference to shallow and deep reasoning can be found in Chapters 2, 3 and 8 of this book. Some of the more important advantages of the DE are:

- Direct experience of specific cases is not required, as it is in the SE, resulting in much more system flexibility. It may deal with situations not encountered before. The corresponding SE would quickly become significantly less reliable as it approached the limits (the 'edges') of its knowledge.
- The causal networks may be modified by the operation of the DE, providing the possibility for learning.
- The logic behind any decision made by the DE can be readily traced, and understood by the operating engineer. It is common practice to provide effective reporting procedures in these systems, in order to permit some degree of monitoring of performance (and instil confidence in the user).

In between these extremes of depth, there are the possibilities of using compiled knowledge bases (Chandrasekaran and Mittal, 1984), which may have the ability to cope with complex problems in certain domains, or to actually link deep and shallow systems together in order to utilise the inherent advantages of both (Abu-Hana and Gold, 1988; Struss, 1988).

5.2.2 Knowledge Representation

The application of AI to engineering problems inevitably involves the selection of appropriate knowledge representation methodologies, of which there are many (accepted) options (Cercone and McCalla, 1987). Some of these methodologies are:

- Logical representations.
- Semantic networks, which are various graph structures.
- Procedural representations, rather than propositions, are argued to be much closer to human knowledge-activity processes and can be included in the semantic networks.
- Logic programming languages, such as Prolog, combine the exactitude of first-order logical representations with the power of procedures.
- Frame-based systems, which cluster the knowledge into chunks. Frames are functionally linked rather than only structurally linked as in the case of semantic networks. In many systems, the semantic network is used as a graphical 'front end' to the underlying frames.
- Production rule systems, which are constructed of pairs of pattern -> action links. This architecture is very effective in shallow expert systems, as the system is merely required to search through the rules until the relevant patterns are identified.
- Knowledge representational languages.

In many ways, knowledge representation is the key to the successful use of AI in any situation. There are appropriate issues such as modelling accuracy (coarse or fine-grained representation), monotonicity, expressive power, etc. The representation problem remains a very active current research topic, and a full discussion of its research aspects, plus some indication of the future of the discipline, can be found in Chapter 8. A more simplified treatment of the subject can be found in Chapter 2.

5.2.3 Modelling

Although the choice of formal representation of knowledge is of paramount importance, in order to understand which is most suitable, the problem needs to be formulated cogently. The major strength of AI is its ability to operate on symbols, rather than numerical data (Charniak and McDermott, 1985). Therefore, it is imperative that the system constructor is able to identify the level of abstraction and symbolism most appropriate for the problem in hand (and any likely extension of it). There may be many and various ways to model system behaviour, but the majority contain two main views. These are the functional and the structural modelling classes (Brady and Connell, 1987); their integration has been the basis of many powerful systems such as the example used in Section 5.5 (Lea *et al.*, 1988, Tezza and Truco, 1988).

5.2.3.1 Structural and Functional Modelling

It is possible to produce models which reflect the structure of a domain, or alternatively, it is possible to model the functionality of objects or entities which make up the structure of a domain. Graphical networks are very often used to demonstrate structure but, when applied to the functional model, they become causal networks (Abu-Hana and Gold, 1988) or influence diagrams. Here the nodes become functional blocks, with the links representing relationships. The influence of one function on another can be implied by a directed search through this network, which can then be traced on the diagram. The system described in Section 5.5 allows the highlighting of parts of the graphical representation of the network by one of its procedures.

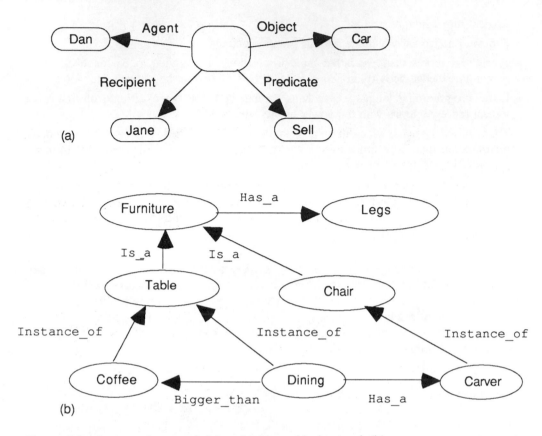

Figure 5.2 A semantic network (a) and a hierarchical network (b)

5.2.3.2 Hierarchies

Network representations can indicate the proximity of a relationship. However, if this is based on a 'flat representation', as in Figure 5.2(a), then the result is limited utility. Figure 5.2(b) shows how hierarchy can facilitate inheritance of properties to specific entities, from more general descriptions. The expert system, which has access to this model, should have the ability to search the hierarchy, to infer that the carver chair, and dining room tables in this example, all have legs. A single hierarchy is seldom sufficient to adequately represent the model in engineering applications. Normally, a set of hierarchies exists, with many shared nodes, each portraying a different view. These may have essentially the same structure, with merely a high-level change as shown in Figure 5.3(a), or could be very diverse with only some of the bottom level objects in common, as in Figure 5.3(b). Wherever the alternate link in the hierarchy exists, this will allow inheritance of other properties by the objects lower in the hierarchy. Hence we could now infer that the dining table has legs, height, but will not fit through the door (conflict in 5.3(b)). A rule would be required, of the form:

'If obj1(void) bigger_than obj2(solid) then fits'.

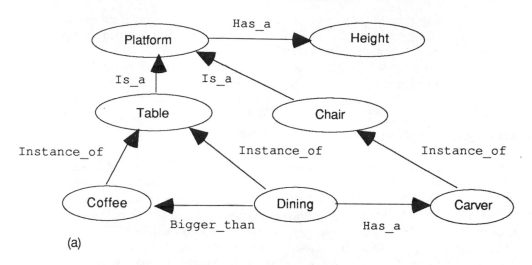

(a)

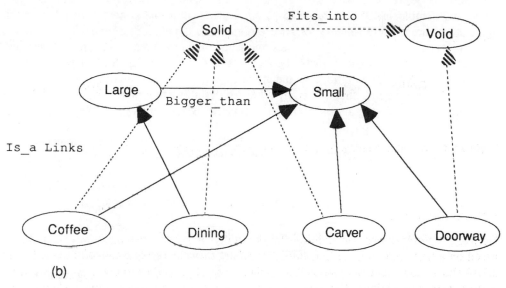

(b)

Figure 5.3 Alternative hierarchical views from that of Figure 5.2(b)

5.2.3.3 Multiple Levels

Hierarchy does not merely indicate relationships between similar artifacts. Its strength lies in the fact that each level of the hierarchy represents a different level of abstraction. As we move 'upwards' in the hierarchy, to coarser granularity (Abu-Hana and Gold, 1988; Zrimec and Mowforth, 1988), a higher level of abstraction is produced, and hence it is more symbolic. This effective consolidation of information results in the expert system operating far more efficiently, but less specifically. This is the reason why hierarchical links are so important; a rule can be used once at a high hierarchical level, and be applied many times to anything attached below that level. As the system must apply itself to specifics at some time, the action of 'travelling' up or down the hierarchy must be carried out efficiently. The system must also be able to deal with rules and information at any level and granularity. This is particularly powerfully illustrated in (Abu-Hana and Gold, 1988) and summarised in Section 5.5.

5.2.4 Operational Considerations

There are many pragmatic factors associated with the application of AI in engineering (Basden, 1984), which can be conveniently subdivided into initial implementation and system operation. However, the hardware and software platforms are of major concern to both of these issues. There is no denying that AI applications can be very expensive, both in necessary hardware, and the not inconsiderable effort required. A full discussion of AI hardware can be found in Chapter 10, and a fuller treatment of available software platforms (and related hardware requirements) can be found in Chapter 3.

Initial and crucial considerations are the choice of the problem in the domain to which a solution is desired, the interface that the developer and user will require and the type, range and complexity of knowledge to be embedded within the system. Not all potential applications are entirely suitable for AI exploitation. It is pointless to expend much effort on problems characterised by simple rules, but a high level of numerical content. Numerical algorithmic processes would probably be ideally suited in such cases. Similarly, care has to be taken not to be over-ambitious, as the application must be capable of being modelled in one of the formalisms mentioned, or any other well established representation. An example of a typical application is marine collision avoidance (Blackwell *et al.*, 1988). They state:

> Although there is a substantial rule book (as in the marine collision problem), it requires skill and experience to interpret the correct action in the majority of complex situations.

In addition to a comprehensive evaluation of the initial problem, the question of what role the system is to play is extremely important. This may include such roles as:

- a consultant, to support the 'non-expert' in making decisions;
- an assistant, checking to ensure consistency, and avoid errors or omissions;
- a trainer making the appropriate expertise available, and reporting on the logical processes of inference on request;
- a 'refiner of the expertise'. It fulfils this role both in the restructuring of general knowledge during elicitation, and in the truth maintenance systems that can resolve and explain conflicts during operation;
- a system controller.

Whatever the role, an attempt should not be made to include everything in the system. As has already been stated, the best solutions are those which employ hybrid methods, in which a mixture of programming paradigms is used within the system when appropriate, i.e. a mixture of numerical and symbolic programming styles. Good examples of the success of this approach abound (Blackwell *et al.*, 1988).

The inference mechanisms for production rule-based expert systems, which make up the greater majority of engineering applications (Adeli, 1988), can be chosen to match the application. There are three main types:

- forward chaining,
- backward chaining,
- mixed chaining (forward and backward chaining being possible within the same system, usually dependent on the type of inference required to solve a particular sub-problem).

A fuller discussion of these issues can be found in Chapter 2, and a treatment of the subject using commercially available systems can be found in Chapter 3. Chapter 3 also deals with another issue important to the successful implementation of expert systems, that of the man–machine interface (also dealt with in terms of knowledge engineering in Chapter 4). Important to the developer is a good programming and system development environment. There are, of course, a number of choices available to the prospective user of expert systems in engineering (reference should be made to Chapter 3). They are:

- buy in a commercially-developed system, ideally suited to the proposed application (there are not many of these);
- purchase a commercial expert system development environment. (These are usually equipped with excellent interfaces, reporting mechanisms and a range of representational facilities);
- purchase an expert system shell. (These are smaller systems, with a limited capacity for representational flexibility, and a less flexible, dedicated user interface (suitable for smaller applications than full environments));
- develop a new system, from 'scratch', using a programming language.

For the latter development strategy, a multiple-window multiple-tasking workstation and a good operating system is usually adequate, so long as the local and bulk memory capacity is adequate to handle the large data and knowledge bases (as well as the shells and tools). Graphics are not only helpful in this kind of development exercise, their availability is becoming crucial in the effective representation of complex relationships (as in semantic networks) on screen. There are many *de facto* standard (Unix/NFS/X-Windows) workstations available, with all these facilities. More simple expert systems can be developed on smaller personal computers, but they can be severely restrictive on all but the more modern, high capacity and performance machines.

One of the most difficult and time-consuming tasks faced by the implementors of expert systems is the acquisition, formalisation and implementation of knowledge and human expertise in a computer system, which cannot possibly mirror the human capacity for general problem-solving. Expert systems operate very successfully in very narrow domains of expertise, but even in such restricted domains, the problems of adequately capturing and automating human problem-solving facilities are immense (see Chapter 4 for a full discussion on the problems and potential solutions to this 'knowledge engineering' process). In addition to the problems related to the actual availability of suitably experienced personnel, there is also the fear, not completely groundless, that the emerging expert system may deprive them of their livelihood. This has resulted in the gain in popularity of learning systems, that can continuously and dynamically develop. Herrmann (1988) describes a system designed to estimate integrated circuit area requirements. the system actively 'experiments' when it is not in use, in order to increase its experience and hence refine its knowledge.

In operation, an engineering expert system (indeed, any expert system) should be easy to use, provide the right kind of support, and should be capable of maintenance. Graphics are useful in the user interface, but the requirements are likely to be radically different to those of the development system. It must also be decided where the expert system is to be most profitably employed. In the more sophisticated engineering systems, the AI component becomes just another module, and has no direct contact with the user at all (Basden, 1984). This is shown in Figure 5.4.

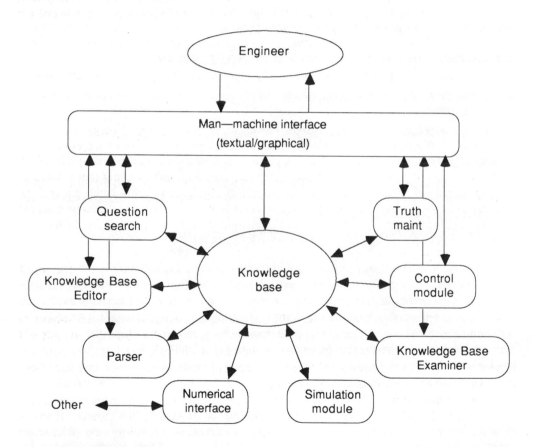

Figure 5.4 An AI system hidden in the application

5.2.4.1 Activity Confinement

These systems rely on performing searches and subsequent comparisons of large quantities of data. Hierarchical abstraction is the prime method of search space limitation, but its effects tend to be arbitrary in that the 'granularity' (information resolution) becomes coarse globally. If perturbations are introduced into a causal model for example, the effect is a propagation throughout the network, well beyond the region of interest, and consequently utilising a potentially massive amount of computing resource. Focusing, on the other hand, as explained by Struss (1988), allows the system to confine itself, at a more detailed level, to a region previously identified as of interest at a more abstract level. A similar action can be seen in the example included in Section 5.5, where the expert system finds probable paths and then constrains search (digraph merging—see Section 5.5) to the devices on this path, avoiding an information explosion.

5.2.4.2 Truth Maintenance

There is always a requirement to make alterations to the knowledge base of an expert system, in a continuous process of updating, after the initial system implementation. Non-monotonic reasoning can be of use in the following ways:

- It can make provisions for 'exceptions' when applying rules;
- It can deal with 'data driven effects', such as the loading of a client's specification file;
- It can provide correction of input errors;
- It facilitates 'what–if?' features;
- It can provide a mechanism for revoking decisions based on weak heuristics.

When the knowledge is 'inexact', or has errors in it, then inconsistencies occur which cannot be resolved. A truth maintenance system can deal with this by using non-monotonic reasoning (Brown, 1985), and/or backtracking (Abu-Hana and Gold, 1988) to find the original cause of the conflict and either request clarification, or correct it. Reference should be made to Chapter 6 of this text for a full discussion on truth maintenance, and Chapter 8 for a comprehensive treatment of non-monotonic reasoning.

5.3 DESIGN

Designing and the design process are different concepts. Designing has been subject to wildly differing views and definitions, from 'Designing is the performing of a complicated act of faith' (Jones, 1970), to 'Design establishes and defines solutions to, and pertinent structures for, problems not solved before, or new solutions to problems that have been previously solved in a different way' (Blumrich, 1970).

It is much more constructive to identify the design process, usually as a set of operations and stages, although opinion differs over how many there are (Jones, 1970; Dieter, 1986). A common view is that the design process is mostly modelled as a serial process, with broad division into:

- *Divergence* The requirements are refined, and a problem specification produced. Constraints are generated for later stages, from the analysis of initial research.
- *Transformation or Conceptualisation* At this stage as many plausible, and pertinent schemes are produced and embodied (with detail). As far as AI is concerned, there is great interest in this domain and the concepts are used to great effect.

- *Convergence* This is the processes of analysis, detailing and evaluation. This domain has the benefit of many existing computer-based tools, which are used to produce a transformation from the 'fuzzy' modelling of the previous stages into numerical models. Errors occurring at this stage are 'expensive' to correct.

5.3.1 Handling Complexity

As engineering systems become more and more complex, the amount of information to be processed in the design process becomes unwieldy, and certainly beyond the limits of individual human capacity. For these reasons, computer tools and automation are essential. Fortunately, the enabling technologies have been developed within the last decade to allow the adequate handling of this type of complexity (AI being one of them). In the realm of VLSI design, the build-up of sufficient computer power at a reasonable cost, and software capable of simulation and database control, has proceeded in parallel with increasing formalism in the design process.

The extended use of hierarchical decomposition has played an important part in the solution: it ranges from layout at the bottom hierarchical levels to system design at the top. Increased abstraction enables adequate understanding of the fundamental functional blocks, and this can be confirmed through simulation. Using this form of hierarchical decomposition of a particular engineering system structure, more detail can be continuously added to the model by further subdivision of these functional blocks. Behavioural models can be used to great effect in the representation of relationships and functional data handling. Using the example of digital electronics, gate level, switch level, transistor level and finally layout level all require appropriate representation, at the most appropriate level of detail. It is this additional information, apparent on descending the hierarchy, that forces decisions to be made and allows the entry of errors. There is often the need to extract the higher level structure as a check, but also for back-annotation and re-simulation, which verifies that the additional detailing has not invalidated previous constraints.

The assumption that top level VLSI design is independent of the implementation at the bottom is naive. This serial process normally requires quite a few iterations in all but the most standard applications. The added detail is usually application specific, in that the application itself determines the final form of the circuit. Also of concern is the hierarchy itself; it is normally purely a functional view at the top, with other, more detailed views and constraints entering at a lower level. This distorts the relationships in the hierarchy, encouraging more errors. What in fact is required is a number of consistent, interrelated hierarchies, each consistent in themselves, but representing different views. The forward-chaining production paradigm used in some systems need other supplementary paradigms (Pepper and Kahn, 1986), in order to more fully support the top–down design and conceptual abstractions inherent in powerful hierarchical schemes.

5.3.2 AI-assisted Design

There has been some notable success in the VLSI domain with 'silicon compilers', for very greatly constrained design in a very short time. This represents pure automation, using mostly algorithmic routines to make various assumptions about the descending route; it produces solutions that may work, but will not necessarily be of the best quality. Secondly, this can be very computationally expensive, even for simple problems. The limitations of this approach are equally as applicable in the other engineering domains. More generally, there has been much work on design methods to assist in synthesis (Jones, 1970; Dieter, 1986). Some of these are effective as an aid to the designer, by reorganising his or her thoughts, and not as a methodology of direct application to AI solutions. Others are merely combinatorial methods which generate all possibilities and abdicate responsibility to the analyser. For AI approaches to the problem (Coyne and Gero, 1986), there are two different schools of thought, exemplified in PRIDE (Mittal *et al*., 1986) and DESIGN (Miller, 1985). The former attempts to model the perceived design processes, whilst the latter models the problem knowledge itself, assuming that the process then becomes implicit; these are synonymous with the DE and SE of Section 5.2.

In reality, many of the most useful AI applications to date have been of the SE consultant or design assistant (Section 5.2), that advises less experienced designers. Companies have tried to speed up design by automating subprocesses, although this is usually in a very specialist area, in which a company may fear the loss of its in-house engineering skills. Notable systems in this area include VT (Marcus *et al*., 1986) which uses plan extensions, constraints and fixes to design elevators, DAA (Kowalski, 1985) describes the set up of a VLSI design system based on the use of a Design Automation Assistant (a classic knowledge-based expert system). The representational system used divides the knowledge into working memory, rule memory and a rule interpreter that is able to answer 'why' and 'how' questions. When looking at many of the simpler design assistants, it can be seen that the drawbacks are usually in the over-simplification of the models, and the requirement to interface to some numerical or algorithmic modules. The next generation of system shells are taking this factor into account.

5.4 MANUFACTURE

The realm of manufacture covers the actual physical production part of the product generation process. This is effectively what goes on in the factory and should include such activities as stock control, storage, distribution, progress, materials handling, process control, maintenance and the many other individual processes that may be more domain specific, such as machining and assembly, or surface and heat treatments. The range is diverse, but a good example, which demonstrates most of the principles, is included in detail within Section 5.5. There is scope for development along the lines of integrated systems however, and this is discussed in the next section.

There are a great many areas of engineering, which are relevant to the application of AI. In addition to the AI-specific areas, such as cognitive science, natural language processing, neural computing, etc., there are more specific issues closely related to engineering. Application areas of enhanced interest include:

- robotics and spatial reasoning (see Chapter 8 for a discussion on spatial reasoning);
- processes, planning and scheduling (see Chapter 6 for a discussion on planning);
- integration and interfacing.

5.5 PRODUCT TEST

This section describes in great detail the work carried out over a number of years by a group of researchers at Brighton Polytechnic. The group has long-standing interests in the development of computer-based systems for use in the engineering discipline of automatic test, and much of this work has resulted in the successful application of AI-based tools in the automatic test equipment (ATE) industry. As such, the main example, which forms the basis of this section, should be regarded as a state of the art exemplar.

5.5.1 Introduction to Test and Test Equipment

This section describes an intelligent, automatic, digital functional test generation system. Testing of digital circuit boards is concerned with the verification of the operation of a digital electronic circuit; therefore, a full knowledge of its logic states is required for any possible operational understanding. Exhaustive testing is a physical impossibility when the number of degrees of freedom in the system increases above 'the mundane'. These days, the combination of all possible inputs and internal states of any useful subassembly would provide an unrealistic requirement on the number of logic states to check. A particularly fruitful avenue pursued by the electronics industry has been to subdivide circuits into increasingly smaller and more manageable functional blocks. Obviously, each 'subcircuit' must be observable and controllable, in order that its state can be ascertained at any time. This kind of access can be gained directly, through the use of a probe matrix (bed-of-nails), called an in-circuit test, or through a method of controlling the circuitry surrounding the object of interest in order that test parameters may be passed 'in and out' through the normal subcircuit interface. This latter mode is referred to as functional test.

The apparently intractable problems related to ever-increasing circuit complexity has been the driving force behind the ATE industry throughout its existence. It is widely recognised (Tsui, 1987) that this complexity, both of electronic circuit boards and integrated circuits themselves, has forced a move within the industry towards functional testing. Lack of internal node access, due to advanced manufacturing techniques (surface mount, etc.) and the continuing integration of circuit function, make an in-circuit, bed-of-nails test on fully populated boards very difficult, although it retains a prominent role in partially completed subsystems. Functional testing becomes even more attractive when it is appreciated that there is no need to 'backdrive' sensitive components, or create much unit under test (UUT) specific hardware. There is also the possibility of highlighting design and performance faults.

To produce a functional test, routes through the circuit under investigation must be identified and validated. Device interaction must then be studied before any devices can be tested, adding a further level of complexity to the test generation process. If the circuit can be modelled at the right level of abstraction, however, it may be possible to utilise existing in-circuit test libraries and/or tests developed at the device level by IC manufacturers. This would greatly ease the generation of a viable test resource; it would be most inefficient to discard the large amount of effort already invested in these libraries. Also, modelling complex devices at the lowest level (gate level) can be virtually impossible, as manufacturers are sometimes reluctant to publish this information.

The generation of functional tests is a non-trivial task. Although a number of designs may contain the same or similar components, these may interact differently due to subtle changes in connectivity. To achieve a high degree of fault coverage, detailed analysis of each board is required and it may take several months to produce a test program for a single design. Furthermore, circuits that are well designed and utilise design-for-test (DFT) techniques (Eichelberger and Williams, 1977; Konemann, 1979; Maunder and Beenker, 1987) in their construction, still require the development of functional tests to capture performance faults generated by device interaction. The circuit partitioning inherent in DFT methodologies makes these faults very difficult to trap.

However, the potential increase in scope and power, provided by a functional test strategy, can only be realised with large and time-consuming test programs. At present there is great demand for suitably skilled test engineers whose lives are complicated by inadequate DFT. To compound the problem, the available 'product window' is steadily reducing due to accelerating technological advancements. Thus, long test development times (typically up to 30% (Schofield, 1985) of the total product cost) must also be cut if commercial viability is to be maintained.

The system detailed in this section is a result of a collaborative venture between Brighton Polytechnic Information Technology Research Institute and Schlumberger Technologies ATE division. Firstly an introduction of the ATE domain, for those unfamiliar with it, is followed by a discussion of the first attempts to elicit a solution to these testing problems using the Automatic Test Pattern Generator (ATPG). The perceived shortcomings have led to the next generation of test philosophy; Intelligent Test Generation (ITG). An overview of the propounded system is followed by more detailed descriptions of the components and their interrelationship, with reference to the premises of Section 5.3. Finally, there follows a discussion on the implications of this work within the ATE domain.

5.5.2 The ATPG—A First Attempt at Solution

In 1966, Roth (1966) and others at IBM (Case, 1981) pioneered an elegant computer algorithm for testing combinational circuits: the technology of that time. The algorithm had great success and reaped a fault-coverage in excess of 95% in most cases. Perhaps a mark of its elegance and initial robustness is the fact that twenty years on, the so-called D-algorithm is still widely used in some form or other in the ATE domain. It is intrinsic to most of the present-day ATPG systems, and has proved popular with test engineers testing combinational logic blocks.

Despite the credence given to the D-algorithm, the prolific advances in technology have rendered it ineffective on digital sequential devices, such as counters and shift registers, that operate in the time domain. The D-algorithm, in its pure form, is incapable of coping with such complexity since it operates in the logic domain only. The advent of LSI and VLSI has further compounded the problem, in that it is unusual to find a gate level description of the device in question. Although most data sheets today give only functional descriptions of LSI and VLSI devices, in some cases it is possible to produce a behaviourally-equivalent gate-level model, which will trap approximately the same percentage of faults. However, the investment in time to perform this task is far too large. This benefit obviously adds to those already detailed in Section 5.5, therefore accentuating the case for abstraction away from the gate level to simplify the process.

The success of any test procedure can be dramatically affected by the detailed design considerations and the final functional implementation. Provisions applied at this stage can simplify testing by many orders of magnitude. The application of an expert system as a 'testability design assistant' is an obvious step towards an improvement in this situation, and has been approached by several researchers (as mentioned in Section 5.3). As the applicability of the D-algorithm principle has steadily diminished with each new advancement of technology, there has been considerable research into methodologies for enhancing the testability of circuits by constraining the designer to produce a design which can be easily tested. Two such methodologies are described in the following.

5.5.2.1 Design for Test (DFT)

DFT is an all-embracing term for maintaining the controllability and observability of the internal nodes of a circuit during the design phase (Eichelberger and Williams, 1977; Carter, 1964). These *ad hoc* techniques, constituting good design practice, and more formal structural design methodologies constrain the designer to a particular design style. Unfortunately, these formalised techniques can derate the performance of the design, add hardware burdens and severely constrain the designer's freedom. Although many would defend formalism as the only effective way to cut the costs of the manufacturing cycle, the aforementioned disadvantages have meant that they are by no means universally applied.

5.5.2.2 Built-in Self-test (BIST)

Certain products operate in areas of high stress or inaccessibility or are required to be particularly reliable on safety grounds. Incorporating the tester in with the circuit can give major advantages of proximity, resulting in greater controllability and observability. The circuitry can also be tested within its own environment at its normal operating speed, thus catering for the detection of performance faults which are inadequately modelled by the normal stuck-at-0 or stuck-at-1 fault (a particular characteristic of digital electronic circuits).

Just as DFT has specific philosophies, BIST also has specific philosophies. The most widely documented is Built-In Logic Block Observation (BILBO)(Konemann, 1979). Although this approach has been successful, the expense of additional circuitry can only be justified in special circumstances. Defence and aerospace applications fall into this category, where the test needs to be carried out periodically and *in situ*.

5.5.3 AI in Digital Test

As we have previously stated, conventional ATPG techniques suffer from exponentially increasing generation times and the inability to operate in the time domain. To overcome these deficiencies, AI-based techniques are being utilised to improve the testing of circuits through formalised DFT features and functional and behavioural modelling. The remainder of this section will briefly review some of the important current developments in this field. The suitability of AI to this problem is due to the abstraction that is facilitated by the modelling, and expert rules applicable to a complex, though not wholly intractable problem. Both shallow and deep approaches are relevant here (Section 5.2), as *ad hoc* and formalised empirical features can be allied to abstracted circuit modelling in the more effective systems. It is important to realise which view(s) of the circuit would give the most relevant behavioural response when used as a basis for abstraction and modelling.

5.5.3.1 Testable Design Expert System (TDES)

This expert system is aimed at improving the testing of VLSI circuits by adding DFT features such as scan path and BIST without altering the normal circuit behaviour (Abadir *et al.*, 1985). The circuit is modelled by a graph representation with nodes as functional blocks and busses, and arcs representing data flow. The functional blocks may be broken down hierarchically into primitives, whose behaviour is based on register transfer logic (RTL). Constraints are provided to the system in terms of the speed of the circuit, fault coverage, number of I/O pins, test time and available external test equipment to guide the selection of the final DFT features.

The DFT rule base contains knowledge which describes the various DFT techniques, their application and overheads. After identifying those circuit features to which DFT could apply, the circuit graph model is searched to identify intrinsic test resources. This implies a deepening of the expert system model. A global view of the circuit is then taken to discover any commonalty of test resources within the constraints specified and thus optimise the DFT features.

5.5.3.2 Alvey KB System for High-Level Built-In Self-test

Although very similar to TDES (described above) in the task, the implementation details are considerably different (Jones and Baker, 1986). This Alvey-funded research project employs the object-orientated paradigm as the underlying control mechanism. The system is intended for use early in the design cycle, starting with a high level description of the circuit before implementation details are finalised. Here we see the value of abstraction in containing the task's complexity.

The circuit description provided in the system is converted to the object-orientated representation through instantiation. Test tokens (a high level abstraction of test vectors which match the circuit abstraction level) are used to explore the circuit by passing them around the objects (devices) and noting if they can be applied and/or propagated. This requires a deeper knowledge base, which understands the causal connections between items in the circuit design. Timing and other test vector constraints have been avoided at this level.

Resources are also identified, that may be used for BIST, and held in a test record along with the additional hardware costs of applying them. The final evaluation of the resource allocation is performed in a similar manner to TDES, by taking a global view of resources with the constraints specified at the onset. Where a number of possible solutions are available, the designer is asked to arbitrate. Finally, the modifications are returned in the high level description, to allow the design to continue, but with the additional DFT structures.

5.5.3.3 Sequential Circuit Test Search System (SCIRTSS)

SCIRTSS (Miczo, 1986) is a system aimed at providing an automatic test generation system for sequential logic circuits. It employs heuristic search techniques borrowed from AI to guide the selection of states. The system requires a gate level model and an RTL description of the circuit. Stuck-at faults are selected in the gate level model and propagated to primary inputs or outputs and memory elements, using conventional ATPG techniques. A sensitised tree search of the state space obtained from the RTL description looks for a set of inputs that will drive the circuit to the required condition. When a memory element is encountered, a propagation search mechanism drives the state machine until the fault appears at the output.

To reduce the search space, the circuit is viewed as data flow and control, where data transfers are not considered to be state changes. In addition to this, a simple heuristic is used to score the nodes in the state space tree to guide the search and reduce the search space. The objective of this heuristic is to indicate the easiest or shortest path to a desired state. Although this system is capable of dealing with the problems associated with synchronous sequential logic, it would appear that it cannot cope with large bus-structured boards and cycle-based devices, since they are difficult to model at the gate level. The authors have concentrated on the containment of the search phase, rather than a true hierarchy that could facilitate abstraction.

5.5.3.4 Automatic Programming Approach to Testing

This is an expert system developed at the MIT laboratory which is capable of writing test programs for digital circuits modelled at the device level (Shirley, 1985). The system itself is a goal-driven rule-based expert system, which encodes knowledge in the form of clichés (known solutions to testing problems), which are applicable to a wide variety of circuit boards. The clichés contain knowledge on the testing, initialisation, traversing, enabling, etc. of the component, providing a route into a deeper expert system with causal understanding. The rules also contain fragments of the test program code and constraints as to how they are combined to produce a test program. The testing clichés are broken down into subgoals having code fragments with the lowest hierarchy level being directly soluble.

In order to produce a portion of test program, rules are applied to a component, containing information about which signals need to be activated and observed to carry out the test. A search is subsequently performed on the circuit model, and each component lying on a possible path is interrogated to determine if it too can be used to propagate the required test vectors. If paths can be built up from the component under test, to the primary inputs and outputs, then the test program is built up from the associated code fragments and constraints, existing as a lower level associated view of the model. This potentially powerful system may suffer at a later date by the locking of analysis, code fragments and constraints within one knowledge base. Diverse tests on more complicated structures may be impossible.

5.5.3.5 Hitest

Hitest (Robinson, 1983) is the only knowledge-based system for test generation to have matured into a commercially available product. The system has a number of features which make it easy to use and help to interpret results. A schematic display showing a 'snapshot' of logic values on the networks, and a timing analyser display, results in a fast analysis of any problems discovered during test generation. Hitest can operate on circuits modelled both functionally and behaviourally, using its own hardware definition language.

Test program generation for this system takes the form of a modified PODEM (Goel and Rosales, 1981) algorithm which can deal with combinational logic testing. The system is also 'clever enough' to stop when it finds sequential logic, at which time it looks in its knowledge base or asks the design or test engineer how to achieve a particular state. This allows test program generation to continue. The knowledge utilised by the system appears to be board-specific, with little or no knowledge retained from one design to another. The system also requires that the test engineer be familiar with the functionality of the circuit, in order that he can provide the necessary interaction. The inherent traceability and explanation paths available to expert systems are enhanced by the display tools in this application. But, as is often the case in the testing world, the test engineer may have little knowledge of how the circuit was designed to operate. This reliance on the test engineer to 'get the system out of a fix' demonstrates the difficulties in defining suitable models and hierarchies to deal with real-world problems.

5.5.3.6 Multi-level Deep–shallow Integration

This is a diagnostic system, rather than a test generator. It integrates shallow empirical rules with a deep simulation-based understanding of the interrelationships and components inherent in the diagnosed system and device models (Abu-Hana and Gold, 1988). These models exist at various levels, and the system recursively descends through the circuit hierarchy, trying shallow knowledge first, then applying deep knowledge as required, until the faulty component is isolated. Once simulation has taken place at a level, the results are incorporated both into the shallow expert for later diagnostic efficiency as well as up the hierarchy.

The shallow expert holds the symptom-to-fault associations and builds a list of suspect components or modules, whilst the deep expert has to refine the process by following the behavioural nets, guided by the structure. This system demonstrates the containment of complexity and effective use of hierarchical abstraction. The deep expert system consists of:

- a knowledge base of the structural hierarchy that models the components connectivity in terms of primitives and modules (containing primitives and other modules);
- an associated behavioural hierarchy that links symptoms and suspects via causal nets which can aid explanation, down to a quantitative description of exact temporal relationships between input and output;
- a multi-level simulator with a range from coarse qualitative modelling down to detailed quantitative modelling, which can generate both healthy and faulty circuit behaviour.

In operation, the system asks the test engineer for probe information as the need arises, to confirm or deny the hypotheses of suspect modules or components. If the system cannot discriminate completely at the top level using the shallow expert, it will simulate through the deep expert. The next approach is to descend down the hierarchy, or to more detailed simulation level, and try the shallow and deep experts again. The simulation results are incorporated into both knowledge bases by construction of new causal nets, and carried back up the hierarchy, so as to aid abstraction for future tests. By this method, the system is learning more about the circuit under test all the time. This system is of interest for several specific reasons:

- The models themselves and their abstractions cover a wide range and yet are well integrated. The concepts of increasing granularity is used to abstract qualitative information and the behaviour is discriminated between responsibilities of a component (what is the intended function) and the side-effects which can be as useful for diagnosis.

These properties aid efficiency greatly and the learning mitigates the high simulation costs.

- Qualitative behaviour alone is not enough for more complex and dynamic systems, although at high levels of abstraction a quasi-static approach may be applicable.

- The developers feel that an object-oriented frame-based system would be more suitable for a large implementation of this system.

- It is assumed that the test engineer may probe at all module interfaces, which suggests total access to the circuit under test.

5.5.3.7 Proof Workbench

The system proposed by Gupta and Welham (1988) uses the fact that test and verification are different. For verification, exhaustive testing is required, but proof only requires us to determine discrimination conditions that show up faults, i.e. only conditions that will detect a fault need be generated. The workbench, implemented in Prolog, attempts to find a condition where the given faulty circuit yields a value different from the correct version. This intentional approach should be simpler, because it avoids unnecessary distinctions between different conditions that satisfy our purpose, nor does it try to ascertain all possible conditions that will highlight the fault.

The workbench provides abstraction to a functional model of the circuit, as do human engineers, hence timing issues are of secondary importance (it is assumed that timing problems such as race hazzards are eliminated at the design stage). The functional behaviour of any circuit module is defined as a Prolog primitive recursive function. Libraries of correct and faulty behaviour are built up hierarchically into one correct and various faulty circuit models. Once the user has defined the correct/faulty circuits, discrimination rules are invoked that work their way down the lattice that represents the device, by partial evaluation (unfolding and simplification). The workbench has various tools to help in this process, which pull out results to the top level, and evaluate or unfold down the model hierarchy. The achieved simplified discriminatory conditions are exported into a list, as the export and action of the discrimination rule are similar in their effect, but this may contain conditions which are internal to the device as a whole and not available to the outside world, if discrimination is performed at a deeper level than the top-level if–then–else.

The system's discrimination rules, which drive the analysis, can be added to by the user, where a syntactic discrimination does not exist, but where the user can see a significant simplifying step towards the discrimination goal. This can be shown to be of relevance for recursive solutions, which need to be called by the user. When this system is complete, it is hoped that libraries would contain:

- behavioural definitions of the primitives, obtained from standard component libraries;

- commonly occurring fault models, available to be chosen by the user; and

- functional descriptions of higher level modules, automatically generated from the structural information in the circuit diagram and the temporal information which may be available from the timing diagrams with a CAD tool to facilitate user interface.

Although the actual test vectors themselves are not produced, other improvements considered by the designers of the system described in this section are:

- automatic specification of the nodes and direction of unfold, which has to be provided by the user at present;
- when and where to apply recursive discrimination;
- addition of bottom–up as well as top–down analysis so that a distinguishable fault can be raised, up to the top level, from a detected faulty module.

All this could be performed with a single comprehensive AI management and control system.

5.5.4 System Overview

The system, used as an example in this section, consists of two heavily interrelated subsystems, as shown in Figure 5.5. KRAFTS* (Lea *et al*., 1988), is an expert-system-based, high-level test formulation and scheduling tool. It utilises a CAD circuit definition, with a component library and functional, generic testing information to build up an associative network, which may be interrogated by the rule base. A report on all tests that may be implemented and observed from the edge connector is then transferred, via the intelligent test planner to DRAUGHT[†], the vector generator. Here, a directed graph merging and backtracking algorithm operates at the behavioural level. The propagation is restricted to the paths specified by KRAFTS, thus minimising an information explosion. Timing violations are reported back to update the formulation and schedule.

The interface between these two major components is not facile, as there is a modelling change and a movement in hierarchical level. This extra information is implemented, or interpreted by a third expert system, which may be considered to be a test scheduler. This entire system is referred to as GIFTS[‡] and is a good example of the use of expert systems in association with purely algorithmic simulations (in this case a merging of interactions in the time domain), i.e. a hybrid architecture.

5.5.4.1 Test Engineering Techniques in the Functional Test Domain

The philosophy behind the system described here is to follow the processes utilised by a good test engineer; processes that have been developed and accepted over a number of years. Adherence to this philosophy results in the logic of the decisions made by the system being readily assimilated by the engineer, who can therefore build up confidence in the system during use. The resulting expert system described in this section was to be capable of producing functional test programs for digital circuit boards which employ *ad hoc* DFT techniques, as well as more formalised methods such as LSSD and BILBO.

The test engineer usually begins with a set of fully engineered circuit diagrams, a circuit board specification and the appropriate device data sheets. To fully understand the implementation details of a design, and thus gain an insight into ways of testing a particular board, the engineer begins by marking features on the circuit diagrams, such as busses, in order to reconstruct a skeletal higher level functional and structural view of the circuit. From this abstracted model, the engineer can more easily study the relationships between devices and the signal flow that takes place on the board.

---

* Knowledge Representaion Automaton for Functional Test Schedules.
† Digraph Representation for the AUtomatic Generation of Hardware Tests.
‡ Generator of Intelligent Functional Test Solutions.

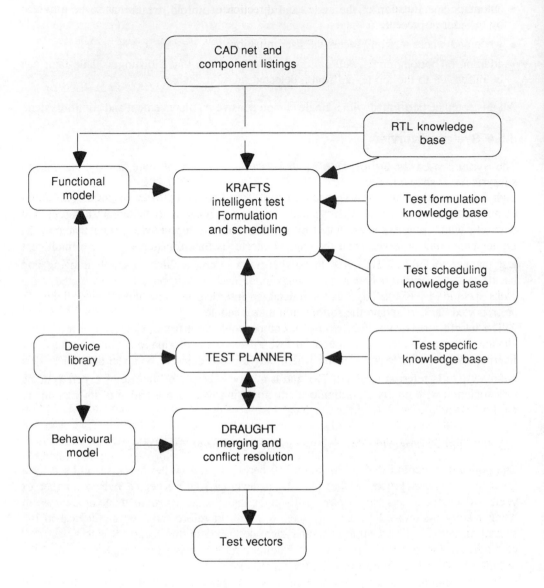

Figure 5.5 System overview

Possible test vector paths are then produced by the engineer, without considering the actual test vectors required, as that is sufficient for this level. These are then validated by a more detailed analysis of the controllability of the devices on the test paths, whilst at the same time identifying devices that may interfere and must be disabled. This is followed by a timing and logic value analysis, carried out either manually or by using simulators. The paths are then exploited to the full when producing functional tests for the devices themselves.

Once the device testability analysis has been performed, the tests are then ordered into a test schedule (the order in which the tests are to be carried out) by the test programmer. This is usually based on some simple heuristics, such as increasing test complexity, both in terms of board coverage and method of testing (e.g. by the edge connectors, then via a microprocessor if relevant, then via a DMA) or by usefulness for utilisation in other tests. Finally, the test program is written to provide the necessary test vectors utilised by the functional tester in order to test the board.

5.5.5 The Test-scheduling Expert System

This system is illustrated in some depth to elucidate the build up of an engineering solution by the application of AI. It shows typical knowledge structures, model abstractions and other attributes common to most solutions, whether in design, manufacture or test.

5.5.5.1 Circuit Model

The output information from CAD packages, consisting of the component and circuit net listings, form the primary input to the expert system. The reasons for choosing component and net listing data as the initial level of representation for the expert system are :

- This is a standard format produced by all electronic CAD packages, from the cheapest schematic editor to expensive top–down design environments. They are also readily available.
- The various hierarchical CAD systems available do not have a standard description of components and connectivity in their hierarchical circuit models. The lack of standardisation introduces problems in adopting higher levels and further abstraction, as an indication of functionality for KRAFTS, although this approach cannot be ruled out completely.
- The presently available (and abundant) libraries of in-circuit tests are defined at the 'device pin level', and thus tie in with this level of representation.
- This is also the initial level utilised by the test vector generator, DRAUGHT, which produces the actual functional test vectors required to test the board. It then descends the hierarchy, and generates time-dependent test vectors.
- The expert system can closely match the test engineers processes, so assisting in the understanding of any reports issued by the decision structures built into the system.
- Although a higher hierarchical view may seem more efficient, it would generate significant extra difficulties in the process of descending down to the level required to validate a test, and provide timing information in the test vectors.

The component listing is used to access the device library to obtain the frame-based symbolic device definitions for the components present on the actual circuit board (Minsky, 1981). The test engineer has only the manufacturer's data sheets containing information about individual device 'pinout', function and timing details (with little or no internal information), which matches his understanding of the circuit components and their function. It is for this reason that the device models used in the KRAFTS expert system are based on the basic pinout of a device, and are the same level of detail present in many schematic capture packages.

The device models are based on directed graphs. Nodes represent the generic name of the device, the actual device name/number and the signal pins. These are linked to a generically based device hierarchy, where rules reside. The device node at the centre of the device model is linked to a leaf node (general concept of that device) in the device hierarchy, via an instantiation link. This allows the inheritance of rules, which describe :

- how to test the device (test formulation);
- how it is to be used in testing other devices;
- how to enable and disable it during testing.

The actual rules themselves are generically based, and thus appropriate for any device of its class, independent of pinout. For example, all microprocessors have address busses, data busses, read/write lines and so on. Although especially true for microprocessors, some other pin-level features of devices do vary, such as types of interrupt and address latch enable. These can be catered for by rules that interrogate the specific device models from the device hierarchy to discover if any additional features are present that need to be invoked. This simplifies and reduces the amount of information stored in the system; detail on individual components is inferred by the expert system when required, after the devices are linked into a specific model. The signal names link to a separate hierarchy, defining pin function and signal flow, with the arcs being the common name associated with the pins.

The device models and the net listing are combined to generate a complete frame-based associative network of the circuit, in the 'specific plane' of the expert systems database. At the same time, inheritance links (instantiated links) are also generated from this specific plane to the fixed general plane (hierarchical model of the domain), where the rules, defaults and device operation information resides in a number of interconnected hierarchies. These are sorted by frame and slot to provide an efficient knowledge search by the inference engine. The more specific rules, etc. reside at or near the bottom of the hierarchies, whilst the more general ones reside at or near the top.

5.5.5.2 Register Transfer Logic Rule Base and Circuit Search

The signal pin hierarchy contains generically modelled rules and information based on RTL. This allows the signal flow and device/signal interaction of specific devices to be obtained, again through inheritance, and forms the basis of the circuit search algorithm invoked by the testability rule base. There has been no attempt to try and model identity mode and transparent mode paths as in the TDES system (Abadir and Breuer, 1985). This is due to the fact that any type of path found on a circuit capable of carrying test vectors, should be tried at some stage in a functional test program, to provide a thorough testing of the board and its features, and not restricted to the few identity mode paths that may exist on a circuit.

The RTL rule base provides one of three responses to the search algorithm in order to define the signal flow structure of the circuit. The algorithm may return:

- Another signal pin node(s) belonging to the same device and indicating a flow of data through it (from input to output (write) or output to input (read)), depending on the search direction). This mapping is independent of any data transformation performed by the device.

- A terminate response. This relation is used for such things as device control signals, whereby a signal enters a device to perform some function but is not directly propagated to an output, although it may cause other modifications to the data.

- The 'no entry' response, used to indicate that a signal cannot enter a device, that is, the expert system is trying to write a signal to an output pin or read a signal from an input.

The nodes are connected via the '**is_a**' inheritance link to form the hierarchy, and the leaf nodes display the inferrable relationships, such as '**prop**' and '**i_prop**' (propagates-to, and -from) that 'could be' inferred for the actual devices existing in the specific circuit model. The rules associated with the '**i_prop**' relationship allow the '**prop**' relationship to be inferred for the board-specific devices. These rules are invoked by a goal enquiry from the search algorithm (Section 5.4).

The reason that a rule structure has been adopted to model the signal flow of devices, is to allow further predicates to be inserted. This is performed in order to investigate other parts of the device and circuit structure before concluding the flow mapping. The alternative would be to build the flow directly into the device model. This preference is illustrated by the fact that a decoder and a demultiplexer are identical devices, but they are used to provide conceptually different functions with regard to specific circuit implementations.

5.5.5.3 The Search

The search algorithm is used to trace pathways through the circuit. It operates under the guidance of the test formulation rule base, responding to requests of the form:

```
<some device>reads_<signal name>_from or <some device>
                                 writes_<signal name>_to.
```

This provides the search algorithm with three essential pieces of information:

- the device from which to conduct the search;
- the type of signal that is to be traced; and
- the direction in which the search is to take place (designated by the '**prop**' relation or the inverse relation, '**i_prop**').

An example of the search algorithm making a '**prop**' inquiry of the pin '**in1_4**' illustrated in Figure 5.6 shows that the inference engine cannot find the relevant relationship. As a consequence, the inquiry follows the **inst** link to the generic concept of the pin '**in1**', where the relation '**prop**' is pointing to '**rule 310**'. In the rule, the '**base_frame**' variable takes on the inquiry value, '**in1_4**'. It is compared in turn against each '**value_frame**' variable, until the premise '**same**' matches the device found by following the inverse input link '**i(_input)**', and the inverse output link '**i(_output)**' from the '**base_frame**' and '**value_frame**' respectively. If they both belong to the same device, then the rule fires and the '**prop**' relation is inferred between the '**base_frame**' and '**value_frame**' ('**in1_4**' and '**out1_4**').

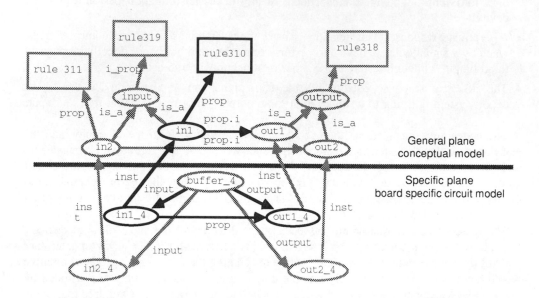

Figure 5.6 A search inference example: Prop structure concluded by rule 310 for **buffer_4**

The reverse signal flow, '**reads_signal_from (i_prop inquiry)**', causes the forward '**prop**' relation to be invoked, and then it traces backwards by the relation '**i_prop**' from the output pin to the input pin. Other responses recognised by the search algorithm are terminate and no entry, used for control structures that do not continue and attempts to drive output pins respectively.

5.5.5.4 Generation of High-level Structure

From the associative network circuit representation and search mechanism, the system is able to identify bus features such as data busses, address busses and control structures. It then generates a higher level circuit representation, superimposed upon the original pin-level representation. The original structure is maintained in order to allow the expert system to drive the test vector generator.

This high-level structure is topographically identical to the signal flow circuit model generated by the test engineer, giving three advantages:

- It allows a high-level test reporting schema to be supported;
- It also increases the efficiency of the test identification knowledge base and test scheduler by constraining the search space, in the same way that abstraction aids the test engineer to focus attention;
- It forms a possible starting point for the analysis of circuits designed using top–down design systems, such as Silvar Lisco's HELIX (Blundell, 1987) and other hardware definition languages (McLaughlan, 1985).

The high-level structure generated by the system is held internally, both as frames within the specific database, and also as a graphical associative network. This enables one to obtain a less complex view of a circuit, as can be seen in Figure 5.7. It should be stated that this diagram is more formalised in electronic engineering terms, but contains identical topological detail to the network.

The isolation of these circuit features (components and buses), along with their respective RTL models, allows the system to conclude testability associations between the separate entities. The system output from Figure 5.7 is a report of all the possible tests and starts, as shown in Figure 5.8. It is combined with a graphical highlighting facility of the paths

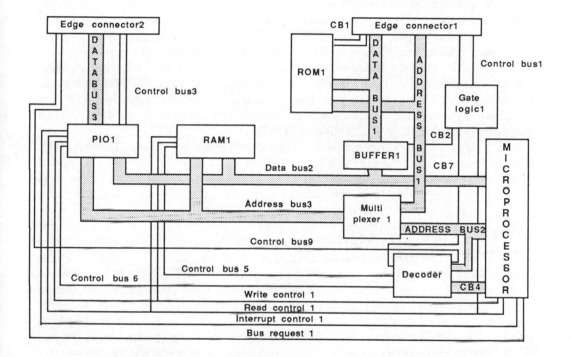

Figure 5.7 Example of a bus-structured microprocessor board as viewed by KRAFTS

microprocessor1 can perform a microprocessor_read_write_ram test upon ram1

ram1 reads_data_from microprocessor1 on the path:
**   ram1   databus2   microprocessor1**

ram1 reads_address_from microprocessor1 on the path:
**   ram1   addressbus2 multiplexer1 addressbus3   microprocessor1**

and so on

Figure 5.8 Output form for the circuit of Figure 5.7

on the frame-based circuit model. This is not the best method of visual representation, and ideally it would be advantageous to drive a schematic capture package directly, so as to highlight the indicated test vector paths.

5.5.5.5 Knowledge Base for Test Identification

The second knowledge base that resides in KRAFTS is the knowledge which provides test identification, and contains rules to elicit from the circuit structure how devices are to be functionally tested: that is, whether the devices present on the board can be tested via the edge connector itself, or indirectly via other devices and then pass the results back to an edge connector. KRAFTS also determines how to control adjacent devices to facilitate the test. This knowledge base operates according to a number of rules containing the criteria that must be satisfied in order to carry out such a test. The rules are based on the supposition that controllability and observability of the various data and control signals lead to device dependency, which directly relates to device testability. The controllability and observability functions make use of the ability to search the circuit and verify that such paths do exist. The rules themselves are independent of circuit structure, but the result of applying this knowledge to the circuit yields, as its output, the circuit substructure necessary to carry out the tests. This is assessed qualitatively as opposed to the quantitative methods of SCOAP (Goldstein, 1979) and CAMELOT (Bennett, 1980).

This follows an alternative approach suggested in outline by Shirley (Shirley, 1985), relating the operation of the expert system to that of the test engineer more closely than has been tried before. It takes the form of first obtaining the testability data flow of a circuit in skeletal form, then fleshing it out with dependencies, timing detail and so on as the circuit is analysed in a top-down hierarchical fashion.

The frames in the generic model of this section of the expert system are organised into three hierarchies, consisting of the test, bus and device hierarchies. These three hierarchies are formed to allow the selective inheritance of knowledge (rules, procedures and user requests for information (asks)) to the specific circuit model, which is then used to solve the functional testing problem. These rules are primarily based in the device and test hierarchies of the generic model. The links that 'could be' deduced by these rules in the specific circuit are shown as '<link name>.i'. The rules themselves can be expressed in a high-level manner, due to the presence of the path-finding algorithm. The algorithm allows the rule base to inquire of the circuit, through any intermediaries and for any particular test, whether signal paths exist. Additional knowledge present in the test hierarchy also checks that the paths can be enabled and that the remaining bus devices tri-stated. This knowledge is procedural in nature, requiring complex search patterns to be invoked, but it is general enough to apply to all tests.

A good example for further examination, as it is one of the simplest to express, is that of testing a random access memory (RAM) directly from the edge connector. The rules for this lie in the device and test hierarchies, and are illustrated in Figure 5.9, along with a list of the rules that appear at these frames in Figure 5.10.

The rule base is called into action in the expert system by an inquiry of the specific plane as to whether a particular device, in this case a RAM, can be tested. This causes the inference engine to take the specific RAM frame and ascertain the presence of a satisfactory test relationship. In this case, there is none (yet), causing the RAM frame to inherit rule 5 from the generic concept RAM in the device hierarchy. Rule 5 informs the inference engine

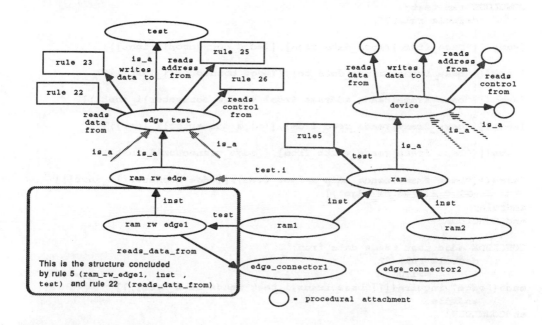

Figure 5.9 Attachment of the 'ram_read_write_edge' test in the generic device and test hierarchies

to check a number of parameters before concluding a positive RAM/edge connector test relation. When firing the rule, the inference engine substitutes the specific instance of the device into the 'base_frame' position of the rule clauses, and then continues as previously described. For further clarification, let us take the first:

```
[ same ([ Λbase_frame ] reads_data_from ],
          [[ edge ] i_inst ])]
```

This clause means: 'Can the base_frame read data from any device representing edge connectors?' If this and the remaining clauses fire, obtaining an answer with a certainty factor greater than 0.2, then the inference engine will carry out the concluding part of this rule :

```
=> a("ram_rw_edge")]
```

This last part of the rule causes a test link between the specific RAM frame and the instantiated 'ram_rw_edge' frame. The test link also takes on the certainty factor value of the lowest clause to fire in the rule (reference should be made to Chapters 2 and 3 for a discussion of uncertainty management, including certainty factors). Once it has been established that the RAM can be tested via the edge connector, it is then necessary to identify the actual edge connectors from which the various data, address and control signals

```
CONCLUDE ram test;
      defrule rule005;
      [
[same([[^base_frame]reads_data_from],[[edge_connector]i_inst])]
      [
[same([[^base_frame]writes_data_to],[[edge_connector]i_inst])]
      [
[same([[^base_frame]reads_address_from],[[edge_connector]i_inst])]
      [
[same([[^base_frame]reads_read_from],[[edge_connector]i_inst])]
      [
[same([[^base_frame]reads_write_from],[[edge_connector]i_inst])]
      [
[same([[^base_frame]reads_chip_select_from],[[edge_connector]i_inst])]
      =>1.0 a("ram_rw_edge")]
endrule;
end CONCLUDE;

CONCLUDE edge_test reads_data_from;
      defrule rule 22;
      [      =>1.0
each("edge",inquire([[[^base_frame]i_test]reads_data_from]))]
      endrule;
endCONCLUDE;
•
•
•
CONCLUDE edge_test reads_chip_select_from;
      defrule rule 22;
      [      =>1.0 each("edge",inquire([[[^base_frame]i_test]

      reads_chip_select_from]))]
      endrule;
endCONCLUDE;
```

Figure 5.10 Rules associated with the edge testing of RAMs

originate. This is achieved by making further inquiries from the newly-formed instance of 'ram_rw_edge', to invoke rules such as rule 22. Thus the network is being built up continuously.

Once all the relevant associated rules have been fired for concluding the RAM's test environment (reads_from, writes_to etc.), a full description of the test is held at the instance of the 'ram_rw_edge' frame. This information drives the search algorithm to extract the paths that lie between the device under test (DUT) (by following the inverse test link), and the participating edge connectors (indicated by the 'reads_data_from' link, etc) along which to apply the test vectors. These paths are further analysed to check for control of the intervening devices in a similar manner. If all is well, the results of the path validation are passed onto the test scheduler, which orders the tests, and then finally to the test vector generator.

All of the certainty factors for these examples are set to 1.0. Certainty factors other than this are used for devices such as a transceiver, that has a direction and enable control. If no control can be gained over one or both of these signals, tests may still take place depending on the logic state of the control, although with less certainty, which then affects the overall test result. The certainty factors assigned also give a numeric value similar to SCOAP for the testability measure, although this is based on a different method of reasoning and numeric model.

Analysis at this stage is similar to the test proposals made by the test engineer and tend to be optimistic, due to the lack of timing and logic value analysis. This level of detail is left to the test vector generator (DRAUGHT) to report back on any problems that occur. After a few tests have been analysed by DRAUGHT, the results obtained allow rapid analysis of those paths that work and those that do not, so that the possible multiplicity of test vector paths of subsequent tests can be re-ordered. This is similar to the test engineer exploiting previous useful test vector paths.

5.5.5.6 Knowledge and Heuristics for Test Schedule

The test-scheduling knowledge base uses the results of the test identification knowledge base as its primary input. The information used is the path information associated with the tests, the devices under test, the types of test that can be carried out on particular devices and the structural associative network model of the circuit. The test schedule is then deduced by using a number of rule-based and heuristic techniques, in order to assess the order dependence of tests on that design. The rule base for scheduling exists in a test hierarchy and establishes an initial group test ordering. These rules contain knowledge of generic testing situations (e.g. how to test with a microprocessor present that cannot be disabled). Heuristics to be found again in the test hierarchy, include:

- a precedence link, which is used to indicate that easier tests should come before 'harder ones';
- a path length heuristic, which evaluates the tests within a test group. (This identifies the shortest test vector paths, indicating the simplest tests. Hence the system can increase the test complexity as the test program progresses, as an aid to later diagnostics);
- a build structure, that weights the use of paths that already have been verified, again for either efficient coverage with the least vectors, or for diagnostics.

5.5.6 The Test Vector Generator and Timing Verifier

This section of the overall system is mostly numeric and algorithmic in nature, but it is of great importance, as it shows the power of a hybrid expert. It interfaces with the AI component via the test planner, whose function is not as simple as might initially seem the case.

DRAUGHT is the component of the Intelligent Test Generator (ITG) system that attempts to define the test vectors to implement a particular test function. Having performed its test scheduling operation, KRAFTS will pass information to the test planner, which mediates between DRAUGHT and KRAFTS. In the case of a microprocessor-based board, where the processor is being used to test a device, the test planner determines which specific instructions need to be implemented by the processor, and how the instructions and data will be conveyed to and from the processor. At the lowest level, this definition of test is a series of read, write and instruction-fetch operations or test primitives. These primitives are passed one at a time to DRAUGHT, along with the bounding subcircuit and any bus data.

DRAUGHT receives a test primitive and re-creates a goal-directed graph model for the specific device operation, incorporating the bus data. The backtracking algorithm is then invoked on the subcircuit and the goal digraph. The algorithm attempts to reconcile the goal digraph with other device digraph models in the subcircuit, by considering the pin activity required and the structure of the subcircuit. When a model digraph can be reconciled to the goal digraph, the two are merged, resulting in growth of the goal digraph. This process continues until the edge connector is reached or an irrevocable conflict occurs. In the latter scenario, a report of the pin and time of the conflict is passed back to KRAFTS via the test planner.

In this way, DRAUGHT builds up a complete picture of the activity at each node of the subcircuit. As well as the input and output test vectors, the system provides valuable internal node activity definitions which can be utilised by diagnostic routines.

5.5.6.1 Test Definition

In order to invoke a test, the test planner has to glean certain information concerning the specific devices involved in the test. If, for example, the scenario is that of a microprocessor testing a RAM, the test planner needs to discern various parameters, regarding both the RAM and the processor, in order to furnish the skeletal algorithm. Information required in this case would be of the form of the start and end address, the size of the data bus and the type of RAM test to be implemented. A subset of the microcode of the processor would also be needed in order to generate the cycle instructions.

Having obtained these pieces of data, the test planner can now apply its algorithm for the particular test in question. This involves passing the relevant information to DRAUGHT, to obtain the required stimulus from the edge connector. Whenever a test fails due to irreconcilable conflict, information is returned at two levels: the test planner can report at which stage the test failed; and DRAUGHT can determine the spatial and temporal location.

5.5.6.2 The Merging and Conflict Resolution Algorithm

This section has already highlighted the problems of generating structural models for complex LSI and VLSI devices. The large cost of producing models for such devices without the necessary information, has led to the investigation of higher levels of model, such as the functional model and the behavioural model. This part of the system uses the behavioural level model as the basis of DRAUGHT because:

• They are defined at the pin level, consistent with the expert system's basic model description.

• The information is abundant and readily available from data sheets, unlike internal gate level information. The timing may also be gleaned from existing in-circuit test libraries.

• Each component has at least one timing model of this sort, but always a finite number of timing models (usually restricted to just a few) describing all the correct behaviour patterns of a device.

- This model defines the correct operation of the device, which is used for backtracking, from which the tests are derived. The emphasis is on validating the board under test, not on an attempt to define all the possible faults which would be a far too big selection.

The vehicle that DRAUGHT uses to model the behaviour of digital devices is the concept of directed graphs or 'digraphs'. Digraphs and digraph theory has, until recently, been a pure mathematical concept. However, its applicability to modelling systems has been recognised in many disciplines. This has generated a great deal of research into all areas of graph theory. Digraphs can portray complex interrelationships as a simple, yet elegant and understandable network, making them a popular modelling vehicle.

One interesting application that bears more than a passing resemblance to the concepts within DRAUGHT, is the project manager's Performance Evaluation and Review Technique (PERT) (Barber, 1987). Here, the vertices model events and the arcs model the interrelationships between those events. In a similar way, the vertices in DRAUGHT model the bounds of waveform edge transitions or pin activity constraints, whilst the arcs model the time delays or the time constraints between two such bounds.

The merging and conflict resolution algorithm is invoked when the test planner requires justification of a test primitive. A goal digraph is re-created by DRAUGHT from the model library of the goal device and the high level data requirements. If, for example, the test planner requires justification of a read operation of a RAM by a microprocessor, then the goal digraph would be one of the RAM's read digraph models with the address of the particular location and data expected, incorporated into the model. DRAUGHT also produces a low-level structural digraph of the subcircuit given by the test planner. In this instance the vertices model the devices in the subcircuit, whilst the arcs model the interconnections between the devices. This model is used by DRAUGHT to determine which devices and pins are driving the device that is currently having activity justified.

The backtracking algorithm is recursive, and stores a snapshot of the current state of the goal digraph in case of unsuccessful merges. After an attempted merge that provides conflict on a particular node in the circuit, the snapshot will be reinstated and the particular model discarded. Another model is chosen to merge with the goal digraph. In order to attempt a merge, a set of events (a subgraph of the goal digraph) is selected and compared, via the structural digraph, with models within the driving device model library. When a possible match occurs, a merge is attempted. This process continues until the driving device is the edge connector, which signifies successful justification, or all possible combinations of merging have been exhausted. The latter case indicates irrevocable confliction. If successful, DRAUGHT has defined windows of time in which edge transitions can occur at all the nodes within the subcircuit. This includes the inputs to the circuit, controllable by the tester.

5.5.7 Test Planner

This essential interface between the two levels of the system performs two major functions. They are:

- To communicate between the scheduler and the test vector software. The planner must pass the sequenced test data defined above, along with the participant circuit substructure identified by the Scheduler. This confines the test generator to a manageable information task. Also, when an unresolvable conflict is detected, a message needs to be returned to the Scheduler for truth maintenance purposes.

- To combine the library held, in-circuit or algorithmically generated test data for the device under test with the appropriate timing information about the participating components during each phase of the test.

In order to understand how the planner carries out these two requirements, it is relevant to identify their function in more detail.

5.5.7.1 Test Planner Function

The test planner operates on a 'per test' basis, when invoked by the Scheduler. It has to perform several actions and procedures in order to succeed in its function. Firstly, it takes the identified subcircuit, including the controlling components of any testing device, and instigates a signal flow tree analysis on the primary test vector routes. The high level results of this analysis (including pruning) are then mapped onto the pin level representation for transfer to DRAUGHT. DRAUGHT then recreates the circuit in digraph representation, and evaluates the path in the time domain, using the test sequences generated. The usefulness of that path is then reported back to the test planner, where it is employed in generating a 'used_in' relationship to guide test path selections from the signal flow tree. Again, there is an analogy with the human test engineer, who will utilise test vector paths that have previously proved fruitful.

Although it may appear tempting to also use unsuccessful vector groups to help guide the signal flow tree analysis, care must be taken to avoid discarding elements that may indeed be useful when used for other test functions. In addition, all test paths need to be assessed at some stage in the test programme, so a naive pruning would prove counter-productive. To utilise this information successfully would require a reinstatement of pruned unviable test paths in case remaining routes also prove unviable. This may well be included at a future stage. The structural details coordinated by the test planner are identified in Figure 5.11.

5.5.7.2 Test Data Manipulation

This second aspect involves the coordination of the test data with the actions that are required to carry out a test in a functional manner. The test data for the DUT is broken down, and then recombined with device specific information from participating components. The sequence of events is then reconstructed, following the test data, enabling DRAUGHT to generate the final test vectors. These sequences can be split into three separate stages:

- Initialisation puts the device into a known start state for that test from an unknown, or otherwise previously defined, but inappropriate state. This is illustrated well in the case of a microprocessor test, by using reset to initialise the program counter.
- Device test is defined by the test scheduler, and applies to the DUT only. The tests may be provided by either an in-circuit test library, which defines the device pin activity in the time domain, or a selection of skeletal algorithms, taking advantage of the iterative generic nature of certain component types (a good example would be the GALPAT test for RAMs), and matching the component models of both the expert system and DRAUGHT. For

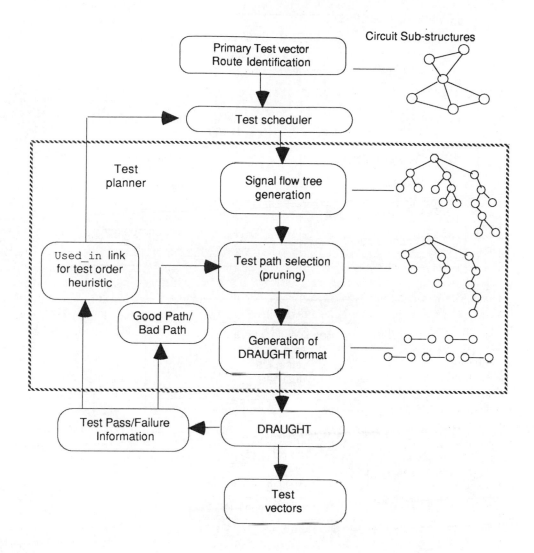

Figure 5.11 The structure and operation of the test planner

testing specific components, the planner fills in the algorithm by inquiring for information, such as address and data bus size, etc., implicit in the expert system's component models.

- Test completion is the stage that ensures that there is minimum interference between tests, and that the final states present on the DUT are known. Failure to do this can mean that the only way of knowing the final state requires a long and time consuming simulation process.

These three stages now are brought together, in order to provide an overall order to the timing cycles of all the participating components. The application of the test patterns

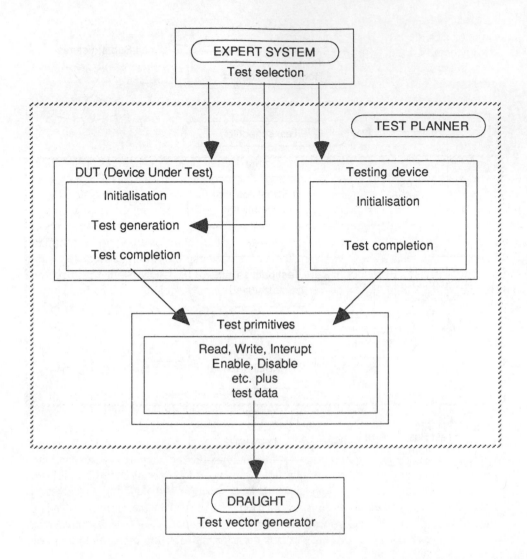

Figure 5.12 The manipulation of test data as carried out by the test planner

themselves can be considered as control and observation phases. Therefore, in the complex case where the DUT is being tested by another (accessible from the edge connector), paths for controlling the testing device are verified first, before analysing the routes between this device and the DUT. These sets of test vectors are added to by considering the complementary requirements of observability through to the edge connector. These test data manipulations are summarised in Figure 5.12.

5.5.8 Conclusions

This test section has highlighted the complexities associated with functional board test, and has propounded a system that is capable of easing the problem, by incorporating several philosophies into the test generation phase. It is quite clear that AI techniques included here are more generally applicable, and will play a significant role in the future in this and other domains. In particular, the role of AI in emulating the thought processes of the test engineer would appear to be essential in the acceptability of these systems to working engineers.

However, it is the circuit designer who will in future be forced to take responsibility for the testing needs whilst designing circuits. The complexities involved make this more integrated approach a prime requirement, where it is the designers' concepts of circuit functionality that are required to guide the testing philosophy. As testing is probably not the designers' specialism, the use of a system, similar to that described here, as an intelligent design assistant would seem to be a natural step. The processes and analyses detailed already do not need to be applied only to existing boards as they use the CAD netlists and various libraries (mostly commonly available), but should be utilised before the implementation has been finalised, to alleviate any unresolvable problems. The implications of the system outlined, on the electronics manufacturing industry in the future, is likely to be profound. By decreasing test generation time, the manufacturer will reap the rewards associated with a shorter time to market. Secondly, even without the intelligent design assistant, manufacturers can be guaranteed a high degree of consistency concerning test quality, which would ensure customer goodwill.

5.6 COMPUTER INTEGRATED MANUFACTURE—CIM

Computer integrated manufacture (CIM) is considered to be the integration of all the functions relevant to the production of a product, in order that it may be observed and controlled more effectively. There are some integrated systems of computer-aided engineering that seem to have advanced a good way towards this within the DMT régime. However, these usually lack integration with the commercial and support systems, and they often have a basically one-directional data flow (down). For observation and control, this flow must be fully bidirectional, with appropriate abstraction and detailing as necessary. The functions needing integration are those that exist at present in the various 'islands of automation': design, manufacture, test, stock control, financial services and commercial. This may be an expensive option, but the benefits are manifold; the more important ones are explained below.

5.6.1 An AI-Controlled CIM Model

In terms of the applications themselves within the DMT domain, the computer-aided simultaneous engineering (CASE) concepts (Ishi *et al.*, 1988) become more attractive and increasingly viable as AI technology develops. The expert CASE could become part of a larger integrated manufacturing (CIM) expert system, based on a central pragmatic knowledge database, which holds various models and views of the associated manufacturing system's behaviour and constraints. A series of modules, some AI to a large extent (but some not), could then interface with the knowledge base and various engineers. A typical system is presented in Figure 5.13.

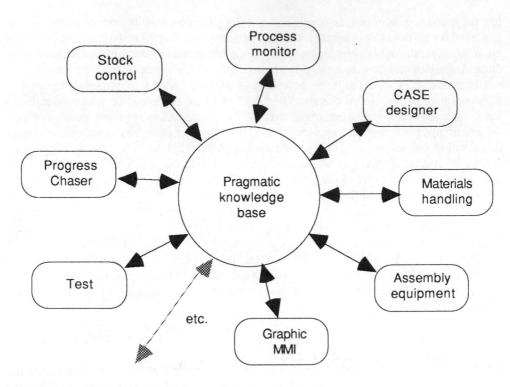

Figure 5.13 A structure for an AI-based CIM system

5.7 FACTORS THAT NEED CONSIDERATION BEFORE DEVELOPING APPLICATIONS

There follows a series of questions, grouped under separate subtitles. If the intention is to implement a system, the first four chapters or Adeli, 1988, give good resumés of the benefits and limitations of expert systems.

(1) *Application*: Is this suitable? Not trying to force consistency when not appropriate? Not too trivial, or over complex? Is it being performed satisfactorily by algorithmic means already? What are you hoping to gain here? What role does the system play?

(2) *Modelling system*: Functional or structural? Both? Mixed mode heuristic and numerical? Can it be represented formally, with hierarchical abstraction?

(3) *System selection*: Is it going to run on suitable hardware? Is it expensive? What interface capabilities does it have? Support? Does it provide other selected functions?

(4) *Representational system*: Will it 'fit' a formal representation? Which is most suitable? Am I being over ambitious? (simple production rule ESs have produced the vast majority of effective engineering applications).

(5) *Inference Engine*: Is it too expensive in memory or computing time? Does the forward or backward chaining (or other) method suit?

(6) *Knowledge*: Where from? How is it acquired? Can it be tested? Can it be maintained? Would machine learning be a viable addition?

(7) *MMI* : What interface is required? Would graphics help and be simple to include? Entry procedures? Should the expert be hidden?

5.7.1 The Future

There is a bright future for AI systems in engineering. We can expect to see the research systems, of which the examples included in this chapter are largely representative, advancing into real-world applications, with the continual decrease in hardware price leading to their increasing cost-effectiveness. Interface standards will be more important, as more systems link the expert in with numerical and procedural modules, with the expert itself better hidden and possibly partitioned. Learning and truth maintenance will become increasingly important as the application knowledge bases become ever larger.

BIBLIOGRAPHY

Abadir, M.S. and Breuer, M. (1985) A Knowledge-Based System for Designing Testable VLSI Chips, *IEEE Design and Test*, 2(4), 56–68.

Abu-Hana, A. and Gold, Y. (1988) An Integrated, Deep-shallow Expert System For Multi-level Diagnosis of Dynamic Systems, In: *Artificial Intelligence in Engineering: Diagnosis and Learning*, Ed. J.S. Gero, Elsevier, Amsterdam, pp 75–94.

Adeli, H. (1988) *Expert Systems in Construction and Structural Engineering*, Chapman and Hall, London.

Andert Jr, E. and Frasher, W. (1988) Compact Expert System for Planning and Control Applications, In: *AI in Engineering: Robots and Processes*, Ed. J.S. Gero, Elsevier, Amsterdam, pp 299–310.

Barber, T.J. (1987) Interactive Critical Path Analysis (ICPA)—Microcomputer Implementation for a Project Management and Knowledge Engineering Tool, *Journal of Microcomputer Applications*, 9, 1–13.

Basden, A. (1984) On the Application of Expert Systems, In: *Developments in Expert Systems*, Ed. Coombs, M.J., Academic Press, New York, pp 59–76.

Bennett, R.G. (1980) Computer-aided Measurement of Logic Testability—The CAMELOT Program, *Proc. IEEE Int. Conf. on Circuits and Computers*, October 1980, IEEE, New York, pp 1162–5.

Blackwell, G.K., Colley, B.A. and Stockel, C.T. (1988) A Real-Time Intelligent System for Marine Collision Avoidance, In: *AI in Engineering: Diagnosis and Learning*, Ed. J.S. Gero, Elsevier, Amsterdam, pp 119–38.

Blundell, B.G. (1987) *An introductory Guide to Silvar Lisco and HILO Simulators*, Computer Science Series, MacMillan, London, pp 97–122.

Brady, M. and Connell, J. (1987) Generating and Generalising Models of Visual Objects, *Artificial Intelligence*, 31, 159–83.

Brown, D.C. (1985) Failure Handling in a Design Expert System, *Computer-Aided Design*, 17, 436–41.

Carter, W.C. (1964) Design of Serviceability Features for the IBM System/360, *IBM Journal of Research and Development*, **8**, 115–26.

Case, P.W. (1981) Design Automation in IBM, *IBM Journal of Research and Development*, **25**(5), pp 631–646.

Cercone, N. and McCalla, G. (1987) What is Knowledge Representation?, In: *The Knowledge Frontier*, Ed. N. Cercone, and G. McCalla, Springer-Verlag, Berlin, pp 1–41.

Chandrasekaran, B. and Mittal, S. (1984) Deep Versus Compiled Knowledge Approaches to Diagnostic Problem-solving, In: *Developments in Expert Systems*, Ed. M.J. Coombs, Academic Press, New York, pp 23–34.

Charniak, E. and McDermott, D. (1985) *Introduction to Artificial Intelligence*, Addison Wesley, Reading, MA.

Coyne, R.D. and Gero, J.S. (1986) Semantics and the Organisation of Knowledge in Design, *Design Computing*, **1**, 68–89.

Dieter, G.E. (1986) *Engineering Design: A Materials and Processing Approach*, McGraw-Hill, New York.

Eichelberger, E.B. and Williams, T.W. (1977) A Logic Design Structure for LSI Testability, *Proc. Fourteenth Design Automation Conference*, June 1977, IEEE, New York, pp 462–468.

Fox, M., Smith, S. and Ow, P. (1986) Constructing and Maintaining Detailed Production Plans: Investigations into the Development of Knowledge Based Factory Scheduling Systems, *AI Magazine*, **7**, 45–61.

Goel, P. and Rosales, B.C. (1981) Podem-X: An Automatic Test Generation System for VLSI Logic Structures, *Proc. Eighteenth IEEE Design Automation Conference*, IEEE, New York, pp 260–8.

Goldstein, L.H. (1979) Controllability/Observability Analysis For Digital Circuits, *IEEE Trans. on Circuits and Systems*, **26**, 685–93.

Gupta, A. and Welham, B. (1988) Functional Test Generation for Digital Circuits, In: *AI in Engineering: Diagnostics and Learning*, Ed. J.S. Gero, Elsevier, Amsterdam, pp 51–71.

Herrmann, J. (1988) Estimations for High-level IC Design: a Machine Learning Approach, In: *AI in Engineering: Diagnostics and Learning*, Ed. J.S. Gero, Elsevier, Amsterdam, pp 359–85.

Ishii, K., Adler, R. and Barkan, P. (1988) Knowledge-Based Simultaneous Engineering Using Design Compatability Analysis, In: *AI in Engineering:Design*, Elsevier, Amsterdam, pp 361–78.

Jones, J.C. (1970) *Design Methods, Seeds of Human Futures*, Wiley, Chichester.

Jones, N.A. and Baker, K. (1986) An intelligent Knowledge-Based System Tool for High-Level BIST Design, *Proc. 1986 Int. Test Conference*, IEEE, New York, pp 743–749.

Konemann, B. (1979) Built-In Logic Block Observation Techniques, *Proc. 1979 Int. Test Conference*, IEEE, New York, pp 37–41.

Kowalski, T.J. (1985) *An Artificial Intelligence Approach to VLSI Design*, Kluwer Academic Press, Boston.

Lea S.M., Brown, N., Katz, T. and Collins, P. (1988) Expert System for the Functional Test Program Generation of Digital Electronic Circuit Boards, *IEEE Int. Test Conference*, Vol 14, IEEE, New York, pp 209–20.

Marcus, S., Stout, J. and McDermott, J. (1986) VT: An Expert Elevator Designer, *AI Magazine*, **8**(4), 39–58.

Maunder, C. and Beenker, F. (1987) Boundary Scan- A Framework for Structured Design-For-Test, *Proc. IEEE International Test Conference*, IEEE, New York, pp 714–723.

McLauchlan, M. R. (1985) *CAD for VLSI*, Van Nostrand Reinhold, New York, pp 167–78.

Miczo, A. (1986) *Digital logic Testng and Simulation*, Harper and Row, New York, pp 373–9.

Miller, A.J. (1985) DESIGN—A Prototype Expert System for Design Oriented Problem Solving, *The Australian Computer Journal*, **17**, 20–26.

Minsky, M. (1981) A Framework for Representing Knowledge, *Mind Design*, MIT Press, Cambridge, MA, pp 95–128.

Mittal, S., Dym, C.L. and Morjaria, M. (1986) PRIDE: An Expert System for the Design of Paper Handling Systems, *Computer*, (July), 102–114.

Pepper, J. and Kahn, G.(1986) Knowledge Craft: An Environment for Rapid Prototyping of Expert Systems, *Pro. SME Conference on Automotive Systems*.

Piaget, J. (1926) *The Language and Thought of the Child*, Harcourt Brace, New York.

Robinson, G.D. (1983) HITEST—Intelligent Test Generation, *Proc. 1983 International Test Conference*, IEEE, New York, pp 311–23.

Roth, J.P. (1966) Diagnosis of Automata Failures: A Calculus and a Method, *IBM Journal of Research and Development*, **10**, (4), 278–291.

Schofield, M.J. (1985) Knowledge-Based Test Generation, *Proc. IEE* **132**(3), 108–110.

Shirley, M.H. (1985) An Automatic Programming Approach to Testing, *Proc. 1985 Workshop on Simulation and Test Generation Environments*, IEEE, New York, pp 55–75.

Struss, P. (1988) Extensions to ATMS-based Diagnosis, In: *AI in Engineering: Diagnostics and Learning*, Ed. J.S. Gero, Elsevier, Amsterdam, pp 3–28.

Tezza, D. and Truco, E. (1988) Functional Reasoning for Fexible Robots, In: *AI in Engineering: Robotics and Processes*, Ed. J.S. Gero, Elsevier, Amsterdam, pp 3–19.

Tommelein, I.D., Levitt, R.E. and Hayes-Roth, B. (1987) Using Expert Systems for the Layout of Temporary Facilities on Construction Sites, In: *Manufacturing Construction Worldwide*, Vol.1, *Systems for Managing Construction*, Eds P.R. Lansley and P.A. Harlow, E & FN Spon, London, pp 566–77.

Tsui, F.F. (1987) *LSI/VLSI Testability Design*, McGraw-Hill, New York.

Woodbury, R.F. (1987) The Knowledge Based Representation and Manipulation of Geometry, PhD Diss., Dept of Architecture, Carnegie-Mellon University, December 1987.

Zrimec, T. (1986) Application of Logic Programming to Task-level Programming of Robots, *Proc. AFCET:Robots and Artificial Intelligence*, Toulouse.

Zrimec, T. and Mowforth, P. (1988) An Experiment in Generating Deep Knowledge in Robots, In: *AI in Engineering: Robots and Processes*, Ed. J.S. Gero, Elsevier, Amsterdam, pp 21–33.

T. Katz
Department of Electrical and Electronic Engineering
Brighton Polytechnic
Lewes Road
Brighton
East Sussex BN2 4GJ

6   Planning and Project Management

P.W. Kuczora

Brighton Polytechnic, UK

6.1 INTRODUCTION

6.1.1 Planning and AI

The topic of planning has been inextricably linked with that of computer-based artificial intelligence (AI), ever since the early attempts to produce computer programs capable of performing as general-purpose problem solvers. During the last three decades, a range of AI planning techniques has been developed, many of which are still used extensively by current planning systems.

Work on planning has involved research into many areas which are of direct relevance to AI. These include:

- search and search space control,
- knowledge representation,
- knowledge elicitation,
- machine learning,
- non-monotonic reasoning,
- explanation and justification of conclusions.

In addition, planning involves a number of problems of its own which are becoming increasingly relevant to knowledge-based systems as they move into use in the 'real world':

- temporal reasoning,
- reasoning with uncertainty,
- perception and vision,
- constraints and resources,
- multiple cooperating agents.

Artificial Intelligence in Engineering. Edited by Graham Winstanley
©1991 by John Wiley & Sons Ltd.

While the notion of planning is closely allied to that of project management in engineering where the term 'plan' generally refers to the *process plan* which defines how an engineering product will be manufactured or constructed, it has a much more specific meaning within the context of AI. The earliest definitions of AI-planning were fairly broad, relating it to the more general concept of problem-solving in a computer system, but it has taken on a more specific meaning in AI circles in recent years. The role of AI-planning is now based around the automatic generation of plans of action to be executed by autonomous robots operating in real-time environments such as building sites, nuclear plants or battlefields. Such applications reflect the fact that most planning research is in fact funded by the military, such as the DARPA Autonomous Land Vehicle (ALV) project. The actual domains in which these systems are eventually intended to operate do not yet exist, given the complex sensor requirements of such applications. It has therefore become the 'norm' to test such traditional AI-planners by applying them to simple robot manipulation domains, such as the classic blocks-world problems, which may easily be simulated on computers.

Here are some descriptions and definitions of AI-planning and the systems which undertake such problem-solving, taken from the literature:

'A Plan is any hierarchical process in the organism that can control the order in which a sequence of operations is to be performed.' (Mille, *et al*., 1960)

'Let us first characterise a problem solver as a program that explores the states that arise from the application of problem-solving operators in search of one that qualifies as a solution to the problem.' (Cohen and Feigenbaum, 1982)

'... systems which can automatically produce plans of action for execution by robots ....' (Tate, 1985)

'There exists a large body of artificial intelligence research on generating plans—linear or non-linear sequences of activities to be executed by some agent (typically a robot) in order to transform an initial world state to some desired goal state.' (Levitt and Kunz, 1987)

'... a *planner* is a program that controls one or more devices capable of carrying out actions in the real world in order to achieve some definite purpose.' (Swartout, 1988)

As successive planners have built upon the techniques developed in previous systems, there has been a progressive 'cross-fertilisation of ideas' between systems. As such systems have been applied to progressively more complex real-world planning problems, the more traditional generalised AI-planning techniques have been supplemented by the use of varying degrees of domain-specific knowledge and traditional procedural computing techniques. While there are no absolute distinctions, the various AI approaches to the generation of project plans and schedules can be divided into the following broad categories:

- Traditional 'strategic' AI Planners,
- Opportunistic and case-based AI planners,
- Advanced AI meta-planners and mixed-AI systems,
- Domain-specific knowledge-based systems,
- Hybrid 'AI-levered' procedural/symbolic systems,

- Non-AI approaches such as game playing and control theory.

These techniques are discussed in detail in the following sections.

6.1.2 Project Management

The field of project management differs in certain substantial respects to that of planning in general, although the two overlap to a certain extent. Some confusion between the two types of system has been caused by attempts to use traditional AI-planners to solve complex project management tasks. The dichotomy between traditional planning work and project management can be viewed from two perspectives:

(1) project management can be seen as a subset of planning, with relaxed timing constraints and working at a higher level of abstraction; and

(2) planning can be seen as a detailed subtasking level within the overall context of a project management system.

Engineering project management is currently carried out in the real world, using a combination of 'paper-based' and computer-aided techniques. The most commonly employed methods are:

- Gantt charts, or bar charts,
- Critical Path Method (CPM) or Critical Path Analysis (CPA),
- Project Evaluation and Review Technique (PERT),
- Scheduling tools, and
- Financial tools.

CPM and PERT were developed around 1960, in order to assist in the running of increasingly large and complex projects in the aerospace, defence and construction industries. Initially, these techniques were used on paper (Fondahl, 1961), but now a wide range of PC and minicomputer-based project management packages are available, offering the facilities of automating the production of project plans in the form of Gantt charts, CPA precedence networks or PERT networks. Other packages, such as ARTEMIS, use OR techniques to perform activity scheduling for complex engineering projects. However, there does not appear to be a standardised overall methodology for project planning and control. This is quite understandable, considering the broad range of projects to which these techniques are applied. Projects vary in many dimensions. Most critically, cost and timescale will affect the techniques used, but other factors are also important, such as the amount of innovation or the location (outdoor projects will need to allow for weather factors).

Surveys which have been carried out in the field of construction project management (Davis, 1974); (Arditi, 1983) have shown that network-based tools are often used as decision support tools for the overall planning of projects, particularly by larger companies in developed countries. However, these surveys also show that such network-based project management tools are less useful for the functions of detailed operational planning and real-time project control. Thus, while modern-day project managers have a bewildering array of tools and techniques at their disposal, it is still normal that when projects become active, problems which occur are not solved using the techniques used to produce the initial project plan.

The question arises as to whether project managers use such an approach, because it is the most suitable, or rather because it provides the only tenable method for manipulating the available data. It seems probable that the problem arises because the systems currently in use do not record the rationale used for original decisions, and so require considerable effort in terms of data re-entry if project plans are to be changed once the project is under way. This point is made by Levitt and Kunz, 1987, who describes how such systems are currently employed within the construction industry:

> 'Existing project management tools are analysis tools that support—and depend upon—knowledgeable managers, who must analyse the project to provide meaningful input data, and who must interpret the significance of the output data.
>
> In planning a project with network-based scheduling tools, a human planner must input activity descriptions and their sequences, either in tabular form or graphically. Next, the human planner assigns durations and resources to activities. The resulting schedule is computed by the CPM or PERT computer system, and can be checked against constraints of total duration, instantaneous or aggregate resource availability limits, weather windows and the like.
>
> Most existing scheduling packages assign activities to their earliest possible start times as a default on the first pass. Where resource-constraints are violated, the manager is able to re-schedule non-critical activities manually to start later. In addition, most packages offer automated *resource-constrained scheduling* using simple heuristics. Resource-constrained scheduling will extend the project's overall duration, where available resources fall short of peak requirements. Re-scheduling heuristics in project management packages have been simple and domain-independent, e.g. '*schedule the shortest activity first*', or '*schedule the activity with least float first*'. As a result, the resource-constrained schedules they generate may be far from optimal.
>
> Except when carrying out this kind of simple resource-constrained scheduling (or when *resource levelling* to minimise periodic fluctuations in resource requirements using similar heuristic approaches), traditional project-scheduling tools are limited to analysis of a project plan and activity durations supplied by the human planner. They do not generate project plans or schedules. Moreover, the knowledge that the planner uses in generating project plans and activity durations cannot be captured or used by these tools. Consequently, each cycle of replanning requires that the planner's knowledge be brought to bear on the problem again.
>
> We gave asserted ... that this requirement for repeated high level input, in order to generate meaningful updates of a project schedule, is the 'Achilles' Heel' of conventional project-scheduling tools (Levitt and Kunz, 1985). In the heat of managing projects, managers appear unwilling or unable to take the time to use their scarce knowledge repeatedly in updating project plans. As a result, although the tools provide valuable decision support during the early planning stages of projects, they often serve primarily an archival purpose thereafter.' (His italics.)

Current research into the application of AI techniques to address this problem is centred around the use of knowledge-based techniques to facilitate the process of project plan generation, and the monitoring and re-planning of projects, once they are under way. As described later, work on the PLATFORM range of decision support systems has shown that AI-based systems which are capable of storing and using knowledge about construction (in this case) and project management, have the potential to overcome the limitations of current network-based procedural techniques.

6.2 PURE AI TECHNIQUES FOR PLANNING SYSTEMS

6.2.1 Strategic and Tactical Planners

Planning systems have been a major research topic within the field of AI ever since the inception of computer-based AI some thirty years ago. However, when the subject is assessed in greater detail, a continuum emerges with regard to the response time which is required of a planning system and the reliability of the information on which the system is required to act. At either end of this continuum are two fundamental types of planning system—*strategic planners* and *tactical planners*. The distinctions between these two types of planning system are demonstrated in the following extract from the DARPA Santa Cruz Workshop on Planning (Swartout, 1988):

> A *strategic planner* is capable of anticipating (some of) the consequences of its actions and using such anticipated consequences to choose between possible courses of action. Strategic planning techniques have primarily been concerned with representing the world in enough detail to allow prediction and efficiently carrying out predictions in order to support decision making. ... Many situations exist in which accurate prediction is either impossible because of a lack of information or useless because of a lack of time. A *tactical planner* is primarily concerned with deciding what to do in situations in which the available information is limited and uncertain. ... situated activity addresses the problem of deciding what to do on the basis of what the robot immediately perceives about its environment. The basic idea is that highly directed behaviour can arise out of interaction with a complex environment to satisfy our requirement that planning ultimately achieve some definite purpose. Purely reactive or purely strategic planners can be considered as extremes on a continuum determined by (1) the amount of time allowed for the system to respond and (2) the availability and volatility of the information potentially useful to the system in deciding how to respond.

For the low-level and time-dependent considerations which apply to tactical planners, it is appropriate to apply conventional control systems theory and procedural programming techniques, and so, in the main, this chapter will refer to systems which can be considered as strategic planners, as this is where the role of AI-planners generally lies. Such systems can be characterised as having the following properties:

- a limited number of actions are required;
- these actions are repeated many times; and
- only a limited knowledge of the world is required.

6.2.2 Traditional (Strategic) AI Planners

6.2.2.1 Planning as Search

Traditional AI approaches to planning began with studies of the topics of general problem-solving and search in goal-directed computer programs some twenty years ago. The earliest planners employed techniques which had been inherited from research into the mathematical treatment of map and network traversals in order to address the twin problems of *search* and *control*. As the study of AI has developed through the 1970s, the topic of search has been pursued in its own right, and a variety of techniques have been developed. These are summarised in Figure 6.1, which is taken from Winston (1977).

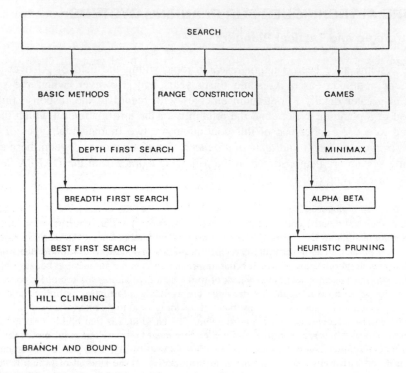

Figure 6.1 The principal divisions of search technology

The earliest basic search methods employed by systems were those of *exhaustive search*. These were developed in order to implement search in graph-traversal problems. *Depth-first search* assumes that one solution is as good as another, and so at each point in the search for a solution, the system makes an arbitrary choice and continues working into the search tree. However, while depth-first search is easy to implement, it has been found to be extremely inefficient in use, due to its tendency to commit the system to the exploration of large numbers of low-level considerations, which are later made irrelevant due to higher-level considerations. For instance, such a planner might spend considerable time exploring the design of a component in, say, aluminium, only to discover later that this material is unavailable or too expensive.

These considerations led to the development of the method of *breadth-first search*, where *all* the alternatives at a given level of the search for a solution were considered before a commitment was made to any of them. Such an approach allowed a planner to avoid the need for extensive computations regarding low-level concerns early on in the planning process, but only at the cost of adopting a careful and conservative search strategy, which in itself was wasteful when dealing with all but the smallest search spaces.

6.2.2.2 Pruning the Search Space

As systems were developed, it soon became apparent that, in any workable system, the size of the search space would be considerable. This, in turn, would lead to computational problems based on a *combinatorial explosion* of the nodes of a search tree for an exhaustive traversal of such a search space. For example, consider a situation where the branching of the search tree is relatively uniform, with the number of alternative branches at each node being b, and the depth of the search tree being n. This means that the number of nodes at the bottom level of the search tree alone is b^n. In a typical game of chess, $b=35$ and $n=100+$, which means that the number of nodes at the bottom of the tree of all possible chess games is of the order of 35^{100}, or 2.5×10^{154}. If the path associated with each terminal node could be calculated in 1 ns, it would take approximately 10^{138} year to compute all the possibilities. Such computational overheads are obviously unacceptable for any application, let alone a real-time one!

With such problems of combinatorial explosion with exhaustive search, it became necessary to develop methods for *pruning the search space* in order for search efficiency to be improved. The first such method was developed from the technique of depth-first search; with the addition of local measurements, this produced the metaphor of *hill-climbing*, where the outcome of each direction of search is compared against some evaluative function, and the path leading to the 'best' next step is taken. Thus, if the search space is envisaged as a mountain, with the solution at the top, search always proceeds in the direction which is 'up' from the present position.

While the classical hill-climbing technique can greatly improve search efficiency in certain domains, it has proved to be easy to cause the method to break down, both in pathological instances such as the 'foothill problem', the 'ridge problem' and the 'plateau problem', and also in cases where there are many dimensions to the problem. Such problems have been addressed by extensions to the classical hill-climbing technique.

In classical hill-climbing, reasoning always proceeds from the previous decision via what seems to be the best descendent of that decision path. A variation on this technique, called *best-first search*, reasons forward from the 'best' node so far, no matter where it is in the partially-developed search tree. The total path length of solutions found using this technique tend to be closer to the optimal than those produced by pure depth-first or breadth-first search, although this is not always the case.

A technique which guarantees to find an optimal solution, if a solution exists, is that of *branch and bound*, where the system searches forward from the least-cost partial path through the search tree. The technique works thus: during a search there will be many incomplete paths through the search space which could be investigated. The system extends the shortest path by one level, creating as many new incomplete paths as there are branches, and these paths are added to the set from which the shortest is again chosen. This process is repeated until the solution is found along some path, which must be the optimal path as it is always the shortest which is extended.

6.2.2.3 State Space Search and Means-end Analysis

By taking the maps or graphs of the simple traversal problems described in the previous sections to be the 'and/or' decision trees of more complex situations, it was possible to apply these search techniques to the area of general problem-solving. The graph traversal techniques have been combined with techniques developed for game-playing on computers (in particular the development of chess-playing computer programs) in an attempt to overcome the problem of combinatorial explosion which has affected exhaustive search algorithms for problem-solving.

Such game-playing programs try to apply legal 'moves' to an initial state in order to arrive at a final state which meets some specified goal criteria, a technique known as *state space search*. In the case of the early planning/problem-solving systems, heuristic evaluation functions were used in order to determine the 'closeness' of any intermediate state to the goal state. An early system which employed such state space search techniques was the Graph Traverser developed by Doran and Michie in 1966. At this stage in the development of AI-planning systems, there was no effective distinction between a planner and a problem-solver, the initial intention of AI being to produce a completely general-purpose 'intelligent' problem-solving computer program.

While Doran's and Michie's work provided one of the first computer programs to address the problem of search and problem-solving, virtually all current traditional AI-planning systems owe their origins to a system first developed some three years earlier in order to undertake theorem-proving in logic. This was the method of *means-end analysis*, first developed by Newell and Simon in 1963 for the General Problem Solver (GPS), and developed over the following years (Ernst and Newell, 1969; Newell and Simon, 1972). As with state space search, means-end analysis works on the basis of differences in initial state and goal state, but reduces the number of intermediate states which are considered by only referring to operators or activities which can satisfy some outstanding goal.

General means-end planners work by defining a set of *literals*, or simple facts, which represent the initial state of the world, and a further set of literals which represent the goal state. These literals are then supplemented by defining a set of possible *actions* which can take place in the planning domain. Actions are defined in terms of their *pre-conditions*—literals which must be true before an action is allowed, and their *effects* on the current world state—which are modelled as additions to, or deletions from, the list of literals which describes the current world state. The system then searches through the actions currently available, and selects one which will reduce the difference between the current state and the goal state. This process is then repeated in order to generate a sequence of actions, which constitutes the *plan*.

6.2.3 Linear and Non-linear Planners

6.2.3.1 Interacting Subproblems, Conjunctive Goals and Constraints

After the problem of limiting search (or pruning the search space), the other major issue which arises in planning systems and general problem solving is that of *interacting subproblems*. This situation arises when a problem has *conjunctive goals*, in that more than one condition must be satisfied in order to arrive at an acceptable solution. In most planning and problem-solving situations, the order in which conjunctive goals should be achieved is not specified, but the order in which a system attempts to solve such goals can have a critical effect on the speed of the search for a solution.

Such interacting subproblems have also been referred to as *constraints* (Stefik, 1981a). These constraints can be deduced from the preconditions of operators in the planning domain, provided the preconditions are explicit. In such a scenario, the production of a plan is seen as a *constraint satisfaction problem*, with the plan requirements expressed as a series of constraints which must be satisfied. Although some work has been done on the development of general-purpose constraint satisfaction techniques, they have generally been found to be inefficient in use, and their use is not widespread.

At worst, interacting subproblems can result in a situation where no plan can be generated at all, but in most cases the effect will be to slow down the overall search for a solution by the system having to carry out extra computations when a premature commitment to an arbitrary ordering of goals proves to generate an insoluble interaction. When this occurs, the system is forced to *backtrack* and change the plan from the point where the choice which caused the problematical interaction was made. Such undoing of reasoning steps is invariably costly in computing terms, and repeated requirements for backtracking in a system will lead to a severe degradation in performance. It is for this reason that successive systems have addressed this problem in increasingly sophisticated ways.

6.2.3.2 STRIPS and Linear Planning

The first planning systems approached the problem of conjunctive goals by effectively ignoring it altogether during the initial phase of planning. A violation of ordering constraints resulted from this approach, and this was dealt with by the system going back and trying to fix the plan. These systems base their operation on a principle known as the *linearity assumption* (Sussman, 1975), which states that: 'sub-goals are independent and thus can be sequentially achieved in an arbitrary order'.

Although the linearity assumption constitutes a huge simplification of the problem of planning, its technique of solving any given goal and then moving on to solve any other goal is often successful, due to the solutions to different goals being decoupled in real-world problems. The use of such an assumption also made it possible to reduce the problem of search to a level where it was computationally tractable on the computer hardware available in the early 1970s, when this work was taking place. The importance of the linearity assumption is illustrated by the following comment from the *Handbook of Artificial Intelligence* (Cohen and Feigenbaum, 1982):

> 'In a historical perspective, this can be seen to be an important heuristic. The number of orderings of problem-solving operators is the factorial of the number of operators, so it is obvious that a problem solver cannot successfully examine *all* orderings in the hope of finding one that does not fail because of interacting operators. The linear assumption says that in the absence of any knowledge about orderings of operators, assume that one ordering is as good as any other and then fix any interactions that emerge.'

The first of the linear planners was the STRIPS system (Fikes and Nilsson, 1971), which introduced a system of representation for plans which has been employed by many later systems. STRIPS employed the technique of means-end analysis developed for the GPS, and was effectively a simplified version of GPS which had been formulated to specialise in reasoning about actions which cause change in the world.

In STRIPS and its derivatives, actions are represented as functions from sets of sentences to other sets of sentences in some appropriate language. This representation may be relatively simplistic, but it offers an efficient means of computing the results of actions performed in the world. The results of actions are represented as changes to a database, so that for an action a there is a database update function $u(a)$ described as a function from sets of sentences to sets of sentences:

$$u(a) : P(L)-> P(L)$$

where L is the language being used to describe any particular planning domain and $P(L)$ is the set of all subsets of L. The representation used by STRIPS associates both an 'add list A(a)' and a 'delete list D(a)' with an action, with the database update function $u(a)$ being given by:

$$u(a)(S) = S + A(a) - D(a).$$

6.2.3.3 Interaction Detection and Correction in Linear Planners

With respect to the detection and correction of interacting subproblems, STRIPS was particularly poorly equipped. The 'STRIPS assumption' was a precursor to the linearity assumption, and assumed that the initial world state is only changed by a set of additions to and deletions from the statements modelling the world state. Interactions between goals were ignored, and no attempt was made to backtrack and repair the plan. Instead, a simple check was made as to whether the conjunction of goals still held, leading to the possibility of mutual goal interference. The result of such interference is, at best, the introduction of redundant actions into the plan and, at worst, the planner entering an endless loop as it repeatedly re-introduces the same goal and tries to satisfy it.

Such limitations were addressed by later systems, with some supporting interaction detection alone, while others have additionally employed a variety of techniques for interaction correction. The first system to move beyond the STRIPS approach was HACKER (Sussman, 1975), where Sussman first predicated the linearity assumption. HACKER was designed to detect 'bugs' in the linearity assumption and to then attempt correction of the plan by backtracking to the point where an alternative goal ordering could be tried. This approach was successful in dealing with some problems, but could easily become inefficient if extensive backtracking was required.

A more sophisticated method of correcting for problematical interactions in a plan is that of *action regression*, where the interfering action which has just been introduced to fulfil some goal is placed at progressively earlier points in the plan until a position is found where the unwanted interaction disappears. Such a scheme was implemented in the WARPLAN system (Warren, 1974) with some success, although the technique can result in the inclusion of redundant actions in a plan, due to the regressed action being unnecessary, or an inferior choice, when it is moved to a new point in the plan.

This problem with the technique of action regression, that it involves the regression of the actual action which satisfies some interacting goal, was solved in planners such as that developed by Waldinger (1975) by the use of *goal regression*. Using this technique, it is the interacting goal itself which is regressed to earlier in the plan when a problematical interaction occurs. This approach has the benefit of avoiding the inclusion of redundant actions in the plan, due to the goal being solved at an earlier point once it has been regressed.

The most sophisticated of the linear planners was INTERPLAN (Tate, 1975), which attempted to resolve interaction problems by the use of analysis, rather than simple regression techniques. With INTERPLAN, Tate introduced the technique of *Goal Structure*, which recorded the link between an effect of one action which was a pre-condition of a later one. This data was recorded in addition to the ordering links between the actions themselves because it is possible for actions to produce effects which are not used until considerably later in the plan. The operation of the system depended on interactions being detected as an interference between a newly-introduced action or goal and one or more goal structure ranges. A data table, which Tate referred to as a *ticklist*, was used to represent the mapping between actions and the goals which they satisfied later in the plan. When correcting for interactions, the system generated the minimum set of goal re-orderings which would effect the necessary modifications, and thus was required to consider fewer re-orderings of the system goals than was required using the regression methods.

6.2.3.4 Hierarchical and Non-linear Planners

The major disadvantage of the early linear planning systems, as epitomised by STRIPS, was that they did not distinguish between actions which were crucial to the success of a plan and those which were simply petty details, and were inflexible in the order in which planning decisions were made. Thus, in the STRIPS notation, all activities are considered as primitive actions which change the world state in some way, with no distinction being made between a minor detail, such as which colour to paint the doors of a building, and a major strategic consideration such as where to locate the building itself. In this representation, a partial plan is simply a sequence of primitive actions which the planner tries to extend by inserting additional actions. In an ideal world, it would always be possible for the system's current partial plan to be extended to form a complete plan; that is to say, a sequence of primitive actions which achieves the goal specification. Unfortunately, linear planners are constantly required to make arbitrary ordering decisions regarding plan actions during the planning process, as they endeavour to keep their current partial plan linearly ordered.

A planning system employing purely linear techniques can therefore easily become 'bogged down' in considering such low-level details at length, only to find later that some higher-level planning goal has been invalidated due to subgoal interaction and that all previous low-level planning must be discarded. Dissatisfaction with the shortcomings of linear planners has been based on their tendency to progress by following the temporal order of the primitive actions which constituted the plan, rather than operating in a top-down manner. A requirement was identified for a successful real-world planning system to be able to abstract away from the primitive actions which linear planners use as their fundamental (and only) representation of a plan, and from a commitment to the precise temporal order in which they should be arranged. This abstraction is based on the precept that all complex plans in the real world are hierarchical in nature, particularly in the structured domains in which it is envisaged that AI-planners will be employed, and that the need for excessive backtracking could be avoided if the planner first developed a complete, but vague, plan before refining the vague high-level goals in greater and greater detail until a fully detailed plan is generated.

This method of *hierarchical planning* has been implemented in a number of planning systems, each of which has employed a subtle variation on the technique. Indeed, the term 'hierarchical planning' seems to be open to a variety of interpretations, as noted at the DARPA Santa Cruz Workshop (Swartout, 1988); 'Hierarchical planning is perhaps the best known and most misunderstood of the approaches to planning...'. The source of the confusion is that *all* plans are hierarchical in nature, but not all planners are hierarchical in their method of operation. To be truly hierarchical, a planner should be able to operate at different *levels of abstraction* with regard to a plan, rather than just traverse up and down a 'flat' representation of the hierarchy of actions which constitutes a plan. That is to say, in addition to being able to model a plan as a hierarchy of goals and subgoals (or tasks and subtasks), a hierarchical planner should also be able to view that plan from a range of perspectives which themselves form a *hierarchy of abstraction spaces*.

To make the issue a little clearer, consider the case of a company which is planning the construction of a new chemical plant. While the plan as a whole will be specified down to the smallest detail, there is a considerable variation in the amount of detail which is required to be presented to personnel at different levels of the business organisation. For example, the managing director will require a highly abstracted overview of the project, possibly of a purely financial and strategic nature, while the technical director will also require high-level manpower and resource data, the project manager for the particular project will require both high- and medium-level data, and team leaders within the project will require both medium- and low-level data regarding the specifics of the jobs at hand. Thus, while there is only one plan, there can be many viewpoints onto it, each of which requires a different level of abstraction if information is to be presented to the user in the most meaningful manner. These levels of abstraction are also seen as corresponding to the way in which complex projects are planned in the real world; first the high-level decisions regarding items such as cost and profitability are made, followed by technical decisions at increasingly fine levels of detail, until the plan is fully specified.

The earliest form of hierarchical planning was implemented in the GPS system (Newell and Simon, 1972) which modelled theorem-proving in logic and introduced the fundamental planning technique of means-end analysis. GPS worked in a single *abstraction space* which was defined by treating one representation of a problem as being more general than others. The original problem space in GPS was defined in terms of four different logical connectives, which were all replaced by a single abstract symbol to produce the abstraction space. A problem could then be solved in the single-connective abstraction space, and the result mapped back into the four-connective space. The first true planning system to employ hierarchical techniques was the ABSTRIPS system (Sacerdoti, 1974), which was developed out of the STRIPS system and shared its notation for representing plans. ABSTRIPS was able to plan in a hierarchy of abstraction spaces by treating some goals in a plan as being more important than others, and, having determined which subgoals were crucial to the generation of a successful plan, it would ignore all other subgoals until these had been satisfied. Non-crucial activities in ABSTRIPS were modelled as *details*, these being subgoals which could be achieved by simple subplan generation, once plans had been developed to achieve the crucial goals which were not details. By considering critical subgoals first and ignoring details, ABSTRIPS was able to reduce the amount of search required during planning, and particularly the need for backtracking to undo premature low-level decisions.

However, while ABSTRIPS planned in a hierarchical manner, it was still essentially a linear planner with respect to its operation within any of the levels of its hierarchy of abstraction spaces. This limitation of being constrained by the linearity assumption was addressed in further work, where partial plans were represented as *partially ordered actions*, allowing the planner to postpone making any commitment to the ordering of actions until it was absolutely necessary. First implemented in the NOAH (Nets Of Action Hierarchies) system (Sacerdoti, 1977), the technique of postponing commitments to the ordering of actions was extended by others to cover commitments to other planning choices, leading to the development of a whole class of non-linear *least-commitment planners*. Given that all current planning systems employ both hierarchical and least-commitment techniques, it is common to find the terms 'non-linear planner', 'hierarchical planner' and 'least-commitment planner' used interchangeably, as are the terms 'linear planner' and 'non-hierarchical planner'. The distinction between the terms is that 'hierarchical' refers to multiple levels of abstraction in the representation of a plan, while 'non-linear or least-commitment' refers to the postponement of commitments to ordering constraints.

6.2.3.5 Hierarchical Search Control and Interaction Detection

While GPS planned hierarchically by abstracting a more generalised view of its problem space, and ABSTRIPS operated by abstracting goals which were crucial to a plan, NOAH was able to achieve its least-commitment strategy due to its hierarchical view of a plan being based on the abstraction of problem-solving operators. By planning initially with generalised operators, which were later refined into the problem-solving operators relevant to the problem space, NOAH was released from the necessity of making *a priori* assumptions regarding the ordering of detailed actions within the plan. Some later planning systems, such as the version of the MOLGEN system developed by Stefik (1981a, 1981b) for experiment planning in molecular genetics, have extended the concept of abstraction to include objects in the problem space as well as problem-solving operators, but many recent planners can trace their parentage directly back to NOAH and the least-commitment approach to planning.

Hierarchical planners attempt to improve the efficiency of search by ordering their goals with reference to different levels of abstraction of the search space. As systems have developed, the methods used to explore the levels of the plan hierarchy have become increasingly sophisticated. Tate (1985) isolates three categories of planners, based on the method employed to explore the levels of a problem's abstraction space:

- strict search by levels,
- non-strict search by levels,
- opportunistic search by levels.

The earliest hierarchical planners, such as ABSTRIPS and NOAH, first generated a solution at the most abstract planning level and then remained committed to it thereafter. This high-level plan formed a fixed skeletal framework into which the low-level details were then placed, with no backtracking to the higher levels of the plan being possible. NOAH's major successor, the NONLIN system (Tate, 1977), took this approach a step further by using the high-level plans simply as a guide to the skeleton solution of a problem. NONLIN was thus able to re-plan or explore alternative solutions at any level of the abstraction hierarchy, either if an overall solution could not be found or if interacting subgoals caused low-level problems which indicated that a previous high-level choice was incorrect. In more recent systems, the abstraction levels are only one of the mechanisms available for the planner to decide on the ordering of goals within a plan. These *opportunistic planners* are able to decide when a particular goal at any level of the abstraction hierarchy is constrained enough to constitute the most preferable goal for the system to pursue. Although they are also hierarchical, these systems exploit a methodology of their own in comparison to pure means-end planners, and are dealt with in more detail in a later section of this chapter.

As with the linear planning systems, the detection and correction of interactions between conjunctive goals is also of paramount importance in hierarchical planning systems. It is only through the system's ability to infer constraints between the various planning operators that a least-commitment planning strategy is possible. While all hierarchical planners are capable of detecting subgoal interactions, they vary in the extent to which they can resolve the situation, this being closely linked to the planner's ability to direct search between the various levels of the abstraction hierarchy.

NOAH utilised a set of procedures called *critics*, which were used to detect interactions between different subgoals within a plan. To assist in the detection of interactions, NOAH also employed a Table Of Multiple Effects (TOME) which was based on the goal structure concept developed for the linear INTERPLAN system (Tate, 1975). However, once an interaction was detected, NOAH could only apply local, limited correction at the abstraction level currently being explored, due to its method of commiting itself to the best apparent choice at each level of abstraction and failing to consider alternatives from that point on. This shortcoming was overcome in the NONLIN system (Tate, 1977), which was modelled directly on NOAH, with the full addition of the goal structure which Tate had introduced in INTERPLAN. This combination allowed NONLIN to generate the minimum set of ordering links in a partially ordered plan which was required to resolve all currently known interactions, with the system also able to consider alternative solutions if a given line of search failed to provide an answer.

Later AI-planning systems have extended their methods of operation in order to address other *constraints* which operate in real-world planning situations. In addition to simple subgoal interactions, these systems have attempted to reason within constraints which are external to the planner itself. DEVISER (Vere, 1983), a development of NONLIN, was able to reason about the timing constraints involved in planning the mission sequences for Voyager space craft. Others planners like MOLGEN and SIPE (Wilkins, 1983) were capable of reasoning about the resource constraints caused by the availability, or lack of it, of scarce physical resources required for a project, such as the aircraft fuel required for SIPE's aircraft carrier mission planning.

6.2.4 Backtracking and Plan Repair

The least-commitment technique adopted by NOAH, MOLGEN and others is a constructive approach to planning, where it is assumed that no planning decision is made until the system is sure that it will not interfere with any past or future decision. This approach is intended to produce a planner which is never required to *backtrack* and undo bad decisions. However, as no system is perfect, even these planners make incorrect decisions at times, and are forced to backtrack and undo conclusions which were formerly thought to be valid. Such retraction of inferences has become an important field of study in its own right within AI, as the computational overheads involved are central to the efficient performance of AI-based systems. In the world of formal logic, such retraction and re-inferencing is termed 'non-monotonic reasoning', while in the field of expert systems it is usually referred to as 'truth maintenance' or latterly 'reason maintenance'. Before going into the details of the backtracking mechanisms employed by AI planning systems, we will consider the topic of non-monotonic reasoning in general.

6.2.4.1 Non-monotonic Reasoning

All conventional logics are *monotonic* in nature. This means that reasoning proceeds in a series of steps, with each new item of knowledge building upon previously established facts. To use a simplistic analogy, monotonic reasoning is a one-way street where all the traffic proceeds in an orderly manner and in the same direction. In human 'common-sense' reasoning, however, matters rarely proceed in such a manner. If a system is to be able to reason with incomplete or uncertain information, it must not only be able to base its conclusions on what is known, but also upon what is not known. Where information is unknown or uncertain, the system will use assumptions or default values to fill in the gaps and so enable it to reason further. The problem arises when, for one reason or another, it transpires that such an assumption or default value is incorrect. When this is the case, it is not sufficient to simply remove or amend the offending datum, as the system can no longer rely on on any information which has been derived on the basis of belief in the datum. The ability to cope with such retraction of belief in an item of information is the ability to perform non-monotonic reasoning. Defined formally, a non-monotonic logic is a logic which allows its theorems to be disproved by later axioms. From this description, it is also possible to see how the subject is closely allied to that of temporal reasoning, which attempts to deal explicitly with the concept of time and time-dependent reasoning. A discussion on the general role of non-monotonic reasoning in AI systems, can be found in Chapter 8 of this book.

The issue may become a little clearer with a couple of examples:

Example 1. The red-haired twins. Suppose we know only one red-haired person, our friend Jane. If we see someone looking like Jane in the crude sense of merely being red-haired, we might assume that that person is Jane, because Jane is the only person we know fitting that description. This inference is non-monotonic, of course, since if we now learn that Jane has an identical twin sister Joan, we can no longer conclude that anyone who looks like Jane is Jane. (Taken from the *Handbook of Artificial Intelligence* (Cohen and Feigenbaum, 1982).)

Example 2. Airline re-routing. The second example, which is attributed to Dov Gabbay, concerns the World Travel Agency (WTA) which offers package tours from New York to Paris and Stuttgart. Every Sunday passengers board a WTA plane in New York; some passengers disembark in Paris, while others continue to Stuttgart. The Stuttgart office arranges for hotels in Europe. At 12.00, the Stuttgart office learns that all flights into Paris are to be re-routed, as the airport is fog-bound. Stuttgart has to decide what to do with the WTA group. They cannot contact New York because all telephone lines are engaged. The only established fact is that the WTA plane left New York, presumably on schedule. The Stuttgart office reasons that the WTA plane will be re-routed to London; the Paris passengers will disembark and the remainder proceed to Stuttgart. However, at 14.00 hours the Stuttgart office manages to establish a telephone link with New York, only to discover that the WTA plane was two and a half hours late taking off. This changes the picture considerably. The Stuttgart office now reasons (in consultation with New York) that the Captain, when hearing of the problems at Paris, decided to return the short distance to New York. The upshot of the newly received information is thus that the passengers will not now be arriving from London, a complete reversal of the previous conclusion.

The need to backtrack and undo previously held conclusions arises in all but the simplest planning systems, with the linear planners often involved in extensive backtracking due to subgoal interaction. This requirement is still present in non-linear planners, despite their use of least-commitment techniques in an attempt to minimise the problem, and a number of increasingly sophisticated techniques have been developed in order to support attempts at plan repair.

6.2.4.2 Beam Search

When a problem is tightly constrained, resulting in a small search space, a method of *beam search* can be employed. This method considers *all* possible solutions to a problem, so that they can be compared, and is usually employed in conjunction with other search methods which employ heuristics of some kind in order to ascertain as to which of the possibilities generated by a beam search is the 'best' choice at that point of the search for a solution. Given that no heuristics are employed and the search is not limited in any way, this technique resembles the 'brute force' algorithms used in conventional procedural programming.

6.2.4.3 Depth-first (Chronological) Backtracking

The first, and simplest, approach to the exploration of alternative solutions when an avenue of planning search failed was that of *depth-first backtracking*, otherwise known as *chronological backtracking*. In the simplest implementations of depth-first backtracking, the current state of the search space (including the possible choices which could be made at this point in time) is saved at each point in the search when a choice has to be made between two or more possible lines of inquiry, and then one of the options is arbitrarily chosen. Search then continues, and if the chosen direction of investigation later fails, the previously saved state is restored and the next alternative choice is made. If the alternatives are exhausted at a given saved point, the system backtracks to the preceding saved state and repeats the process of choosing and exploring alternatives. Such backtracking could be implemented using simple stack-based programming techniques, and this was the basis of the backtracking mechanism employed by STRIPS and similar planners.

As planning systems were developed further, and considerations such as search control entered into the general development of AI and knowledge-based systems, it was found that such simple implementations of chronological backtracking could suffer from severe deficiencies when extensive backtracking was required in order to arrive at the solution to a problem. These deficiencies became apparent when planning systems were applied to problems which exhibited large amounts of low-level planning detail, resulting in the need for almost incessant backtracking in the search for a solution. In many real applications, there may be a number of largely unrelated parts to the overall solution, with work carried out on one part of the solution having no effect on other parts of the solution. When a planner restores a previously saved state during chronological backtracking, it is quite possible that the system will be forced to abandon a large amount of reasoning work which is not directly affected by the need to backtrack, but simply happens to have been carried out more recently than the reasoning work which requires changing. These limitations were to be addressed by the development of systems which recorded the *data dependencies* between different conclusions, intended to allow more efficient updating of a system's database during backtracking.

6.2.4.4 Dependency-directed Search and Truth Maintenance

The term 'Truth Maintenance System' (TMS) was first used by Jon Doyle in a paper published in AI Magazine(Doyle, 1979), although Doyle now prefers to use the term 'Reason Maintenance System' on the grounds that it is more indicative of the nature of the system. Doyle's original TMS was actually an attempt to formulate a non-monotonic logic which addressed the problem of resolving conflicts produced in a forward-chaining type of reasoning system. With a forward chainer, the system is first loaded with whatever 'facts' are available, and it then reasons forward to see what possible conclusions can be reached from the given data. Given such a mode of operation, it is quite possible that the system will produce a logical inconsistency, such as concluding both that a fact is true and also that it is false, the two conclusions having been arrived at via different chains of reasoning.

Doyle's TMS was intended to reason around such contradictions by employing a technique known as 'Dependency-directed Backtracking', which had been proposed by Stallman and Sussman (1977). Here, each item of information in the reasoning system's database is tagged with a pointer to the information upon which it immediately depends for its inference. When the TMS is triggered, it backtracks along the dependency links until it finds an item of data whose retraction will remove the contradiction. The TMS must then also work forward and update any other items of information which might depend on the changed datum. With the reasoning system's database restored to consistency, the system can continue reasoning until either another contradiction is encountered or it can reason no further.

Doyle's early work, and his later work with McDermott, was of a theoretical nature, in that it was an attempt to formulate a non-monotonic logic rather than build a functioning TMS in software. Other workers have also attempted to formulate logics which might be applicable to such reasoning tasks, notably Reiter's 'Default Logic' (Reiter, 1980), McCarthy's 'Circumscription' (McCarthy, 1980) and Moore's 'Autoepistemic Logic' (Moore, 1980). While dependency-directed backtracking is a theoretically sound technique, attempts to implement working TMSs using this method have been found to suffer from a number of efficiency problems, just as chronological backtracking had done. If we pause to consider the memory overhead involved in maintaining a data dependency network of any degree of complexity, along with the computational workload involved in the dependency directed backtracking itself, it is possible to appreciate the scale of the problem. In addition, it should be noted that a TMS is only capable of identifying a set of choices which have been implicated in a planning failure, it does not address the problem of which of the implicated choices should be changed in order to rectify the situation. Some systems attempt to apply heuristics in an attempt to rank the possibilities in some way, but systems will often simply choose the conclusion to be changed on an arbitrary basis, as was the case with chronological backtracking.

6.2.4.5 Multiple Belief Spaces

While such justification-based TMSs operate by attempting to preserve the overall consistency of a knowledge-based system's database, a second class of non-monotonic reasoners operates in a radically different fashion. Rather than keeping track of data dependencies, these systems allow the reasoner to hold mutually contradictory items of information in its database concurrently, so that contradictions do not have to be reasoned around. Systems which have employed variations on this 'partitioning' technique include Data Pools, Contexts, Worlds, Viewpoints and de Kleer's Assumption-based Truth Maintenance System (ATMS) (de Kleer, 1986a,b,c).

This ability to reason in *multiple belief spaces* can be used to implement hypothetical reasoning systems where different possibilities can be explored simultaneously. This can be a useful, possibly essential, feature for planning systems, and is a highly desirable facility in any advisory expert system. (For an analysis of expert system types see Hayes-Roth *et al*. (1983).) It is also possible to use this type of mechanism to model, say, an industrial process at different points in time, and hence allow a system to reason explicitly about time-dependent relationships.

Johan de Kleer's ATMS (1986a,b,c) probably constitutes the first truly workable TMS system to be developed. The ATMS is designed to be used alongside a reasoning system and to allow it to reason with assumptions which may later be retracted, then maybe reinstated, and so on. As opposed to dependency directed backtracking, where every datum is tagged with pointers to the information upon which it immediately depends, the ATMS tags all data with a list of the set of assumptions for which it holds true. This enables the ATMS to reason about all possible world states simultaneously, there being 2^n possible states for a system reasoning with n assumptions. In order to make this system computationally feasible, de Kleer has not only developed the system theoretically, but has also detailed the ways in which the system's operation must be optimised so that it does not constitute the prime user of computational resources within the overall reasoning system. For example, as the system spends most of its time performing set membership tests, de Kleer uses bit-vectors to represent the assumption sets, which reduces the set membership test to the ORing of the relevant bit-vectors.

The ATMS is similar in operation to the facilities offered by the major AI workstation development environments, such as ART's viewpoints, KEE's worlds and Knowledge Craft's contexts. By the time of the development of the ATMS and these other systems, the use of pure means–end planning systems had been extended by the use of the symbolic processing power of AI workstations such as the Symbolics and TI Explorer to provide a knowledge-based solution to planning problems. The use of such 'AI-leveraged' techniques will be discussed at greater length in Section 6.3.

6.2.5 Script-based and Case-based Planning

An alternative approach to the use of means–end analysis to achieve planning tasks has been based on observations of how some human planning activities are carried out. Most project management departments will have on file the details of previous, hopefully successful, projects. When a new project comes in, managers will often try to find a previous project which was broadly similar to the project at hand. Given that most companies operate within fairly well-specified sectors of any given industry, it is to be expected that there will be a good degree of commonality between the details of different projects. When a similar project is found, it is used as the basis for plans for the new project. Such an approach can be efficient, in that the project manager does not have to constantly 're-invent the wheel', and hopefully will produce more reliable plans, as future plans can be based on the experience gained in previous projects.

Such an approach to planning has been implemented in a number of *case-based planners*, which can also be referred to as *script-based planners*, a name based on work carried out in natural language understanding where stored *scripts* were used to provide top-down expectations of the course of a story. In such planners, the system chooses a skeletal framework from a store of such high-level templates, onto which the specific details of the new project will be refined. The refinement process proceeds in a hierarchical manner, similar to that of the least-commitment means-end planners, but in this case the subplan fragments are retrieved from a store of such fragments and are then ranked heuristically for their applicability to the section of the project in hand, rather than generated afresh by the use of means-end analysis of the planning problem.

Such planning systems work on the premise that the most important aspect of generating new plans and debugging old ones is the organisation of previous experience, that is, the storage of previous plan fragments and their subsequent retrieval and transformation for use in new plans. Given this concentration on storage and retrieval, these case-based planning systems involve extensive consideration of the issues of machine learning and the organisation of knowledge bases.

One of the earliest planners to employ such storage and retrieval techniques as a form of learning was HACKER (Sussman, 1975), an early attempt at automatic computer program generation. HACKER was a conventional procedural program, but was supplied with libraries of plans, debugging techniques and domain axioms. Memory organisation techniques in HACKER were simplistic and retrieval was achieved by using standard pattern-matching techniques, but the system was able to demonstrate two of the important facets of case-based planning:

• that planning consists of using and debugging known plans; and
• that once a planner has constructed a useful plan, it should store the results for later use.

Other planning systems have taken up this approach, particularly in areas where an archive of previous project data is already available. For example, the case-based version of MOLGEN produced by Friedland (1979) was able to build on data gathered from experiments in molecular genetics which had been carried out at Stanford University.

6.2.6 Opportunistic Planners

Another method of human planning, which has been described as *opportunistic* (Hayes-Roth and Hayes-Roth, 1978), has also been implemented as a technique for AI-planning systems. In human opportunistic planning, a person who is required, say, to journey into town in order to collect an item at one shop might well opportunistically plan to carry out any other shopping which is required, either immediately or in the near future, at the same time as the original errand in order to save a wasted journey at a later date.

Unlike AI systems which employ a fixed search strategy, either data-driven (forward chaining) or goal-driven (backward chaining), opportunistic search allows a more flexible approach to be adopted. With these systems, the current focus of the search varies according to which of the operations currently under consideration is the most highly constrained, and hence is likely to obtain maximum benefit by being included in the plan at this point in time. This flexible control strategy is supported by a control structure called a 'blackboard', whereby the various problem-solving components of the system exchange information. This blackboard acts as a 'clearing house' for suggestions about planning steps, which are posted by the various autonomous planning *specialists* which are each designed to make a specific type of planning decision.

Planning in an opportunistic planner does not therefore take place in the top-down manner of a hierarchical planner, but is driven in a bottom-up manner by the opportunities which arise to include detailed problem-solving operators into the developing plan. Rather than including detailed problem-solving actions as late as possible, as with top-down refinement where the planner tries to develop a plan fully at each level of abstraction before proceeding further, opportunistic planners develop 'islands' of low-level planning actions independently, working in an asynchronous manner as the plan grows organically out of clusters of problem-solving operators.

Some systems have taken the concept of autonomous planning specialists within a single planning system a stage further, allowing the sources of problem-solving expertise and knowledge to reside in separate software modules or even physically separated hardware systems. Such *distributed planners* allow fully-distributed planning to take place, with subproblems being passed between different specialised planning experts. The control of such a system will vary according to the physical arrangement of the computer system(s) being used to implement the system. Where planning is distributed between software modules running on a single hardware platform, control is normally carried out by using a centralised blackboard data structure, together with a scheduling executive similar to the operating system of a conventional multi-user programming environment such as UNIX. If the autonomous units are physically distinct, with information shared via communications links of some kind, a more distributed system of control via pair-wise negotiation can be used.

6.2.7 Mixed-AI-planners

A number of recent AI-planning systems have made use of large amounts of problem-specific knowledge combined with the classic technique of means-end analysis. Such domain-specific data is usually employed to guide the process of correcting for interacting subproblems by the use of analysis techniques. The first planning system to employ such techniques was NOAH, which was also the first of the least-commitment planners. Many of the procedures used by NOAH to solve any particular problem were specific to that domain. Such knowledge was represented using functions written in an extension of QLISP, and called SOUP functions as they defined the Semantics Of User Problems.

Other systems have also used similar amounts of problem-specific data to increase the efficiency of primary search or backtracking and plan repair after a planning failure. Such systems are called 'mixed AI-planners', from their combination of generalised problem-solving techniques with items of domain-specific knowledge. In particular, opportunistic planning systems have tended to employ a mixed-AI approach, as they often require the application of efficient local heuristics to guide the search for the currently most-constrained goal, which they will attempt to satisfy.

6.2.8 Meta-planning

As AI-based planning systems and search control strategies have developed, a distinction has been drawn in some systems between *planning*—the generation of the basic actions which constitute the output of the system—and *meta-planning*—which is explicit reasoning regarding how the planning system goes about generating the plan. This distinction mirrors that made in knowledge representation between *knowledge* and *meta-knowledge* (knowledge about the structure of other, more specific, knowledge). While previous hierarchical AI-based planning systems have generated plans by the successive expansion of plan actions at varying levels of abstraction, meta-planners are concerned with the order in which the different planning operations are carried out in the planning system itself. Planning operators and data structures are partitioned into layers, with operators within any given layer controlling the execution of operators in the layer immediately below via a message-based interface. In this respect, such systems can be seen as early examples of the use of an object-oriented paradigm in system design.

The term 'meta-planning' was coined by Stefik (1981b) in his description of the operation of the MOLGEN system, developed for planning experiments in molecular genetics. MOLGEN generates plans by dividing the problem into sub-problems and then explicitly studying the interactions arising between those sub-problems. This reasoning is carried out using an extension of the concept of means-end analysis, augmented by the use of a *constraint posting cycle* to formulate, propagate and satisfy constraints. The execution of the constraint posting cycle is controlled by a four-layer planning architecture, with the operators of the upper layers being more general than those of the lower layers. The four control layers in MOLGEN are summarised in descending order as:

- the interpreter (which executes strategic operators),
- the strategic layer (which creates and executes design tasks),
- the design layer (which creates and executes laboratory tasks),
- the laboratory layer (which models modifications to physical objects).

The benefits of using such an explicitly-layered control structure are noted by Stefik (1981b) as being that: 'The approach ... exposes and organises a variety of decisions, which are usually made implicitly and sub-optimally in the planning programs with rigid control structures.' Operator-like representations of the different operations which a planning system can perform have also been used to implement meta-planning in systems for natural language understanding(Wilensky, 1981) and a number of other opportunistic planners. However, while meta-planners such as MOLGEN and the others can distinguish between basic plan actions and planning meta-actions, a distinction which is not made by traditional means-end planners, they are not capable of *strategically* planning the execution of such meta-actions. This limitation has been addressed by work carried out at Carnegie-Mellon University, Pittsburgh on the representation of process plans for construction and manufacturing engineering, resulting in the development of the PLANEX (Zozaya-Gorostiza *et al.*, 1989) architecture.

PLANEX is a domain-independent, knowledge-based process planning system architecture, which explicitly generates both *process plans* (which define how the desired product will be manufactured) and *operator plans* (which specify the problem-solving steps which will produce that process plan), and can thus be classified as the first *strategic meta-planner*. The problem-solving behaviour of PLANEX incorporates both strategic and opportunistic elements, and the system provides an explicit set of control operators which, in turn, generate operator plans. The user is able to modify not only the system's knowledge base, but also the search control strategy used by the system in generating a process plan. PLANEX has been used to develop two prototype process planning systems: one in the field of construction engineering, and the other configures for wiring harness design. CONSTRUCTION PLANEX is used to plan the construction of mid-rise concrete and steel-frame office buildings, while HARNESS PLANEX plans the manufacturing operations required for electrical wiring harnesses in the automobile industry.

Such a system represents the current state of the art in AI-based planning systems, and has produced viable planning results within the framework of academic research, but there may be problems in migrating such technology into the engineering workplace in the short term. The CONSTRUCTION PLANEX and HARNESS PLANEX systems were implemented in Common Lisp and Knowledge Craft on a Texas Instruments Explorer AI workstation, with the associated graphics sub-systems written in C on a Silicon Graphics IRIS workstation. Such high-powered dedicated hardware and software is not to be found in any but the largest engineering institutions, and probably in few academic engineering departments outside the United States. For AI-based planning systems to be taken up in mainstream engineering applications, it will be necessary for such systems to be capable of acceptable speeds of operation on engineering hardware platforms. This fact has been recognised by the developers of PLANEX, and a version of the system written entirely in Common Lisp (albeit minus the graphical schedule displays and animation sub-system) has been ported to hardware platforms such as the Sun Workstation and some of the more powerful members of the IBM PC family. Such migrations are bound to increase as, with the distinction between high-level PCs and workstations fast disappearing and developers offering a choice of two plug-in Lisp processors for the Apple Macintosh II, a generation of desk-top computers which are capable of handling the computational loads imposed by symbolic or functional processing comes into (relatively) common usage.

6.3 KNOWLEDGE-BASED, 'AI-LEVERED' AND NON-AI SYSTEMS

Since the development of traditional AI means-end planners in the 1970s and the development of mixed AI techniques in the 1980s, a number of other AI-related techniques have been applied to the domain of planning problems. The methods employed include domain-specific expert systems, hybrid 'AI-levered' project management systems and various non-AI approaches such as control theory.

The need to explore beyond the bounds of traditional AI-planning techniques has been generated by attempts to apply such planners to the solution of real-world problems of increasing degrees of complexity. Most AI-planning literature deals exclusively with simplistic low-level domains such as the 'blocks-world' block-stacking problems, where generic preconditions can be applied repeatedly by the planning system each time that it considers moving a particular block, or similar action. However, most real-world engineering projects require planning to be undertaken at a level where there are many distinct individual tasks, involving little repetition and significant amount of specific knowledge in order to determine which action should be performed next. In such high-level plans, the precondition for each unique activity tends to be the completion of another unique activity. For example, when NONLIN was applied to the problem of house construction (Tate, 1976), the system specified 'brickwork done' as a precondition of starting the activity 'roofing and flashing', while 'brickwork done' was an effect of the completion of activity 'lay brickwork'.

The list of activities, preconditions and effects which acts as input to the NONLIN system can be considered as an unlinked list of activity successors which is compiled into a linked list of activities by the planner, giving an explicit sequence of activities which satisfies the plan objectives. In real-world projects where the precondition for each action tends to be the completion of other actions, the only possible sequence of activities is already implicit in the unlinked listing supplied as input to the planner. In compiling the final linked list of activities, the means-end planner is only making explicit a set of sequential relations which were already implicit, and thus no new knowledge is actually generated.

6.3.1 Domain-specific Expert Systems

With the advent of expert system shells and knowledge-based development environments, the ability to create bespoke knowledge-based applications has been made directly available to engineers and other domain experts. This has resulted in the development of systems which employ solely domain-specific knowledge in order to carry out planning tasks in a variety of narrow domains. The two main areas of planning and project management to which such systems have been applied are schedule interpretation/generation, particularly job-shop scheduling, and constraint-based planning and design.

A number of expert systems have been developed in recent years, aimed at dealing with problems within the project management domain, which require the representation of activity, state and resource. In 1986 the Alvey community club PLANIT was established to attempt to develop an Interactive Planners' Assistant (Drummond, 1986) capable of reasoning with project models and serving as a test-bed for various experiments in project planning and scheduling. Experiments are envisaged in plan representation, plan construction, and plan execution monitoring. Previous to PLANIT the ISIS project (Fox *et al*., 1981; Fox and Smith, 1984) developed a system for factory scheduling which reasoned about the activities involved in the production of orders in a job-shop.

More recently, the Lockheed Software Technology Centre has looked at resource allocation in Software project management (Bimson and Burris, 1986), implementing a project-modelling scheme based upon Fox's developed for ISIS. In this latter system, models of the project are manually constructed (including details of activity, product, state, aggregation, sequence and physical connection) and are then assessed to determine suitable resource allocations to effect a reasonable schedule.

The development of the EXCAP system (Wright, 1988) has been a long-running project undertaken at UMIST in Manchester, aimed at applying AI techniques to mechanical process planning. Its development is based on the premise that there is an increasing demand for process planning on the shop floor, due to a combination of the shortening of product lifetimes and the fall in the supply of experienced shop-floor personnel with increasing automation. With requirements of easy knowledge engineering by on-site personnel and knowledge-base maintenance by non-specialist personnel, the system has evolved through a number of distinct development phases. Early prototypes were constructed using proprietary expert system shells, but these were found to be too inflexible. Flexibility was obtained by the use of pure Prolog for later systems, but only at the cost of great difficulty in the building and modifying of knowledge bases. The final version of the system, which is being implemented in industry at ICL Kidsgrove, has been developed using the Poplog development system (which offers Prolog, Lisp and Pop11 languages) and employs a custom shell combined with knowledge-based tools such as an interactive rule editor and graphically-based knowledge-base maps.

Just as the distinction between planning and problem-solving is not completely clear-cut, the process of design in an industrial context can be considered as having much in common with the requirements of a pure planning system. In both cases the problem-solving process can be seen as having the properties of a constraint satisfaction problem. In a planning system, the requirement is to resolve the constraints imposed by the specified planning goal(s) and any external constraints (such as DEVISER's time windows), while the design process involves resolving design constraints such as materials, costs and fabrication techniques. For example, there would obviously be a great deal in common between systems intended to: (a) design an engineering component; and (b) plan how to manufacture that component. It is due to this commonality that the first-wave techniques of computer-aided engineering, which are referred to as CAD/CAM (Computer Aided Design/Computer Aided Manufacture), are being replaced with the concept of CIM (Computer Integrated Manufacture).

6.3.2 Hybrid 'AI-levered' Project Management Systems

Although domain-specific expert systems have proved useful in a number of cases, often because no computer-based solution to a problem existed previously, the use of purely knowledge-based computing techniques to resolve a problem inevitably brings about difficulties of its own. At the present time, expert systems and AI techniques are the 'flavour of the month' and there is a tendency to regard them as a panacea, which they patently are not. Such systems tend to have drastic shortcomings in comparison to procedural programming techniques with regard to issues such as mathematical calculations, particularly floating-point arithmetic and trigonometrical functions, and handling graphics.

In many real-world cases, there are likely to be any number of functions required which can best be implemented using procedural techniques or a knowledge-based system might be required in a situation where traditional procedural software is already in use. A strand of recent research into computer-aided project management has been the development of systems where AI techniques are used as extensions to conventional procedural programming, such systems having been termed 'AI-levered systems' (Levitt and Kunz, 1987). Work on such systems has been carried out at Stanford by Levitt, who has used AI workstations and proprietary development environments to assist with project management tasks, and at Brighton Polytechnic, where project management tools have been constructed around a custom development environment designed to run on engineering workstations.

6.3.2.1 The PLATFORM Experiments at Stanford

Part of the recent work carried out at the Center for Integrated Facilities Engineering at Stanford University (Levitt and Kunz, 1987) has resulted in classifying the cognitive tasks associated with project management in two dimensions. These two dimensions are those of the various *functions* of the project management task, and the different *levels of abstraction* at which project management is carried out. Levitt then uses this classification to propose which of the different types of knowledge processing and/or procedural computer techniques to aid decision-making for each of the various subtasks of the project management process.

Using Levitt's analysis, there are considered to be four different *project management functions* which make up the stages through which the project management task progresses as a project proceeds in time. These project management functions are, in order of execution, *project objective-setting, project-planning, project-scheduling* and *project control*. A number of these terms are used in different ways by different AI researchers, and so Levitt starts by defining each of these terms as follows:

Objective-setting consists of the decisions which are made during the first conceptual stages of a project. This is the phase of a project when decisions are made regarding trade-offs between a project's location, size, cost, overall duration and performance levels, resulting in the definition of a set of goals for a project. This phase of project management is sometimes called conceptual design, and its outputs, which cannot be specified in enough detail to carry out any planning of separate activities, are usually referred to as design criteria, project goals or project objectives.

Project-planning is the stage of a project where a list of detailed project activities is generated, along with their sequential relationships. These are the activities which must be executed in order to achieve the set of project objectives specified in the objective-setting phase of the project. The output of the planning phase of a project is termed the project plan by Levitt.

Project-scheduling consists of assigning durations and resources to the various activities which constitute a project plan, calculating the start and finish times of activities, and computing the utilisation rates for project resources. Two heuristic methods of re-scheduling activities to satisfy resource constraints are currently used in planning systems. *Resource-constrained scheduling* re-schedules activities so that they can be executed within the limits of available resources, while *resource levelling* operates by minimising fluctuations in resource utilisation rates. Levitt calls the output of this scheduling phase a project schedule.

Project control is the phase of a project which takes over once work on the project has actually commenced. It consists of a number of actions aimed at monitoring the progress of the project and ensuring that it is brought to a successful conclusion. These actions include:

- measuring actual activity progress and resources consumption,
- interpreting past performance data,
- forecasting the duration and resource consumption of each remaining activity,
- comparing predicted durations and resource consumptions against the project schedule,
- eliminating or reducing any slippages from the project schedule,
- maintaining or increasing any favourable deviations from the project schedule.

Levitt notes that these actions may require re-scheduling activities, replanning or even revising project objectives in the light of large variances from the project schedule, or of externally imposed changes to the project specification.

With regard to the varying levels of abstraction from which a project can be viewed, Levitt notes that there is a continuum of detail ranging from one-sentence project objective statements to project plans consisting of thousands of activities. He defines three levels of project planning which correspond to discrete organisational and management responsibility levels for many real-world projects. These are termed the *executive level*, the *work package level* and the *task level*, which are defined as follows:

- The *executive level* corresponds to a high level corporate officer who has overall responsibility for each project. This person would typically manage the project at a level of detail involving about 50 to 200 activities. Activities at this level of abstraction tend to be unique and involve little or no repetition.

- The single manager who is generally made responsible for each of the plan activities at the executive level carries out this task by breaking the executive level activity down into its discrete component activities, usually numbering from ten to a few hundred. This action generates the *work package level* of the plan, at which level there may be some repetition of activities.

- At the final *task level*, each of the activities at the work package level of the plan is broken down into a task level plan of some 5 to 50 activities, under the direction of a supervisor such as a foreman. At this level of the plan, there is a small number of different activities and a significant amount of repetition.

The breakdown of the project management tasks associated with house construction, plotted along these two dimensions of project management function and level of abstraction, results in a three by four matrix of project management tasks, as shown in Figure 6.2. Based on these observations, Levitt has constructed a number of prototype AI-leveraged project management decision-support systems, which go under the generic title of PLATFORM.

PLATFORM I has been developed to support the task of project control, based around interpretation of interim project status, replanning and re-scheduling. Developed using the Intellicorp Knowledge Engineering Environment (KEE) running on a Symbolics AI workstation, the system employs a PERT network (Lockyer, 1964; Moder *et al.*, 1983) to model the sequential and temporal connections between the activities required to build a concrete gravity type of off-shore oil platform in the North Sea. It integrates this PERT network with knowledge-based techniques to reason heuristically about the risk factors which might affect an activity. PLATFORM I represents and reasons with planners' knowledge in two ways—*contingent subnetworks* and the concept of *knights and villains*. A project may have some activities whose construction will depend upon variables the values of which cannot be predicted in advance. Such activities can be treated as a set of *contingent subnetworks* which the planner defines, each containing the tasks that would have to be performed for any one state of the unknown variables. The appropriate contingent subnetwork will be activated when the uncertainty regarding any variables is settled as the project proceeds. The concept of *knights and villains* is based on a heuristic evaluation of the impact of various risk factors, such as weather, on the duration of an activity, which it is hoped will improve on the assumption that activity durations are either deterministic (CPA) or vary independently of one another (PERT) and on the use of purely statistical techniques for risk analysis. Risk factors having a positive effect on activity durations are considered as knights, while risk factors having a negative effect on activity durations are considered as villains. The durations of future activities are revised downwards if the activity is impacted by one or more knights and upwards if the activity is impacted by one or more villains. Since any villain can delay an otherwise productive activity, it is assumed that villains will always override knights. With this approach, PLATFORM I can interpret schedule performance on a partially-completed project, diagnose potential risk factors as knights and villains, and provide revised forecasts for the duration of remaining activities.

The PLATFORM II system combines knowledge-based programming and interactive graphical interfaces to support project planning and scheduling. While graphical editors to provide decision support for project planning exist, they can only act as passive

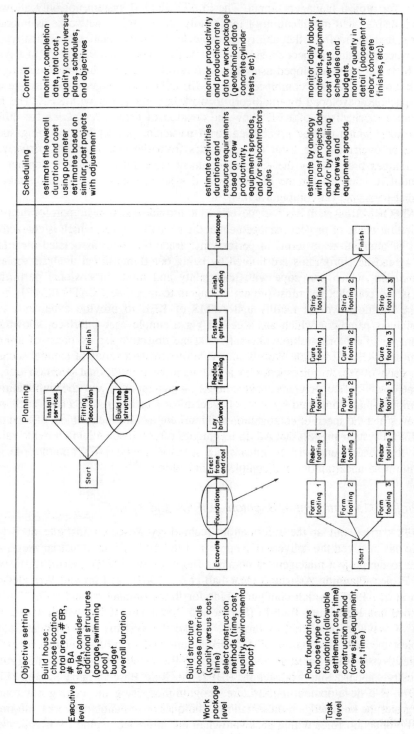

Figure 6.2 A taxonomy of project management tasks for house construction

'sketchpads' for project managers' input. The PLATFORM II system not only allows for the creation, display and modification of graphically based project networks as interactive Gantt charts, but also offers the ability to evaluate a project plan automatically using stored heuristics regarding project construction methods. As with the PLATFORM I system, PLATFORM II has been developed using the KEE development environment running on a Lisp workstation, in this case combined with the SIMKIT knowledge-based discrete event simulation package developed by Intellicorp to work in conjunction with KEE. The frame and inheritance mechanisms of KEE allow the creation of *project templates* for different types of project, which can be specialised for a particular project by providing specific project data or overriding the local default values from the template. The *active values* and *active images* provided by the KEE environment have allowed the development of a rich and intuitive interface for the PLATFORM II system, ideally suited to operation by non-specialist personnel (in computer terms).

The PLATFORM III system has been developed to provide decision support for the project objective-setting phase of project management at the executive level, which is characterised by analysis of alternatives in terms of performance trade-offs with associated uncertainty. Computer spreadsheet programs are limited by being one-dimensional analysis tools, and *decision analysis*, which can cope with uncertainty and multi-dimensional performance attributes, is too complex for most project managers to use. The PLATFORM III system employs the Multiple Worlds facility and ATMS of KEE to provide extensions to the decision analysis paradigm which are accessed via a simple user interface, allowing the project manager to explore multiple alternatives at the objective-setting phase of a project. The system uses KEE's Multiple Worlds and ATMS to create a series of parallel scenarios, or *worlds*, each of which corresponds to a leaf node of a traditional decision tree. The ATMS treats each choice or chance outcome as an assumption, and the implications of these assumptions are computed by a set of procedures attached to objects in each world. Levitt believes that the speed of computation and intuitive feel for the problem offered by the PLATFORM III system suggest that ATMS techniques offer the potential to provide valuable extensions to decision analysis or spreadsheet scenario analysis for complex decisions involving multiple uncertainties and complex computations.

6.3.2.2 Project Management at Brighton—PIPPA and XPERT

Recent work carried out at the Information Technology Research Institute at Brighton Polytechnic has involved the delivery of a number of hybrid IKBS to industrial users, which are specific to the project management domain. These are the PIPPA system (Professional Intelligent Project Planning Assistant) (Marshall *et al.*, 1987a, b; Lim and Wrigglesworth, 1987; Lim *et al.*, 1987), which configures bids for flight simulators, and XPERT (eXpert Project Expedition Reasoning Toolkit) (Barber and Boardman, 1988), which assists in the expedition of wayward construction projects. Unlike the systems developed at Stanford, which exploit the dedicated power of AI workstations and Lisp-based environments such as KEE, these systems are based around a custom IKBS development environment running on an industry-standard Sun workstation. RBFS, the Rule Based Frame System (Eklund *et al.*, 1987), is a development architecture, programmed using the Poplog environment, which integrates the knowledge representation techniques of semantic networks, information frames and production rules with a backward-chaining inference engine (Winstanley *et al.*, 1989). VEGAN, the Visual Editor for the Generation of Associative Networks (Kellett and Esfahani, 1988), provides an interactive graphical editor which allows the knowledge engineer to generate and manipulate associative networks on a computer workstation screen in a natural and accessible way, while simultaneously maintaining the underlying frame base on which the RBFS inference engine operates. Further reference should be made to Chapter 2 for a discussion on the structure of the RBFS system in the context of representation.

Project planning in a bid proposal phase within the flight simulator industry is an important activity, which transforms the existing range of products and available resources to meet a specific customer's requirements. It can be defined as an act of preparing a coordinated set of detailed information, which consists of a product document, master production schedule and product tree structure, together with a set of activity, material, time and cost estimates. The principal aim in this project, was to design and develop an intelligent planning system for assisting sales engineers and project planners in respect of:

- configuration of flight simulators;
- identification of non-standard requirements; and
- specification of a set of activities, cost and time estimates to satisfy different customers' needs, the available resources and present market conditions.

A hierarchy, which covers all the key parameters was regarded as fundamental to the concept of a 'project'. Members of the hierarchy were termed 'actor', 'action', 'object', 'resource' and 'activity'. (Later 'actor' and 'resource' came to be regarded as subclasses of a more generic concept 'object', renaming the original parameter 'object' as 'component' in the process.) Activities were regarded as composites of actions upon objects, in a sense similar to actions being verbs and objects being nouns in sentences that expressed the activity itself. Thus, the object hierarchy is depicted as a cone whose apex is the product, (the overall objective of the project), whilst the action hierarchy is depicted by another (inverted) cone whose apex expresses the motivation of the project: see Figure 6.3. At the connection

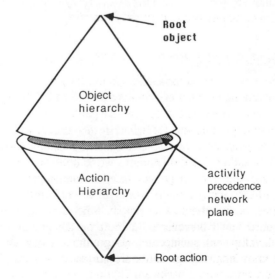

Figure 6.3 Emergent view of object and action hierarchies

of the cones lies an *activity plane* into which notions of sequence and precedence can be read.

A prototype product configuration system has been developed and installed which, in its present stage, is capable of preparing a bid summary report and a product tree structure of the main installations of a simulator, given a customer's specifications. The second system to be developed—XPERT—has been aimed at assisting the project manager in the expedition of wayward project, i.e. the pulling of 'slipped' projects back on schedule.

For a project manager to postulate the most appropriate strategy for expediting an activity, he needs two abilities: firstly, his experience, which gives him a rich pool of knowledge about which strategies exist and how they apply in a range of situations, and secondly, an insight into what exactly is involved in particular activities, in order that the best strategy may be found for the given situation. The project expedition system must possess a knowledge of the strategies and when they are applied, and of specific information concerning the context for activities.

Work to date has developed a hierarchy of abstract expedition strategies, shown in Figure 6.4, which describes the range of strategies. The knowledge about which strategy best fits a given situation has been gleaned from the domain experts and encoded in a knowledge base in the form of production rules. The XPERT system as a whole is a hybrid system: a combination of the IKBS and a heuristic procedural algorithm. The heuristic algorithm performs the actual expedition, taking as its input data the set of activities which the IKBS has identified as being good contenders for expedition via a hierarchy of expedition strategies.

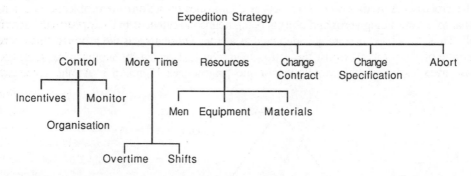

Figure 6.4 A hierarchy of generic expedition strategies

The first area, to apply more control, is primarily concerned with making people work harder and more efficiently. Incentives in this context means paying workers according to their productivity, the aim being to get more work done in the time available. Better organisation leads to improved efficiency and closer monitoring is concerned with ensuring that workers are actually attaining a reasonable level of performance. More-time relates to obtaining more working time without using project time, the two principal methods being overtime and shifts. There will however be a limit on the amount of overtime that can be worked in a week, and consequently a limit on the amount of time expedited. The same applies to shift working, although shifts are superior since they may potentially use twenty-four hours in a day.

Resources suggests that extra resources may be added in order to get the task performed more quickly. There are three subdivisions, manpower, equipment and materials. There will be practical limitations as to the amount of resources that can be added. Change-contract applies to those activities which have been subcontracted out, or to purchase contracts. It might be possible to change the contract for expedition purposes, but usually at more cost. Change specification might enable the activity to be achieved easier and more quickly. It might be that the quality requirements are reduced or just altered. For example, if an activity was the delivery of a particular material which was proving difficult to obtain, the client might accept a more readily available alternative. Abort is the final and most drastic measure. If an activity is currently impossible to achieve it might be possible to remove it from the jurisdiction of the project, treating it as a completely separate scheme.

XPERT also estimates a time–cost trade-off function associated with the selected expedition strategy, so that it may be used by the heuristic algorithm. For some of the strategies previously outlined, the estimation process is quite automatic, using the procedural operators available within the Pop11 language. For the strategy of working overtime on a particular activity, the information detailing the number of men working on the activity, rates of pay, duration of the activity and so on, is used to estimate the cost function based on standard company costing procedures. Fortunately, it is the strategies which most commonly occur, like overtime, shift-work and additional resources, which can be estimated automatically. For strategies which cannot be dealt with automatically, such as changing a contract with a subcontractor, the cost function is entered by the project manager when prompted by the system. The project network can be displayed in an Activity-On-Node (AON) representation, or as a timescaled network chart similar to a Gantt chart.

6.3.3 Non-AI-planning in Uncertain Domains

As was mentioned in the earlier discussion on strategic and tactical planning systems, there is a continuum with regard to the response time which is required of a planning system, and the reliability of the information on which the system is required to act. Traditional AI planning lies at one end of this continuum, where it is assumed that there is complete knowledge of the world and the effects of possible actions on the world. At the other end of the continuum lies traditional control theory, where there is virtually no advance planning and the control system reacts to differences between its required state and the observed state of the world. Between these two extremes lie a number of techniques for dealing with degrees of uncertainty in the planning process, these being, in order of increasing reactivity, conditional planning, automatic programming, plan monitoring and re-planning, deferred planning, game-playing and operating systems. This predictability–reactivity continuum is shown in Figure 6.5, taken from the DARPA Santa Cruz Workshop on Planning (Swartout, 1988).

Traditional AI-planners and many scheduling systems are characterised by the degree of predictability which they assume about their environment. In general, all the relevant aspects of the initial world state are assumed to be known, and the effects of actions are assumed to be correct. There are few real-world applications where this degree of predictability is available, but such systems have been successful in man-made domains such as machinery, remote spacecraft control and factory scheduling. Within these limitations, there are still many areas of interest for future research, including search control in potentially large search spaces, truth maintenance, temporal reasoning, the 'frame problem', interaction detection

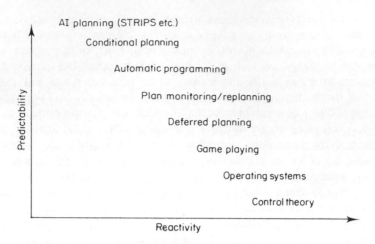

Figure 6.5 The predictability–reactivity trade-off

and resolution, goal representation and constraint handling. However, if planning systems are to be applied to more uncertain domains, they will require the use of techniques from other areas of the predictability–reactivity continuum.

If information about the world is missing, or operators with non-deterministic multiple outcomes exist, then traditional AI-planning methods will not work. If the amount of uncertainty is relatively small, the technique of *conditional planning* can be employed, which involves generating alternative plans for each possible outcome, along with tests to choose between the different alternatives at the appropriate time. Levitt's contingent subnetworks operate in a similar manner, but the plan is assembled manually by the project manager, assisted by critiques from the computer-based decision support system. The main problem with this approach, just as with the early exhaustive search systems, is the possibility of a combinatorial explosion of possible world states. For conditional planning to be successful, it is necessary to ensure that most of the conditional branches re-join, so that the world state remains largely the same, whichever choice is taken at any given point. The use of these conditional branches effectively removes all uncertainty about the planning situation, because every branch in the plan is completely determined. The only uncertainty in the system is that regarding which of the various possibilities will actually execute at run-time.

Although *automatic programming* is generally considered to be a separate area of research from planning, the two disciplines have many features in common. In the case of automatic programming, the execution environment is a particular computer or target programming language which is entirely predictable in its behaviour, rather than the uncertain real world where a planning system must execute its actions. Thus the *actions* of traditional AI-planning systems equate to simple programming commands such as arithmetic and assignment operators, and conditional planning equates to the programming 'if' statement. As with conditional planning, the overall system is completely deterministic, in that the program must be able to cope with all possible alternatives generated by the input conditions, the only uncertainty is that regarding which of the possibilities will execute at run-time. The major difference between automatic programming and traditional planning is that, in programming loops, the world only changes as dictated by the program itself, while in iterative planning loops, the system may well be required to cope with effects caused by unpredictable external events. While there are also other minor differences between such systems, all such differences are in no way fundamental, and it is believed that both disciplines could benefit from a cross-fertilisation of ideas.

While early traditional planning systems have been mainly concerned with the generation of plans to achieve given goals, real-world project management has a strong requirement for *plan monitoring*, where the actual progress of a project is checked against expected progress and any mis-matches discovered. In the robot control domain where most traditional planners are designed to operate, such feedback would be generated by sensors and modelled in the planner as a *sensor operator*, similar to the standard type of action operator, except that generates information rather than altering the state of the world. For high-level project management systems such as PLATFORM, the feedback would be supplied as user input during the project control phase. *Re-planning* is required whenever new information is discovered, which invalidates previous planning assumptions, or a situation which has not been planned for arises during the execution of a plan. The only difference between re-planning and planning itself is the trade-off between speed and optimality, which is fundamental to the problem of exploring search spaces for possible solutions to a problem. The initial planning of a project can be seen essentially as an off-line process, where there is the option of spending a considerable amount of time looking for an optimal, or near-optimal, solution to a problem, while the need for re-planning occurs on-line during project execution, due to unforeseen circumstances. Given that the need to re-plan indicates some degree of crisis in project execution, it is usually considered more important to produce a corrected plan quickly, even if it is of a suboptimal quality.

If large amounts of detail about the world are missing during the initial formulation of a plan, so that conditional planning is impractical, a planning system can insert a call to itself in the plan at any point where information is currently uncertain. During plan execution, the planner will be called at the appropriate time, the assumption being that the necessary information to make a decision will now be available, a technique termed *deferred planning*. If most low-level planning detail is deferred in this way, the result is a high-level skeletal plan which defines an overall structure, with the exact nature of detailed actions being decided during plan execution. In effect, this technique, which is considered to mirror closely the methods employed by human planners, extends the concept of least-commitment planning to cover both the plan generation and plan execution phases of a project.

The fundamental difference between planning and game-playing is the degree of unpredictability involved in the latter, due to the uncertainty regarding the other player's responses. This uncertainty is even present in chess, which is a *perfect information game*, with the range of possible responses making detailed long-range planning impossible. In this type of domain, the planning exercise is reduced to making immediate moves in response to those of an opponent, based on a look-ahead which, in the case of chess-playing programs, may only be four to five moves deep. Multi-agent planning, involving multiple cooperating agents (Koo, 1986), can be seen as a kind of cooperative game, with the planner being either one of the agents or some centralised dictatorial agent in a hierarchical control structure.

The field of *operating systems* theory also has some degree of overlap with elements of planning, in particular the scheduling of processes in an operating system bears similarity to the scheduling of tasks in a planning system, although the timing constraints involved obviously differ. Given the degree of opportunism involved in the working of an operating system's scheduler, there is potential for cross-fertilisation of ideas between the fields of opportunistic planning systems and multi-tasking operating systems.

At the far end of the continuum lies traditional *control theory*, where the system is completely reactive, responding to perceived changes in the world state with fast pre-compiled stimulus–response type reflexes. In situations where there is a critical need for rapid responses, the tendency for a highly reactive, short cycle-time system to go into oscillation due to over-reaction can be avoided by the application of control theory techniques.

The trade-off between predictability and reactivity shown in Figure 6.5 has a number of consequences which should be addressed in the search for an optimal planning solution in any given circumstance. Each of these consequences is, in itself, a trade-off which is an aspect of the fundamental predictability–reactivity continuum. In the first instance, there is a trade-off between reactivity and response time; the more reactive the system, the shorter the time available to plan a response to any given occurrence and the faster the feedback cycle required. A further trade-off is that between plan length and predictability; the more predictable the environment, the greater the likelihood that a plan consisting of a long chain of activities will succeed. Thus automatic programming can expect to produce a plan (the program) which may be millions of steps in length with a high likelihood of success, due to the complete predictability of the computer as an environment. There is a trade-off between sensor information and predictability, in that highly reactive situations will usually require large amount of sensor information if planning is to succeed, and between reactivity and hierarchical planning in that the higher the level of planning becomes, the more predictable the environment. Finally, there is a crucial trade-off between correctness and robustness in the design of a plan. While traditional AI-planners have always been designed around the concept of producing a plan which is provably correct, this method is only really applicable to highly predictable domains. In order to deal with the inevitable uncertainty involved in real-world planning issues, it is more important that a plan is robust and can therefore deal with unexpected eventualities with some degree of success. For example, a route-planning program might produce a 'correct' travel plan, only to find that it involves tightly-timed airport connections which any seasoned traveller knows to avoid, due to any slight delay meaning a missed connection which constitutes a planning failure.

6.4 DEVELOPMENTS IN ENGINEERING PROJECT MANAGEMENT

The results of recent work in planning for engineering project management point the way towards the development of commercially viable real-world systems in the very near future. Based on the project management systems previously described, areas of current project management research include assessing the lessons learned from the PLATFORM experiments at Stanford, work on graphical interfaces for such systems by a number of researchers and the development of an integrated project management system at Brighton Polytechnic.

6.4.1 Lessons Learned from PLATFORM

Levitt and Kunz (1987) draw the following conclusions regarding the success of the various PLATFORM systems, and the lessons learned from their application:

- The use of a knowledge-based system (referred to by Levitt as a Knowledge Processing System, KPS) in tandem with project management algorithms and database access tools provides enhanced capabilities for project control. PLATFORM I can generate reasonably intelligent updates of project schedules using small amounts of stored knowledge and a few heuristics.

- A hybrid development environment provides smooth integration between project management analysis tools and knowledge processing systems. The procedural analysis and database tools currently employed for project management require expert knowledge in order to generate their input and understand their output. An integrated, object-oriented development environment such as KEE allowed the combined development of Lisp-based analysis and KEE-based inferencing for the PLATFORM systems.

- The combination of an ATMS (multiple-worlds) system and decision analysis gives enhanced decision support for the objective-setting phase of project management. PLATFORM III demonstrated that an ATMS capability can provide leverage to traditional decision analysis concepts for project objective-setting by facilitating the generation of alternative scenarios, evaluation of outcomes, and comparison of alternatives for both decision and state variables.

- Interactive graphical displays and their *active* images, programmed in the AI development environment, provide excellent decision support for project planning and scheduling. This leveraging of conventional interactive graphics with knowledge processing functionality opens up a whole new range of possibilities for interactive decision support tools for project management and other domains.

Having made these observations regarding the particular lessons derived from the three PLATFORM systems, Levitt goes on to draw conclusions as to the most beneficial ways that AI leverage can be applied to the four different project management tasks isolated in the taxonomy of project management tasks shown in Figure 6.6.

To support *project objective-setting*, it is proposed to use an ATMS in conjunction with concepts from decision analysis or spreadsheets, while *project planning* will require different types of support, depending on the abstraction level at which the planning is taking place. Means–end planning techniques are best suited to the lower *task level* of project planning, and mixed-AI or domain-specific systems should be used at the higher *work package* and *executive level*. The use of interactive graphical displays for data display and entry, in conjunction with knowledge-processing systems for reasoning about the data, has the potential to provide new types of decision support for *project scheduling*. For *project control* at all levels of abstraction, Levitt proposes the use of hybrid systems combining traditional network-based algorithms with knowledge processing capabilities.

6.4.2 Open Architectures and Graphical Interfaces

The success of the use of graphical network tools for knowledge engineering was demonstrated during the development of the XPERT system at Brighton, when the VEGAN semantic network produced by the knowledge engineer was used as a basis for discussion with the domain expert, a project manager with no experience of AI-based tools. As development of the system progressed, the domain expert began to use the VEGAN editor

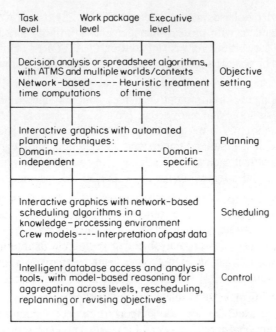

Figure 6.6 Guidelines for using AI and procedural techniques for project management decision support

to modify the system's knowledge base as a natural part of the knowledge engineering process, something which would have been impossible with a system based on, say, a raw Prolog implementation.

This experience, and the recommendations of Levitt and Kunz regarding the benefits of interactive graphical interfaces and hybrid systems, are reinforced by similar findings reported by others, such as the developers of the EXCAP process-planning system (Wright, 1988), developed at UMIST and now being implemented at ICL Kidsgrove. The development cycle of the EXCAP system provides salient pointers to the potential problems which must be considered when developing systems for real-world use. The first prototype was developed using a commercially available expert system shell, but the fixed control structure of the shell was found to be too restrictive to fulfil all the requirements of the proposed system. A second prototype system was then developed in Prolog, which provided the necessary expressive power, but it was found that such a system would be impossible to maintain by on-site personnel with no direct AI expertise. This resulted in the development of the third and final version of EXCAP, which is implemented using the Poplog environment on Sun workstations and makes extensive use of graphical user interfaces.

A point of particular interest in the development of EXCAP is the rationale behind the adoption of graphical user interfaces for the system. Despite the fact that ICL is a major industrial company, the developers of EXCAP have stated that it is economically impossible to support the development and maintenance of the system's knowledge base by external specialised consultants. At the very best, it is envisaged that 50% of the initial development of the knowledge base might be carried out by specialist AI consultants, with the remaining development and all maintenance being carried out by in-house personnel. This economic factor raises the need for user-maintainable knowledge-based systems to an imperative, along with the need for the use of graphical user interfaces. In this respect, many of the benefits conferred by the use of interactive graphical interfaces to access network-based project models can be seen more as an Information Technology phenomenon, rather than an AI phenomenon.

6.4.3 Integrated Applications for Project Management

The PIPPA and XPERT systems at Brighton have shown that both knowledge-based and hybrid programming techniques can be successfully applied to the domains of product configuration and project expedition. The insight into the nature of activities comes in the form of an improved project model: whereas in standard CPA procedures, activities take attributes of just time estimations, they now include a whole set of more descriptive attributes, such as human dependency and weather susceptibility. It has been found empirically that project engineers talk in terms of these kind of attributes when describing activities and any associated problems. The next stage in the application of these techniques to project management as a whole is the development of an integrated project management system which will assist the project manager in the areas of:

- plan generation,
- risk analysis and plan critique,
- project monitoring,
- project expedition and plan repair.

An analysis of the requirements of such an integrated project management system has highlighted the need to generate a synthesis between the concept of *top–down planning*, where the plan generation process is concerned with a consideration of high-level goals and constraints, and *bottom–up control*, where the project manager is required to assimilate a large amount of detailed information generated by a project that is actually under way.

This has resulted in the adoption of a bipolar control structure, based around the information base of the Rule-Based Frame System, as shown in Figure 6.7. Here there is a bi-directional flow of information to and from the information base. Top–down planning goals and constraints are used by the intelligent planning system to produce a project plan which is stored in the system's information base. During the duration of the project, the project manager is able to access the information base via an interactive graphical monitor, which allows the system to be informed of the progress of the project. If and when a slippage is detected, the system passes control to an intelligent expedition system, which updates the information base as required.

Thus in Figure 6.7, general project requirements are seen as filtering down the system from the top–down planning process via the intelligent planning system to the system's information base. At the same time, detailed project feedback travels up the system from the bottom–up control level via the project monitoring system into the information base. The system is able to compare its model of the current state of the project, derived from the monitoring system, with the original project plan so that it not only reports actual slippage of activities, but also isolate activities where slippage is likely to occur if remedial action is not taken.

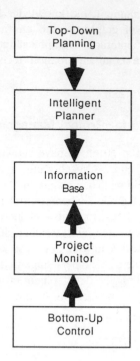

Figure 6.7 A bipolar control architecture

The proposed structure of the Integrated Project Management System (Kuczora *et al.*, 1988) is shown in Figure 6.8. The project manager will bring some form of project specification as input to the system, and can choose whether to generate a project plan manually with an editor such as VEGAN, or automatically via a knowledge-based system such as PIPPA or its derivatives. Once generated, the project plan can be handed over to a risk analysis module, which will look for potential flaws in the plan.

This hybrid system will use a modified version of the XPERT knowledge base to isolate those activities which are 'bad contenders' for expedition should slippage occur in the project, or carry some other form of risk, combined with procedural heuristics to assess the impact of such information on the overall project plan. A project plan which has a high percentage of un-expeditable activities on its critical or near-critical paths will obviously incur a high degree of risk, and may lead to the project manager deciding to modify the plan. Such modifications will initially be carried out manually using a graphical editor such as VEGAN, with a future possibility being the development of a knowledge-based plan repair module.

Once the project is under way, control passes to the project monitor module. This is a purely procedural program with a graphical interface, again using the same 'house style' as other modules in the system, which accesses the project information base. For the project manager, the monitor module will function as a graphical project 'diary' which employs either Activity-On-Node (AON) or Gantt chart representations, and which can be updated as required. The monitor module will also allow the user to set and monitor arbitrary project 'milestones', display detailed activity data and overall resource loadings and perform CPA

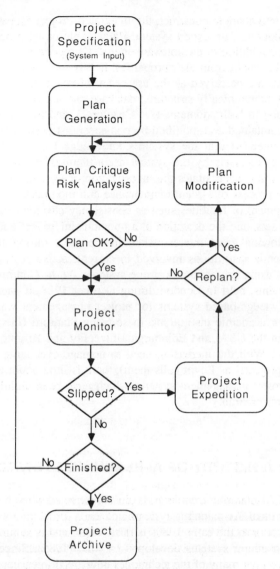

Figure 6.8 An integrated project management system (IPMS)

analysis of the project. It is also intended that the monitor will be able to present project plans at varying levels of abstraction, to suit the needs of users at different managerial levels of the organisation.

The resulting system can be said to be integrated from two standpoints. In the first instance, the system offers the user an integrated working environment, in that it consists of a set of tools which are built to a common ergonomic design, share a common information base and allow the user to move freely between them. The system is also integrated in terms of the underlying methodologies and techniques which it employs. Just as the development of RBFS involved drawing the distinction between knowledge and information and then integrating the two into a single construct, the Integrated Project Management System goes beyond the simple notion of an expert system. Rather, rule-based programming techniques are seen as a valuable addition to the software engineer's repertoire, to be used as and when the application would benefit from their properties. In this context, the main benefits of such rule-based techniques are perceived as the use of an inference engine to allow access to a complex, abstracted, hierarchically structured database, the use of rule-based inferencing to synthesise information in 'soft' domains (see Chapter 2), and the relative ease with which rule bases can be maintained and modified by non-specialist personnel.

Given that it is intended that the system's knowledge bases will be maintained and modified on an on-going basis by the system users themselves, rather than by specialist knowledge engineers, this necessitates the investigation of techniques which will maximise the reliability of the process of system maintenance and modification. Work is currently in hand on the development of facilities such as consistency checking for semantic networks and production rule sets, and the detection of ambiguous inferences and network structures.

Surprisingly, following great interest taken in the topic during the 1970s, there are relatively few academic institutions involved in current, active research into planning or project management using symbolic processing systems. At the Conference on Knowledge-based Planning Systems, held in London during October 1988, it was noted that research into the use of knowledge-based systems for project management was only being carried out in earnest at four academic institutions, these being Stanford University and Carnegie-Mellon University in the USA, and Edinburgh University and Brighton Polytechnic in the UK (Trimble, 1988). With the increasing need to manage ever larger and more complex projects (or 'mega-projects' as Levitt calls them), there is little doubt that the requirements for planning and project management systems which have been highlighted still provide a fertile area for future research.

6.5 APPENDIX: A HISTORY OF AI-PLANNING SYSTEMS

The development of AI-planning techniques is closely reflected in the historical development of the various individual AI-planning systems and early problem-solvers. The lineage of planners developed prior to the early 1980s is fairly clear, and is summarised in Figure 6.9, a taxonomy of AI-planning systems developed by Tate (1985). Since that time, planners have tended to incorporate many of the techniques developed in earlier systems and so their lineage is less clear-cut. There follow details of sixteen of the most important AI-planning systems to be developed over the past twenty years.

6.5.1 Graph Traverser (Doran and Michie, 1966)

One of the first computer programs which was intended to act as a general purpose problem-solver was the Graph Traverser system developed by Doran and Michie in 1966. Graph Traverser operated by employing state space search to apply legal moves to an initial state, aimed at achieving a given goal state. Heuristic evaluation functions were employed in order to rate the intermediate states produced by Graph Traverser in terms of their closeness to the required goal state.

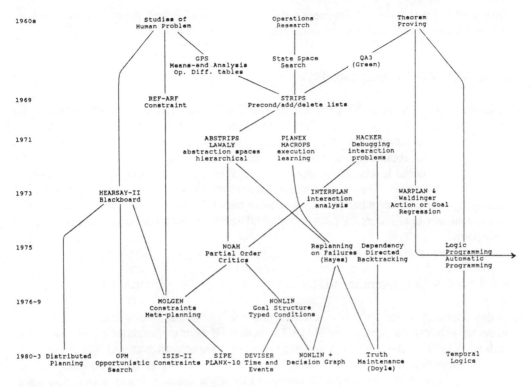

Figure 6.9 A taxonomy of AI-planning systems

6.5.2 General Problem Solver (GPS) (Ernst and Newell, 1969)

The first major step forward in early planning systems came with the development of the technique of means–end analysis for use in the GPS. This system was developed between 1963 and 1972 by Ernst and Newell in order to undertake theorem-proving in logic. As well as introducing the fundamental technique of means–end analysis and the concept of differences between an initial state and a goal state, GPS was also the first problem-solving program to act in a hierarchical manner. By only considering operators which could achieve some currently outstanding goal, means–end analysis offered greater efficiency of search than earlier systems.

6.5.3 STRIPS (Fikes and Nilsson, 1971)

Building upon the techniques introduced in GPS, the STRIPS system is widely recognised as the first true AI-planning system to be developed. Intended to provide low-level control for robot applications, the system was a specialisation of GPS, combined with ideas taken from situational calculus, and was aimed at solving problems regarding actions which cause change in the world. A linear (non-hierarchical) planner, employing the 'STRIPS assumption', the system ignored subgoal interactions during the generation of a plan and only attempted a limited form of plan repair by the use of stack-based depth-first backtracking if interactions were found and the conjunction of goals failed to hold.

6.5.4 ABSTRIPS (Sacerdoti, 1974)

A further development of STRIPS was designed to overcome the limitations imposed by the linear nature of this system. The first hierarchical AI-planner, ABSTRIPS planned in a hierarchy of abstraction spaces, so that plans were refined in a top–down manner. Also intended for robot control, ABSTRIPS carried out a strict search of the hierarchy by increasingly detailed levels, using means–end analysis in a linear manner at each level of abstraction of the plan. The system ignored subgoal interactions at each level and attempted a limited form of plan repair by the use of depth-first backtracking, the intention being that the hierarchical nature of the search would keep the need for such backtracking to a minimum.

6.5.5 HACKER (Sussman, 1975)

A development in the field of automatic program generation which became relevant to planning systems was that of the HACKER system, developed by Sussman as a model of skill acquisition in the generation and debugging of computer programs. It was in this work that Sussman first predicated the linearity assumption which is one of the mainstays of AI-planning theory. Although HACKER operated in a linear manner, it was designed to detect bugs in the use of the linearity assumption by their generation of subgoal interactions, and to attempt to repair such interactions by the use of a case-based method of selecting debugging procedures (written in conventional procedural code) from a library of such routines.

6.5.6 INTERPLAN (Tate, 1975)

The most sophisticated of the linear planning systems was the INTERPLAN system developed by Tate in 1975. Intended as a completely general-purpose planning system, INTERPLAN was designed to detect subgoal interactions and to correct them by analysis. The technique of goal structure was used to record links between actions and their effects on other subgoals in a plan, and goal structure ranges (originally called Holding Periods) were used to construct a data table called a ticklist, which could suggest the minimum set of re-orderings required if an interaction was discovered during plan generation.

6.5.7 NOAH (Sacerdoti, 1977)

The next major advance in planning systems came with NOAH (Nets Of Action Hierarchies), where Sacerdoti first developed the concept of the non-linear least-commitment approach to planning. NOAH, which was implemented in order to model engineering assembly tasks for use as an intelligent assistant supervising apprentice mechanical engineers, allowed planning steps to be developed in parallel until the necessity arose to linearise them. NOAH used procedures called critics to search for subgoal interactions within plans which were represented as procedural nets, and using a Table Of Multiple Effects (TOME), based on the concept of goal structure, as a data structure to aid the search. Although NOAH was a hierarchical system, it allowed only a limited correction to be applied if interactions were detected, due to its being limited to exploring the hierarchy of abstraction spaces via a strict search by levels. Many of the procedures used by NOAH to solve any particular problem were specific to that domain. Such knowledge was represented using functions written in an extension of QLISP, and called SOUP functions as they defined the Semantics Of User Problems. This large amount of problem-specific knowledge, combined with classic means-end analysis, has led to NOAH being termed a mixed-AI planner.

6.5.8 NONLIN (Tate, 1977)

A combination of means-end analysis, the least-commitment approach of NOAH and the Goal Structure interaction analysis technique of INTERPLAN allowed Tate to produce what is perhaps the definitive general-purpose non-linear planning system called, appropriately, NONLIN. Employing a declarative partial order representation, called a 'Task Formalism', NONLIN has, to date, been applied to the areas of house-building, electrical turbine overhaul and naval logistics. Interactions were detected and corrected by analysis, with hierarchical search being non-strict by levels of the abstraction hierarchy. A system of one-then-best backtracking allowed NONLIN to backtrack to a higher level in a plan if choices further down the hierarchy failed. A further development of NONLIN, the Decision Graph system (Daniel, 1983), has also allowed the system to exploit dependency-directed backtracking.

6.5.9 MOLGEN (Friedland, 1979)

There have been a series of AI systems developed at Stanford University in order to assist with experiment planning in molecular genetics, all of which have carried the generic name MOLGEN. The version of MOLGEN developed by Friedland was a case-based planner which used libraries of items of script-based domain knowledge, placing it in the class of early mixed-AI-planning systems.

6.5.10 MOLGEN (Stefik, 1981)

Probably the best known of the MOLGEN systems was that developed by Stefik in the early 1980s. A hierarchical, least-commitment (non-linear) planner, this version of MOLGEN was one of the first planning systems to employ a form of opportunistic search, both locally and within the levels of the abstraction hierarchy, in addition to straightforward means-end analysis. Interactions were detected and corrected by analysis, using domain-specific knowledge, making Stefik's MOLGEN another mixed-AI planner. The system was also able to reason about resource requirements, supported dependency-directed backtracking for plan repair and had a knowledge of meta-planning, that is, a knowledge about the generalities of planning techniques.

6.5.11 SIPE (Wilkins, 1983)

Developed for aircraft carrier mission planning, the SIPE system (Wilkins, 1983) was a least-commitment planner which also employed mixed AI techniques, in a generalisation of the work carried out in the experiment planning domain with Stefik's version of MOLGEN. The system employed classic means-end techniques to control the search for a plan, but moved beyond the limits of the STRIPS notation by using a declarative partial order representation of plans, called the Notation, which was similar to the Task Formalism developed by Tate for NONLIN, and a Plan Rationale which was performed the same function as NONLIN's Goal Structure. Interactions were detected and corrected by analysis, with the resolution of constraint propagation being carried out using domain-specific knowledge. The system supported a rich representation of the planning domain, using frame-based inheritance hierarchies as a knowledge representation and being able to reason about constraints and objects which were treated as scarce physical resources.

6.5.12 DEVISER (Vere, 1983)

The DEVISER system (Vere, 1983) was a development of the NONLIN system, and was designed to carry out mission sequencing for Voyager spacecraft. Like its predecessor, DEVISER was a hierarchical general-purpose least-commitment (non-linear) planner where interactions were detected and corrected by backtracking based upon information stored in the goal structure which was inherited from NONLIN. In addition, the system was able to reason explicitly about events in time, such as time windows, and consumable resources such as fuel. With regard to temporal reasoning, DEVISER was able to detect and correct for temporal displacements caused by the time constraints imposed on a goal or action which had just been introduced into a plan. It was also able to plan within externally imposed timing constraints, such as the orbits of planets and their satellites.

6.5.13 BB1 (Hayes-Roth, 1985)

The pioneering work in the development of opportunistic AI-planning systems has been carried out by Barbara Hayes-Roth, based on earlier work on a cognitive science model of human planning developed in conjunction with Frederick Hayes-Roth. Developed following earlier work on the HEARSAY speech recognition system, the BB1 architecture (Hayes-Roth, 1985; Hayes-Roth et al., 1986a,b) has been designed as an AI platform for planning-related applications, offering dynamic opportunistic search control via a blackboard architecture which allowed modules to exchange data and control information. As with other opportunistic planning systems, BB1 employs a mixed-AI approach, using domain-specific scheduling knowledge to guide the search for the currently most-constrained goal, which it will attempt to satisfy as the next step in the plan. Employing multiple knowledge sources at three levels of abstraction—'strategic', 'focus' and 'heuristic'—BB1 used a combination of local heuristics, means-end analysis, constraint propagation and other problem-solving techniques to propose Knowledge Source Activation Records (KSARs) for execution via the system blackboard.

6.5.14 O-PLAN (Bell and Tate, 1985)

A further development of NONLIN is the O-PLAN system (Open Planning Architecture), produced by Tate and Bell (Bell and Tate, 1985) as a more sophisticated general purpose least-commitment (non-linear) planner than its predecessor. Working in a hierarchical manner, O-PLAN is able to reason explicitly about time and resource constraints, representing them as a numeric minimum–maximum pair which defines the permitted range of activity duration, activity start and finish times, and delays between activities.

6.5.15 COMTRAC-O (Koo, 1986)

The COMTRAC-O system developed at Stanford (Koo, 1986) employs a de-centralised mixed-AI approach to the planning of movements and tasks by autonomous robots. These 'multiple intelligent agents' use means-end analysis to plan their own work and then use domain-specific knowledge to communicate selectively between themselves, sending messages asking for commitments to perform required predecessor activities in order to achieve outstanding goals. Work is currently proceeding on developing this system of pair-wise negotiation to address the issue of asymmetry between different agents, which would model the hierarchical nature of human planning teams, and in the development of conflict resolution strategies to allow for the sharing of scarce global resources between different activities.

6.5.16 Knowledge Intensive Planner—KIP (Luria, 1987)

KIP, the Knowledge Intensive Planner (Luria, 1987), has been implemented in order to provide user assistance for the UNIX operating system, with regard to operations such as the writing to or deleting of files. KIP is a mixed-AI system which operates by using concerns, these being goals whose probability of failure are inferred as being significant using domain-specific knowledge. These concerns are only injected into the plan as subgoals to be achieved once their probability of failure is inferred as exceeding a specified threshold value. Where concerns are perceived as having a high likelihood of success, they are ignored in current planning decisions. It is hoped that this method of limiting a planner's search space can be applied to more complex domains.

6.5.17 PLANEX (Zozaya-Gorostiza et al., 1989)

PLANEX (Zozaya-Gorostiza et al., 1989) is a domain independent, knowledge-based process planning system architecture, used to develop process planning systems at Carnegie-Mellon University, Pittsburgh. Written mainly in LISP, with graphics sub-systems written in the language C, PLANEX has been used to develop two prototype process planning systems in the fields of construction engineering and wiring harness design. CONSTRUCTION PLANEX is used to plan the construction of mid-rise concrete and steel-frame office buildings, while HARNESS PLANEX plans the manufacturing operations required for electrical wiring harnesses in the automobile industry. PLANEX explicitly generates both *process plans* (how the desired product will be manufactured) and *operator plans* (the problem-solving steps which will produce the process plan), and can thus be classified as the first *strategic meta-planner*. The problem-solving behaviour of PLANEX incorporates both strategic and opportunistic elements.

REFERENCES

Arditi, D. (1983) Diffusion of Network Planning in Construction, *Journal of Construction Engineering and Management*, **109**(1), 1–12.

Barber, T. J. and Boardman, J. T. (1988) Heuristic Optimisation Techniques in Knowledge-Based Project Control, *Proc. Twelfth IMACS Con. Paris, July* 1988.

Bell, C. E. and Tate, A. (1985) Using Temporal Constraints to Restrict Search in a Planner, *Pro. Third Workshop of the Alvey IKBS Programme Planning Special Interest Group*, Sunningdale, Oxon., UK, April 1985, Alvey Directorate.

Bimson, K. D. and Burris, L. B. (1986) Knowledge Representation in Software Project Management: Theory and Practice, A Lockheed Software Technology Centre Report, Austin, Texas. Presented at the Forum on Artificial Intelligence in Management, Richmond, Virginia.

Cohen, P.R. and Feigenbaum, E.A. (1982) *The Handbook of Artificial Intelligence*, Vol. 3, Addison-Wesley, Reading, MA.

Daniel, L. (1983) Planning and Operations Research, *Artificial Intelligence: Tools, Techniques and Applications*, Harper and Row, New York.

Davis, E.W. (1974) CPM Use In Top 400 Construction Firms, *ASCE Journal of the Construction Division*, **100**(CO1), 39–49.

de Kleer, J. (1986a) An Assumption-Based TMS, *Artificial Intelligence*, **28**, 127–162.

de Kleer, J. (1986b) Extending the TMS, *Artificial Intelligence*, **28**, pp. 163–196.

de Kleer, J. (1986c) Problem Solving with the TMS, *Artificial Intelligence*, **28**, 197–224.

Doran, J.E. and Michie, D. (1966) Experiments with the Graph Traverser Program, *Proc. Roy. Soc.*, A **294**, 235–259.

Doyle, J. (1979) A Truth Maintenance System, *Artificial Intelligence*, **12**, 231–272.

Drummond, M. (1986) A Design for the Core of the PLANIT Club Interactive Planners' Assistant, Planit Alvey Community Club Working Paper 4.4, June 1986.

Eklund, P. W., Barber, T. J. and Teskey, F. N. (1987) An AI Environment for Knowledge Based Systems Development, *Proc. KBS '87 Conference*, London, UK, Online Publications, Pinner, Middx, pp 127–35, June 1987.

Ernst, G. and Newell, A. (1969) *GPS: A Case Study in Generality and Problem Solving*, Academic Press, New York.

Fikes, R. and Nilsson, N.J. (1971) STRIPS: A New Approach to the Application of Theorem Proving to Problem Solving, *Artificial Intelligence*, **2**, 189–208.

Fondahl, J.W. (1961) A Noncomputer Approach to the Critical Path Method for the Construction Industry, Technical Report No. 9, Stanford University, November 1961.

Fox, M. S. and Smith, S. F. (1984) ISIS—a Knowledge-based System for Factory Scheduling, *Expert Systems*, **1**(1), 25–49.

Fox, M. S., Allen, B. and Strohm, G. (1981) Job Shop Scheduling: an Investigation in Constraint-based Reasoning, *Proc. IJCAI-81*, Vancouver, Canada, Morgan Kaufmann Publishers Inc., Los Altos, CA.

Friedland, P. E. (1979) Knowledge-based Experiment Design in Molecular Genetics, Report No. 79-771, Computer Science Dept, Stanford University (doctoral disseration).

Hayes-Roth, B. (1985) A Blackboard Architecture for Control, *Artificial Intelligence*, **26**, 251–321.

Hayes-Roth, B. and Hayes-Roth, F. (1978) Cognitive Processes in Planning, Report No. R-2366-ONR, Rand Corporation, Santa Monica, California, USA.

Hayes-Roth, B., Buchanan, B. G., Lichtarge, O., Hewett, M., Altman, R., Brinkley, J., Cornelius, C., Duncan, B. and Jardetzky, O. (1986a) PROTEAN: Deriving Protein Structure from Constraints, *Proceedings of the AAAI*.

Hayes-Roth, B., Garvey, A., Johnson, M. V. and Hewett, M. (1986b) A Layered Environment for Reasoning About Action, Report No. KSL-86-38, Computer Science Dept., Stanford University, CA. Hayes-Roth, F., Waterman, D. and Lenat, D. (eds) (1983) *Building Expert Systems*, Addison-Wesley, Reading, MA.

Kellett, J. M. and Esfahani, L. (1988) VEGAN and KET, Graphical Interfaces for Knowledge Engineering, ICL/Ergonomics Society Con. on Human and Organisational Issues of Expert Systems, Stratford-upon-Avon, UK, May 1988.

Koo, C. (1986) COMTRAC-O: A Communication Language for Multiple Intelligent Agents, unpublished paper, Stanford University.

Kuczora, P. W., Kellett, J. M. and Boardman, J. T. (1988) An Integrated Project Management System for Engineering Planning and Control, Submission to the Eighth Alvey Special Interest Group on Planning, University of Nottingham, UK, November 1988.

Levitt, R.E. and Kunz, J.C. (1987) Using Artificial Intelligence Techniques to Support Project Management, Working Paper No.1, Center for Integrated Facilities Engineering, Stanford University, CA.

Lim, B. S. and Wrigglesworth D. (1987) A Prototype Intelligent Planning Assistant in the Manufacture of Flight Simulators, *Proc. IEE Colloquium on Expert Planning Systems—A New Application for Control Theory*, London, May 1987.

Lim, B. S., Marshall G., Kuczora P. W., Boardman J. T., Murray P. M., Wrigglesworth D. and Whitehouse M. (1987) A Foundation for a Knowledge Based Computer Integrated Manufacturing System, *Fourth European Conf. on Automated Manufacture, Birmingham, England, May 1987*.

Lockyer, K. (1964) *Critical Path Analysis and other Project Network Techniques Solution Manual*, Pitman, London.

Luria, M. (1987) Concerns: A Means of Identifying Potential Plan Failures, *Proc. Third IEEE Conference on AI Applications, Orlando, Florida, USA, February 1987*.

Marshall, G., Boardman, J. T. and Murray, P. M. (1987) Project Formulation and Bidding in the Flight Simulation Industry Using Knowledge-Based Techniques, *Proc. of the Royal Aeronautical Society International Conference on The Acquisition and Use of Flight Simulation Technology in Aviation Training*, April 1987.

Marshall, G., Kellett, J. M., Lim, B. S. and Boardman, J. T. (1987) PIPPA: Expert Project Planning System in Manufacturing Engineering, *Proc. KBS '87 Conf.*, London, Online Publications, Pinner, Middx.

McCarthy, J. (1980) Circumscription—A Form of Non-monotonic Reasoning, *Artificial Intelligence*, **13**, 27–39.

Miller, G.A., Galanter, E. and Pribram, K.H. (1960) *Plans and the Structure of Behaviour*, Holt, New York.

Moder, J. J., Phillips, C. R. and Davis, E. W. (1983) *Project Management with CPM, PERT and Precedence Diagraming*, Van Nostrand Reinhold, New York.

Moore, R.C. (1980) *Reasoning From Incomplete Knowledge in a Procedural Deduction System*, Garland, New York.

Newell, A. and Simon, H.A. (1963) GPS: a Program that Simulates Thought, In: *Computers and Thought*, Eds E.A. Feigenbaum and J. Feldman, McGraw-Hill, New York.

Newell, A. and Simon, H.A. (1972) *Human Problem Solving*, Prentice-Hall, New York.

Reiter, R. (1980) A Logic for Default Reasoning, *Artificial Intelligence*, **13**, 81–132.

Sacerdoti, E. D. (1974) Planning in a Hierarchy of Abstraction Spaces, *Artificial Intelligence*, **5**, 115–135.

Sacerdoti, E. D. (1977) *A Structure for Plans and Behaviour*, Elsevier-North Holland, Amsterdam.

Sacerdoti, E. D. (1979) Problem-solving Tactics, *Proc. IJCAI-79*, Tokyo, Japan, Morgan Kaufmann Publishers Inc., Los Altos, CA.

Stallman, R. M. and Sussman, G. J. (1977) Forward Reasoning and Dependency-Directed Backtracking in a System for Computer-Aided Circuit Analysis, *Artificial Intelligence*, **9**, 135–196.

Stefik, M.J. (1981a) Planning With Constraints, *Artificial Intelligence*, **16**, 111–140.

Stefik, M.J. (1981b) Planning and Meta-planning, *Artificial Intelligence*, **16**, 141–169.

Sussman, G. J. (1975) *A Computer Model of Skill Acquisition*, Elsevier, New York.

Swartout, W. (ed.) (1988) Workshop Report—DARPA Santa Cruz Workshop on Planning, *AI Magazine*, (Summer) 115–131.

Tate, A. (1975) Interacting Goals and Their Use, *Proc. IJCAI-75*, Tbilisi, USSR, Morgan Kaufmann Publishers Inc., Los Altos, CA, pp. 215–18.

Tate, A. (1976) Project Planning Using a Hierarchic Non-linear Planner, Research Report No. 25, Dept of Artificial Intelligence, University of Edinburgh, UK, August.

Tate, A. (1977) Generating Project Networks, *Proc. IJCAI-77*, Boston, MA, Morgan Kaufmann Publishers Inc., Los Altos, CA.

Tate, A. (1985) A Review of Knowledge-based Planning Techniques, *The Knowledge Engineering Review*, **1**(2), 4–17.

Trimble, G. (1988) Expert Systems for Project Management—An Overview, *Proc. Conf. on Knowledge-Based Planning Systems Day Two, London, UK, 27 October 1988*, IBC Technical Services Ltd., London.

Vere, S. (1983) Planning in Time: Windows and Durations for Activities and Goals, *IEEE Trans. on Pattern Analysis and Machine Intelligence*, **PAMI-5**(3), pp 246–267, May 1983.

Waldinger, R. (1975) Achieving Several Goals Simultaneously, SRI AI Center Technical Note 107, SRI, Menlo Park, California, USA.

Warren, D.H.D. (1974) WARPLAN: a System for Generating Plans, Dept of Computational Logic Memo 76, Artificial Intelligence, Edinburgh University.

Wilensky, R. (1981) Meta-planning: Representing and Using Knowledge About Planning in Problem Solving and Natural Language Understanding, *Cognitive Science*, **5**, 197–233.

Wilkins, D. E. (1983) Representation in a Domain-Independent Planner, *Proc. IJCAI-83, Karlsruhe, West Germany*, Morgan Kaufmann Publishers Inc., Los Altos, CA, pp 733–740.

Winstanley, G., Kellett, J.M., Best, J. and Teskey, F.N. (1989) IDEAS—For Expert Systems, In: *POP-11 Comes of Age: The Advancement of an AI Language*, Ed. J.A.D.W. Anderson, Ellis Horwood, Chichester.

Winston, P.H. (1977) *Artificial Intelligence*, Addison-Wesley, Reading, MA.

Wright, A. (1988) The Use of Expert System Techniques to Automate the Selection and Sequencing of Manufacturing Processes, *Proc. Conf. on Knowledge-Based Planning Systems Day One, London, UK, 26 October 1988*, IBC Technical Services Ltd., London.

Zozaya-Gorostiza, C., Hendrickson, C. and Rehak, D.R. (1989) *Knowledge-Based Process Planning for Construction and Manufacturing*, Academic Press, London.

P.W. Kuczora
Information Technology Research Institute
Brighton Polytechnic
Lewes Road
Brighton
East Sussex BN2 4AT

7 Computer Vision in Industry

R. Thomas

Brighton Polytechnic, UK

7.1 INDUSTRIAL CONSTRAINTS AND VISION SYSTEM PERFORMANCE

7.1.1 Introduction and Overview

Computer vision is a truly multidisciplinary field of study. The term itself serves to associate the machine world of digital computers, electronic engineering and robotics with the biological world of human vision and understanding (in common with the discipline of artificial intelligence (AI) in general). This simple view encapsulates the dream of intelligent automata, which has captured the imagination of the general public via the media.

On close inspection, the 'computer' aspects of computer vision encompasses both the hardware elements of optical sensors, parallel processing architectures, computer graphics and displays, and the software elements of data manipulation and calculation. The 'vision' aspects of computer vision encompasses the three main elements of the human visual system, namely the eye, optic nerve and cerebral functionality. The human system is highly adaptive and also non-uniform; its performance depending upon both the physical characteristics of the eye itself, and on the complex interaction and interrelation of the eye and brain. It is easy for a human to select an apple from a bowl of fruit; the process involving image data capture, interpretation of the shape, size, texture and colour information from all the fruit in the bowl, and producing the resultant information to control physical movement. In attempting to define a viable computer vision system, which emulates the essential functionality of the human system, a decision must be made on which characteristics of the human system must be, or should be, integrated into such a complex system. In addition to this, how can we identify and implement those higher cognitive aspects which are readily performed by the human brain as opposed to the human eye. Moreover the human system readily perceives objects within an image, even though the image itself may have been subjected to changes in scale, orientation, distortion and colouring, and in situations when objects are only partially visible or poorly illuminated, or perhaps even camouflaged. The human vision system operates instantaneously, whether or not there is relative motion between the human and the image of the world that the human perceives.

Before the industrial revolution, most agricultural and manufacturing activities were

Artificial Intelligence in Engineering. Edited by Graham Winstanley
©1991 by John Wiley & Sons Ltd.

labour intensive. The industrial revolution saw the introduction of mechanisation, with a consequent increase in productivity. This process was the beginning of the long road to automatic production and assembly techniques which are currently in use, and the introduction of computer-vision-based equipment is but a further step along this same path. Therefore it is worth remembering that the driving force behind the changes to date is one of economic viability. A computer vision system will be introduced into industry only when it can carry out a task that cannot be performed manually, or when it can perform the task more accurately, reliably and efficiently than an employee.

Over the last fifty years or so any industrial operation requiring significant visual input was performed by suitable trained human operators. Since a great many such tasks can be identified, the introduction of industrial robots has been restricted mainly to repetitive mechanical tasks of machining and movement, and these have been installed in 'tightly controlled' operating environments. However, with the introduction of computer vision systems able to equal or better the human visual performance, albeit in task-specific application, the spread of robots into the manufacturing industry as a whole has been increasing. Computer-based machine vision is both better and worse than human eyesight. In certain circumstances it can be faster; it will work continuously, provided all appropriate parts are available and the system is correctly set up. It can work in adverse environmental conditions which would be very unhealthy or dangerous for a human operator. It inherently produces objective measurements, as opposed to the subjective assessment routinely performed by trained human operators.

An overview of manufacturing industry readily identifies tasks and subtasks being implemented either in parallel or in the appropriate sequence. A similar scenario occurs when an industrial vision problem is examined for a possible computer system solution. The vision problem is broken down into a number of more simple vision functions, some or all of which may be readily implemented with a computer-based system. Some of these vision functions can be identified here as the capture of an image, the recognition of certain objects within this image, and the initialisation of subsequent action taken as a result of the recognition and in order to complete the task in hand. Figure 7.1 shows the results of the application of computer vision to a sample image. Figure 7.1(a) shows the (computer) unprocessed image emerging from a typical, relatively low-cost television camera (a vidicon camera). Figures 7.1(b) and (c) demonstrates the effects of some computer-based processing on the image of Figure 7.1(a). Some of the processes involved in producing such transformations will be discussed later in this chapter, in the context of image understanding (and how AI can help in this), but for the purposes of illustration, suffice it to say that certain attributes of a 'standard' image can be highlighted through the application of appropriate computer-based processes.

In discussing the human vision system as the model for a computer-based vision system, it must be remembered that the human system has evolved to operate within a three-dimensional world and readily interprets all input data accordingly. All computer monitors and television displays inherently present a two-dimensional image; this two-dimensional image being the resultant projection of three-dimensional data onto a two-dimensional plane. Such a process can be explained mathematically, but the converse process, that of constructing three-dimensional models from two-dimensional image data, is not so easily performed. It is the machine implementation of this process that is essential before fully automated, vision-controlled robots can become a reality. Such robots will require the ability to recognise their three-dimensional surroundings, in order to 'navigate' around them, and

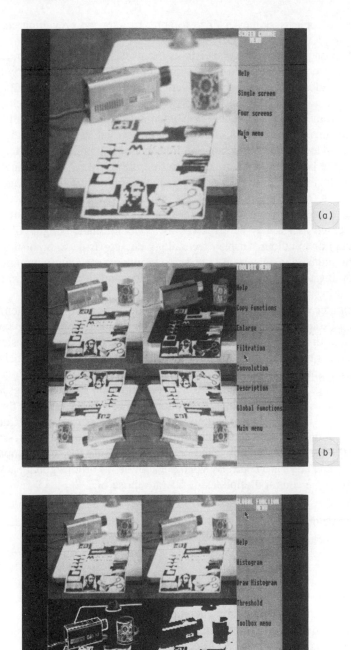

Figure 7.1 Computer vision processes. (a) Digital image acquired by vidicon camera; (b) captured image and three transformed images—negate, invert and translate; (c) captured image and three processed images—smoothed, threshold and edge enhanced

'see' a variety of objects as they happen to appear under the prevailing conditions, at the time.

The introduction of computer-based vision systems has therefore been limited to operation within controlled environments, in which some of the more simple vision functions have been separately identified and a solution 'engineered'. Images are usually captured as a square matrix of numbers, typically 256, 512 or 1024 pixels square. Each pixel may take on a range of values to represent the light intensity at that point, typically 1 byte (8 bit) is used to store this information, thus allowing for 256 values, resulting in a typical image containing 1 Mbyte of data. If the captured images need to undergo a variety of processing operations, before being presented to a human in order to make some critical decision, then this amount of data will require processing within the frame time of the image monitor being used (40 ms for a frame rate of 25/ps). If the captured image is in colour, then three image planes will have been captured and stored. Regardless of the original input image data, the resultant processed image may be displayed as monochromatic (single colour), achromatic (shades of grey) or in colour. Display technology encompasses some of the non-linearities inherent in the human vision system in order to produce a meaningful and pleasing display. Digital computing techniques offer extreme precision and flexibility but may be relatively slow in computing speed and expensive to purchase.

As the computer vision system is expected to tackle more complex problems and integrate more human vision characteristics, then the major problem confronting computer vision engineers is that of relating the three-dimensional objects and two-dimensional images. In short, a solution is required to the problem of recognising three-dimensional objects from a two-dimensional image. Since this process in humans is clearly a cognitive one, the academic disciplines of image understanding and AI merge with those of image processing and pattern recognition from within the computer vision domain. This has resulted in the integration of image noise removal, finding edges within an image, recognising patterns from the resultant edge images, building up representations of the real world from the processed image data, and incorporating a degree of knowledge or intelligence into the total computer system.

Computer Vision (CV) is a relatively young discipline, its origins going back some twenty-five years. The three central disciplines of CV are those of:

- Image Processing (IP),
- Pattern Classification (PC), and
- Scene Analysis (SA),

each of which may be thought of as an 'enabling technology' for CV. However, none of these disciplines serve to individually encompass the central CV issue, i.e. the development of a symbolic scene description from images taken of that scene, in order that the appropriate interaction with the scene may take place (Horn, 1979).

Image processing—the science of producing an improved version of an original image— was the first discipline to undergo 'redevelopment'. It has well defined mathematical roots in linear systems theory, and the use of one-dimensional signal processing, originated in the 1940s, in connection with the enhancement of Radar signals. However, the availability of digital computers in the early 1960s saw the first experiments with two-dimensional signal enhancement for chromosome analysis, and the study of crystal structure by the processing of X-ray diffraction patterns. Researchers at NASA's Jet Propulsion Laboratory

(JPL) subsequently applied this technique in 1964 to television camera images transmitted by the USA's Ranger series of moon probes (Sheldon, 1987).

It was known that images taken by American space probes were seriously degraded by unavoidable, but predictable, noise patterns and geometric and photometric distortions. Thus, provision could be made to eliminate these effects by post-processing the received images, utilising simple subtractive filtering techniques and by the application of calibration data obtained from the cameras whilst still on earth. The use of image processing techniques expanded rapidly, and their contribution to the success of each space-flight became greater as the probes were dispatched farther afield, when the intrinsic data contained within the returned images became increasingly valuable.

During this period, the development of image processing was very much driven by the needs of space exploration. Evidence of this is the amount of early work performed on image compression to raise both the rate and reliability of data transmission. For the first time, these techniques allowed images to be received at a greater rate than they could be processed. In turn, they provided the stimulus for the beginning of research into more efficient image processing algorithms and architectures. Figure 7.2 summarises the historical trends in image processing over the last twenty years.

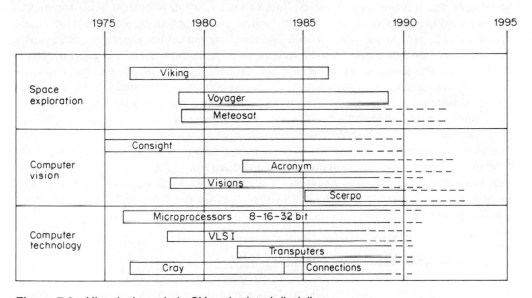

Figure 7.2 Historical trends in CV and related disciplines

The problem of signal propagation delay results in the situation where it becomes almost impossible to maintain interactive control of an unmanned space probe or robot vehicle, even whilst within the limits of our solar system. Therefore, the desire to explore the outer reaches of space has necessitated the development of pattern classification and scene analysis techniques, in order that automata could make its own decisions about course changes, sample collection and terrain mapping, etc.

Pattern classification is the ability to recognise objects within a captured image. Such recognition requires the extraction of features from the image, such as edges or regions and the classification of these features to one of several prespecified categories. Ideally this should be performed in order that no decisions are incorrect. However, in applications where this cannot be achieved, the probability of errors should be minimised. In this case, the problem of classification becomes a problem in statistical decision theory, where the decision problem is posed in probabilistic terms and all the relevant probabilities are known. Pattern recognition is mainly concerned with images of two-dimensional objects.

Scene analysis is concerned with the transformation of simple image features into abstract descriptions relating to more complex objects in the scene. Scene analysis is often used synonymously with image understanding, in which the three-dimensional nature of the image necessitates the use of a wide variety of descriptions, or representations. Image understanding systems have a strong AI content, being heavily concerned with symbolic processes for representing and manipulating knowledge within the CV domain.

The first application domain to receive the benefits of 'spin-off' technology from space exploration was that of medical imaging, since the time when workers at JPL began to apply their techniques to the processing of microscope and X-ray images of the human body. Their results were so impressive, that in 1968, funding was given by the American National Institute of Health to JPL to set up the Life Sciences/Biomedical programme. In the early 1970s, the use of pattern classification for the automatic classification of white blood cells was a major area of interest. More recently, the development of AI techniques have resulted in proposals for an intelligent medical image analysis system. These intelligent systems will be able to use a prior knowledge about the human body, encapsulated as rules derived from a domain expert, to achieve improved results from scene analysis systems.

In recent years, developments in digital device technology and, correspondingly, improved computer architectures, have allowed the storage, processing and manipulation of full colour images at full broadcast television (TV) standard and frame rates. This technology is in very great demand world-wide from the entertainment TV industry, which has now become the arch-exponent of such 'real-time' technology, and thus exerts a major new influence on the direction of CV research, specifically in the area of image processing. Developments in the field of telecommunications to apply image understanding to the achievement of superior data compression of colour TV imagery (for the purposes of videoconferencing), may perhaps be seen as a complementary research thrust in the direction of pattern classification and scene analysis.

7.1.2 CV as an Industrial Discipline

The application of CV within the industrial environment can be readily identified with tasks such as component part inspection, and the guidance and handling problems associated with manufacturing assembly processes. (see the discussion on manufacture in Chapter 5). Developments in these broad areas are intimately associated with the development of industrial automation robotics.

From 1960 onwards, researchers at Stanford Research Institute (SRI), Stanford University and Massachusetts Institute of Technology (MIT) undertook a series of investigations into robotics and CV (Nitzan et al., 1979). The SRI Vision Module was the first complete vision system suitable for industrial use, providing the ability to recognise parts on the bases of their size and shape, regardless of position or orientation, the ability to train the system by showing various views of the part in question, and finally being able to utilise a library of subroutines that could be applied to the development of specific application programmes.

Several commercial manufacturers have adopted some of the general principles of the Vision Module, and made various improvements or modifications for specialist applications. This development has had a substantial impact on the generation of a CV industry, and especially on robot vision applications.

Current applications of CV encompass a broad spectrum of activity, including automatic guidance of welders and cutters, object acquisition by robots for assembly tasks, inspection of printed circuit boards, screening medical images, remote sensing and satellite exploration, document readers, as well as military applications (Wallace, 1988; Hunt, 1985).

7.1.2.1 Existing Industrial CV

The use of automation within industry is a strategy which, when properly applied, is likely to bring about a significantly increased demand for the products of that industry. In the short term, this will be due to the ability to reduce the cost of manufacture, and thus the selling price, whilst increasing 'responsiveness' to market trends. In the longer term, the enhanced reputation for the product quality and reliability should ensure a greater share of the market. Taken in isolation, these factors suggest that all industry should be able to benefit from automation and thus from the application of CV. However, the situation is not as clear cut as this.

To date, the majority of applications for industrial CV has existed within the automobile manufacturing industry (Wallace, 1988; Hollinghum, 1984; Geisselmann et al., 1987; Loughlin, 1987) as illustrated in Table 7.1.

Table 7.1

Task	Manufacturer
Identification of motor car bodies by outline	Ford UK
Automatic wheel body assembly	Ford Germany
Automatic tyre to wheel realignment	Ford Germany
Part classification prior to paint spraying	Ford Germany
Windscreen alignment and placement	Austin Rover
Application of body seam sealant	Austin Rover
Casting identification, orientation and sorting	GM Canada
Inspection of brake assemblies	Volkswagen

This represents the classic example of mass production, having a very long history of automation, beginning with the mechanised production line concept invented by Henry Ford I at around the turn of the century. This industry operated on an enormous scale, with a correspondingly high capital investment in production machinery. The product is manufactured in large numbers, has a high unit cost and is of a relatively large and manageable size.

Because of its importance as a provider of income and jobs, the automobile industry has been heavily supported by funding from national government sources. The availability of adequate funding, coupled with the clear need to increase efficiency and maintain competitiveness, must have provided an obvious opportunity to rationalise production methods and exploit automation. The costs incurred can be readily amortised over the large

Table 7.2

User	Productivity	Task
Advel	1000 million rivets	Inspection of break stem rivets
Green Giant	121 million cobs	Inspection and measurement of corn cobs
Kodak	51 million metres of film	Inspection of photographic film
DHSS	40 million books	Sorting and reading postcodes
United Glass	101 million bottles	Inspection of glass containers

numbers of vehicles capable of utilising that system during the life of the particular product line, and the techniques developed are sure to be of benefit to a subsequent product which will differ only in detail, thus requiring only relatively minor implementation modifications.

Considering other applications, typified by Table 7.2, none of these share the special advantages revealed for the automobile industry, and they cannot all be classified as production industries, by any means. However, the one attribute that they all do have in common is the scale on which they operate.

Thus it seems clear that the productivity of a process is a major factor in deciding the feasibility of a problem solution, based on a typical CV system. It is reasonable to deduce that the numbers have to be very large for the system to be affordable, as the development and implementation costs of those systems are likely to have been very high indeed. It is also interesting to note that in each case listed in Table 7.2, it must have been relatively easy to justify the use of the CV system in preference to human labour, primarily on the grounds of sheer speed of operation.

Batchelor has indicated that Automatic Visual Inspection (AVI) is not generally applied to production in batches of less than one million (Batchelor et al., 1985). This contrasts strongly with the statistic that approximately 75% of all manufactured goods are produced in batches of less than 50 items. From this, it would appear that CV has to undergo a considerable change of direction before it is able to address this large market sector.

Another related problem facing the application of CV is exemplified by a major international company which currently produces 45 000 different product lines, each of which has an average 'half-life' of approximately three years. This very short product lifecycle is becoming ever more typical in modern manufacturing industry, because of the rapidly changing 'tastes' of the consumer societies, and the desire for more economically efficient and responsive production methods. However, it presents a very great problem to any manufacturer operating in this domain who needs to exploit automation in general, or CV in particular, in order to achieve the desired production efficiency. With such a rapid turnover in product lines, with each being manufactured in relatively small total quantities, it is not surprising that manufacturers are unwilling to invest in very costly technology that must necessarily be dedicated to its chosen function in order that it is at all practicable. This will be true, even though the particular company in question operates on such a scale that it should be well able to organise its funds to support such an investment on an evaluation basis, if it wished to do so.

In the case of a more typical manufacturer, with a much smaller range of products, this situation is compounded by the lack of corporate finance necessary to under-write the risk of investment in new technology, resulting in the small scale manufacturer being faced with an uncomfortable dilemma, namely the knowledge that the introduction of automation is required in order to survive and prosper in the increasingly competitive markets being built on short-life products, tempered by the fear that investment in the wrong technology, and at too early a stage in its development, could prove fatal.

Automation is considered to be the key to dealing with a wide product range, made in small batches, within one factory. The dominating concept is that of flexible manufacture, which provides what is effectively 'time-division multiplexing' of the manufacturing facilities, so that a variety of products can be made on a single Flexible Manufacturing System (FMS), in batch sizes that ideally would be controlled by the arrival of customers orders. As the name suggests, the FMS has to be adaptable, and here CV has a major role to play. The use of CV systems, well adapted to the tasks in hand, but of sufficiently general purpose nature will eliminate the need for much expensive and cumbersome fixturing in a 'FMS workcell'. Provided that the CV systems can justify themselves in terms of cost and flexibility, they may turn out to be to be the 'pivot', upon which the viability of FMS and the consequent health and well being of the majority of manufacturing industry depends.

7.1.3 Performance Criteria

Recent reports indicate that approximately 90% of all industrial vision systems are being used in inspection, which has hitherto been one of the most difficult tasks to automate (White, 1988). Typical examples included a four-camera vision system to measure the flatness of hot steel plate as it leaves the rolling mill, and visual inspection within automated printed circuit board assembly cells to help improve quality and yield. Since the application of CV technology to industry is a relatively new phenomena, with few vision suppliers (in the UK) being in business for more than five years, potential users really need to have an awareness of optics and sensors, together with computing, electronics and engineering. Furthermore, this relative immaturity necessitates that both suppliers and users of the vision system resolve the sometimes conflicting views of academics, who are making significant advances in image understanding and scene interpretation, and the industrialists, who are forced to make severe compromises in terms of viable functionality and basic economy.

Figure 7.3 (Wright and Bourne, 1988) contrasts these industrial and academic perspectives; an industrial vision system must accommodate the 'desire' to effect a global understanding of a scene with that of performing very accurate measurements on particular objects within the scene.

When considering the possible installation of a CV system in an inspection role, it would be unreasonable to expect the system to perform any task impossible for a human agent. Although this appears to be an obvious statement, it is this type of over-expectation that can 'dismay' industrialists.

There are some technological barriers associated with the processing of image data:

- the resolution of the sensor,
- the image acquisition time,
- processing generally utilises the two-dimensional projected image of a three-dimensional original,
- the ability to process colour, and
- the response time to process the image and then make a decision.

Cameras are available that are able to capture images up to 1024 × 1024 pixels in resolution, and the adoption of 'array sensors' or 'linescan sensors' can affect the resolution and acquisition time parameters. Binary and grey-scale images can be processed with current vision systems, whilst the processing of colour images (with the inherent increase in data), remains too expensive for the majority of industrial applications. The analysis of two-dimensional projected images of a scene can produce shortcomings when objects are partially occluded during an inspection process; such limitations may be overcome by the use of more than one camera, or the utilisation of different camera angles, all of which add to the cost and time in processing and merging the data. Image-processing speed limitations are generally

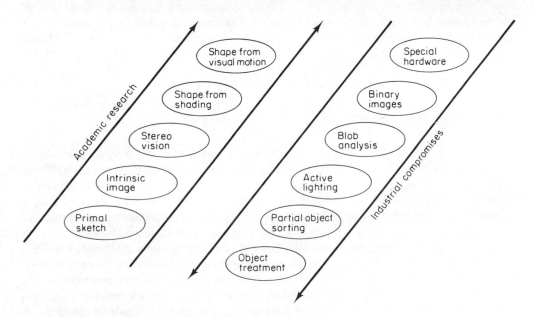

Figure 7.3 Industrial *versus* academic perspectives for CV

1 Is the object in question readily identifiable?

2 Is the background pattern easy to control?

3 Is the objects's colouration known and regular?

4 Is the lighting easy to control?

5 Can multiple views of the object(s) be collected?

6 Can the camera be manipulated around the object?

7 How diffucult is it to maintain the camera's position?

8 Is the required accuracy within the capability of the camera system?

9 Are there special optical effects that must be considered in order to get good images?

1 0 What is the best spectrum of light to get the clearest images?

1 1 Can the object be readily treated with a foreign substance to improve the image?

1 2 How difficult is it to interpret the visual information once it is retrieved

1 3 Are there reasonable constraints on the task requirements?

1 4 Are there less expensive sensors that can do the job better?

Figure 7.4 Reference guide to vision checklist (after Wright and Bourne, 1988)

overcome by using parallel-pipeline architectures and special purpose integrated circuits, but with the obvious cost penalty.

Before any manufacturing company purchases a CV installation there are many factors to consider. If the company already has substantial computer networking facilities then the CV system can easily be integrated into the existing network as another item of peripheral equipment. For inspection applications it is beneficial if the CV system is integrated with the company CAD database. Moreover the company research and development policy will influence its investment in the purchase of a general purpose solution or the development of a company specific solution. Reference to a check list such as that given in Figure 7.4 (Wright and Bourne, 1988) may help in deciding the practicality of any system.

7.2 VISION SYSTEMS AND SYSTEMS ENGINEERING

7.2.1 Industrial CV Taxonomy

Within the context of this chapter, the application of CV to industrial systems can be subdivided into two broad categories (Wallace, 1988):

- visual inspection and
- visually controlled robotic guidance.

The first category involves the comparison of an image, or image subsection, against a predefined standard, and has been used to inspect products in the automobile, textile, food- and metal-processing industries. The second category consists of trajectory planning, clash avoidance, adaptive positional control, and the direction of the robot to a precise position in three-dimensional space relative to a particular object. Examples include assembly of mechanical parts or printed circuit boards, and adaptive control to perform paint spraying or welding tasks.

An industrial CV system provides three basic functional units of image acquisition, analysis and interpretation. These can in simple terms be allied with the three fundamental disciplines referred to initially as image processing, pattern recognition and scene analysis.

The majority of vision systems rely on two-dimensional image processing, and assume that the three-dimensional world is constrained in some way by prior knowledge. Three-dimensional industrial vision systems are being constructed, however, using time-of-flight sensors and laser triangulation and, to a lesser extent, stereo vision.

7.2.1.1 Visual Inspection

Automated visual inspection consists of the examination of a known part, or assembly, to ensure that it meets a specified standard. The inspection task has two stages: the first test phase defines and quantifies specific image features, and the second test phase classifies the part or assembly and, hence, accepts or rejects it on the basis of the acceptable specified standards.

In general, the system designer has to consider alignment problems (manual or automated positioning for location of key features), and optics and illumination problems (control and structure of lighting, colour and type of light source). Requirement of high resolution over a large area may necessitate a scanning system in two or three dimensions. Data compression is often used to reduce the data storage requirements of high resolution images.

Three principal categories of automated visual inspection are identified:

(1) dimensional inspection of natural and fabricated items;

(2) Surface inspection of natural and fabricated items; and

(3) Electronic circuit and layout inspection.

7.2.1.2 Robotic Guidance

The three principal categories for CV applications to robotic guidance and control can be identified as:

(1) automated assembly and manufacture;

(2) electronic assembly; and

(3) bin-picking.

Many flexible manufacturing systems use robots in repetitive assembly-line operations. Here there is no interactive operation, and the robots are constrained to operate within a precisely controlled environment. Because of the controlled nature of these operations, the resolution requirements are less stringent than those for visual inspection. The use of adaptive control is less widespread, owing to the unconstrained nature of the problem. A high proportion of industrial applications use binary imaging techniques to provide the speed of processing, comparable with the speed of the robot servo-control system. To exercise fast and reliable control, it is necessary to provide a highly constrained environment, coupled with the use of appropriate computing hardware. Visual control systems may rely on two- or three-dimensional sensing, dependent upon the particular application. In most visual control systems, there is some element of inspection, since the vision system must first determine the nature and orientation of the part before grasping and manipulating it.

7.2.2 A Systemic Approach

Section 7.1.2 indicated that the use of vision systems in industry was heavily concentrated in the manufacturers of automobiles and electronics, with a relative lack of penetration into small- and medium-sized companies across the product spectrum. Considering industrial computer vision in general, there are several reasons behind any installation, including increased flexibility, lower cost, increased reliability and accuracy, and greater speed and serviceability.

From a systemic viewpoint, two approaches emerge which may contribute to improving still further the market penetration of CV systems. The first step in a systemic solution is the adoption of 'soft systems' thinking (see Chapter 1), which identifies the nature of the goal and the human factors that prevent its achievement. Such an approach must be extended to include consideration of a strategy which attempts to manage the future impact of human factors upon the problem. Thus, the 'top-level' managers in industry must be subjected to a subtle process of education, in order that they can actively consider themselves as potential users. Such soft-systems thinking may suggest the need to utilise existing standard computer technology, and therefore produce systems which are computationally efficient. In addition, such soft-systems thinking suggests a need to lower the development costs and thereby decrease the risk associated with selecting a vision-equipment solution. Finally, such thinking suggests the need to prevent the continued use of piecemeal (dedicated) solutions and hence stimulate the production of widely applicable technology.

Any soft-systems strategy which is adopted will make use of 'hard systems' methodologies to structure the approaches to be taken. A suitable hard-systems approach to complex problems is to decompose them and thus form a hierarchical, or a distributed, system consisting of submodules. If the result is to be a true system rather than a simple aggregate, each module must be designed and developed with the system in mind. Each one must provide a defined communications interface with the modules around it in the system; this communication interface ensures that relevant detail about the overall goals of the system is propagated throughout the set of subsystem modules, in order that they are each able to make a wholly beneficial contribution to the achievement of those goals. Exploitation of this property of 'interdependence' ensures that the resulting system has the potential to be 'near-optimal' in terms of efficiency.

The level of flexibility which is necessary may be achieved by the adoption of a generic modular structure for the particular application domain; a complete range of subsystem modules, sufficient to provide all the functions essential to problem-solving in that domain, must be identified. Each subsystem module must be allowed the freedom to perform its function in the way best able to realise the overall goals of the system. The term 'ordered liberty' describes this kind of control structure operating within a system architecture (see Chapter 1 for a discussion on systems thinking, and its implications for AI).

A generic model of a CV system is shown in Figure 7.5. Inspection of this model reveals the scale and multidisciplinary nature of the task of implementing a design strategy for an industrial vision system, which stretches from optics and design of structured lighting at the the one end of the model, to AI and robotics at the other end. Few industries will employ personnel with appropriate expertise in all aspects of the system. Furthermore, as the development costs for both hardware and, perhaps more importantly, software increase, there has been a tendency for industrial companies to seek to purchase a ready-made solution to their vision tasks. Any general purpose system still requires refinements so that it becomes efficient (dedicated) to the particular task in hand; such action may well have contributed to the relatively slow penetration of CV systems across the spectrum of industry, and in particular to speculative equipment investment.

The main functional blocks identifiable from the generic model in Figure 7.5 are as follows:

Scene constraints This is arguably the more important element, since the ability to acquire a good quality image, appropriate to the task in hand, greatly influences subsequent operations. Within the industrial environment many factors, e.g. production line equipment, safety rules, lighting conditions, dust and dirt, hazardous chemicals, operatives interaction, all conspire to make the installation of

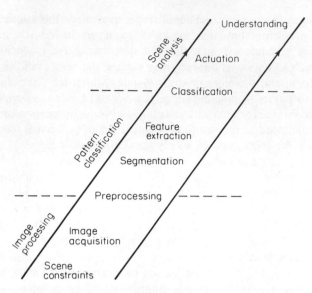

Figure 7.5 A CV hierarchy

cameras and special lighting fraught with difficulties. In essence the objects under consideration have to be presented to the camera photosensors in order to produce the optimum optical image, making use of additional special purpose and controlled lighting techniques.

Image acquisition This is the process of translation from light stimuli on the photosensors to a stored digital value, within the computer frame store memory, producing a digitised picture up to 512 by 512 pixels resolution, with each pixel representing binary, colour, of grey-scale values (up to 8 bits per pixel).

Preprocessing This process seeks to prepare the digitised image in a form that is suitable for subsequent operations. Typically correction for sensor distortion, contrast enhancement and adjustment, filtering to remove noise and improve quality, may all be necessary during this stage. These operations only change the digitised pixel values of the digitised image within the frame store, or subsequent display storage elements.

Segmentation This is the initial stage for any recognition process, whereby the acquired image is 'broken up' into meaningful regions or segments. The segmentation process is not concerned with what the regions represent, but only with the process of partitioning the image. For example, in a natural scene to be segmented, there may be regions of sky, clouds, ground, building and trees. Typically, segmentation utilises thresholding, edge-based or relaxation approaches.

Feature extraction This is the final process prior to classification, whereby the inherent characteristics or features of the different objects within the image are identified. Moreover, decisions are made about how these features are computed from the raw data. Typical features include square, ellipse, line, area, perimeter, centre of area. After these operations, the features no longer reside as an image array, but as a list of descriptions or feature vectors within the computer.

Classification This is the process, in which the pattern recognition or image classification takes place, the result being a description of the scene under investigation. Various techniques exist for the classification stage, including template matching, various statistical approaches and syntatic methods; *a priori* knowledge is often used in the training and test phases of the classification function.

Actuation From an industrial viewpoint, any CV system must contribute to the efficiency of production, inspection or test functions within the company. Hence, the actuation module identifies any subsequent action that the vision system makes, either directly through explicit robotic interactions with the original environmental scene, or indirectly as implicit interactions (as in manual intervention following vision system directives).

Clearly, the functional blocks identified within the generic model all act upon the data they receive, before passing this modified data to the subsequent module. The CV system can be regarded as containing several subsystems which, though complete in themselves within the enclosing subsystem boundary, cause interaction between modules across the functional spectrum. Moreover, it is clear that the level and degree of complexity of computing during the preprocessing stage will relate intimately to the quality of the image captured during the acquisition stage. This illustration of subsystems interdependence can both assist and obstruct in the hardware and software designs of the vision system itself, particularly if the engineering interdisciplinary nature of the system is considered, i.e. the scene constraints can impinge upon the domain of the production engineer, the analysis and actuation areas can impinge upon the manufacturing and test engineers, and the rest of the system can be influenced by the hardware and software design engineers. In addition to this, the speed of operation and accuracy of any system will necessarily be governed by the financial limitations to which the system is subjected.

Consideration of the generic CV model of Figure 7.5, in the light of a systems philosophy, has indicated the clear presence of subsystem boundaries, together with a degree of modular interdependence. Each subsystem would be a candidate for application of the latest technological developments, and *a priori* knowledge within the constrained industrial environment, should produce 'emergent properties' from the vision system, that are both predictable and beneficial (see Chapter 1 for a discussion on emergent properties within the systems engineering context). In short, the generic model can be utilised, but a systems approach to optimality should be adopted, rather than the alternative option of optimising each module individually for the task in hand.

The performance of a vision system can be evaluated by consideration of many factors such as cost, speed, reliability, competence, flexibility and user friendliness. The competence of any system is limited by the representations it uses to describe the acquired images and the knowledge available for manipulating and transforming them (Horn, 1979) (see Chapter 2 for a full discussion on the representation problem, and Chapter 8 for an advanced treatment of the topic). Ideally a vision system should be capable of representing the environment scene at many levels:

- the 'pixel images' and resulting pictorial features (edges and regions);
- the more complex intrinsic surface characteristics (texture and scratches);
- three-dimensional surfaces and objects; and ideally
- space maps and symbolic representation to allow for intelligent interpretation of the scene in a way similar to the human viewer.

Current industrial systems tend to rely on very basic representation, using two-dimensional representations, obtained within a well structured environment, and utilising *a priori* knowledge of a limited number of detailed objects.

It has already been shown in Figure 7.3 that current areas of academic research highlight developments in intrinsic image description, stereo vision and three-dimensional shape definition using a variety of techniques. Thus, the academic 'push' towards the intelligent vision system (with human capabilities), encompassing the AI domain, is of restricted use within the broad spectrum of industrial vision systems equipment and applications (as yet). However, a systems overview of both the industrial problem, and possible CV solutions, may contribute to enhanced adoption of computer-based vision equipment across the broad spectrum of manufacturing and service industries.

7.3 VISUAL INSPECTION

7.3.1 A Two-dimensional Perspective

When considering the generic model of a CV system, it is clear that, the more control is exerted over the environmental constraints, the simpler is the subsequent processing solution. In short, if the vision problem can be fully controlled and specified , the vision task becomes one that a human operator, and hence CV system, can easily perform. Most visual inspection applications require reference to be made to stored data, either within a CAD system or as a set of stored parameters. Inspection of dimensions, features, shapes or regions can usually be made by ensuring that the component parts are presented to the image acquisition system in a suitable orientation, and under suitable illuminations, to produce the highest quality digitised image. Thus, the three-dimensional world can be constrained using prior knowledge of the problem, in order to produce a two-dimensional problem and hence a simpler solution. Referring to Figure 7.5, much of the visual inspection applications can be accomplished by utilising essentially the low-level portion of the CV spectrum (essentially image processing). Moreover, the inspection 'throughput rates' can usually be controlled to suit the equipment available, and hence there is usually no processing hardware limitation. However, market forces and industrial aspirations will accelerate the trend into faster hardware, utilising the latest developments in parallel processing.

Visual inspection applications can therefore be regarded as essentially a two-dimensional image processing problem.

7.3.1.1 Scene Constraints

It is important not to forget the constraints of the industrial environment on the scene under consideration. For any vision system, the quality of the lighting is a key parameter. However, the most appropriate method of lighting the scene may not be the obvious one. In some cases, striped or polarised light sources may be advantageous. Laser scanning techniques may be appropriate, and the surface finish of the objects under inspection must be taken into account, i.e., whether shiny, matt, opaque etc. Typically, the vision system designer is required to consider a number of scene constraints, including the output level of the light sources, the spectral response of both optical components and image sensors within the acquisition system, the spatial and temporal distribution of light, the operating environment itself, and finally the type of sensor to be used. In essence, the system should be designed to reduce the demands on the image processing subsystems to an insignificant level, by acquiring the ideal image to analyse; the system software should be capable of dealing with non-optimum images and hence be reliable and consistent (Batchelor et al., 1985).

7.3.1.2 Image Acquisition

Industrial CV systems employ a variety of linescan and array cameras in the acquisition of digitised images. Linescan cameras are capable of building up two-dimensional pictures, by virtue of the relative movement of the object in front of the camera, as in a conveyor belt inspection system, or by the relative movement of the camera or lightscan across the object under consideration. Figure 7.6 shows an example of this.

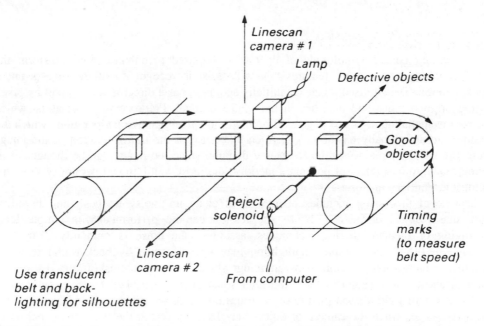

Figure 7.6 Two-dimensional image acquisition using relative motion of object and camera. (Reproduced by permission of IFS Ltd.)

Although TV quality Vidicon cameras are capable of producing the very best high quality high resolution images, they tend to be fragile and susceptible to damage from 'optical overload'. Solid state, charge-coupled device (CCD) cameras are now in widespread use, owing to the robustness, reliability and the photolithographic technique used for solid state devices. Typically, a high resolution image will be made up of 512 × 512 pixels, with each pixel having an 8-bit grey-scale resolution, i.e., 256 grey levels. Reducing the grey-level values to 16 (4 bit) does not seriously degrade the image for the human viewer, and the reduction to a two-level binary representation (1 bit) is often quite adequate for industrial tasks. Similarly, image resolution can be reduced from within the context of the task in hand. Linear arrays are well suited to linescan applications, whereas two-dimensional arrays occupy a similar market niche as the TV camera.

One important advantage of the reduced resolution and memory array storage capacity is the 'speed up' obtained in computation. This may be an important consideration for 'real-time' vision systems, in which the image 'update rate' is to be less than the 'frame time' of the monitor (40 ms for standard television receiver). If real-time systems are required at the highest resolution, then the use of parallel array architecture becomes necessary. Such parallel architectures are often appropriate to perform the local and global processing operations at the higher levels within the vision hierarchy. Moreover, special array processor architectures have been developed specifically for the image processing domain (Kittler and Duff, 1985; Fountain, 1987).

Although most cameras are capable of producing grey-scale and colour images, it is often the case that such images are reduced to a simple binary image for subsequent processing in many industrial applications.

7.3.1.3 Image Preprocessing

Preprocessing covers a multitude of operations required to enhance and transform the original image, and identify features that will assist in recognition of the objects under consideration. The digitised image, as initially acquired, has a direct relation, pixel by pixel, to the original scene and thus lies in the spatial domain. Preprocessing operators, which utilise the spatial domain, are in general pointwise or pixel operators. Operations which are required to act globally on the image, often make use of the transformation of the image from the spatial domain into the frequency domain, since the image can be thought of as being made up of a series of periodic functions; the most well known operator of this type being the Fourier transform.

Low level processing includes operations to clean up 'noisy images', and highlight particular features of interest. Noise suppression can be performed with simple local averaging or median filtering techniques, assuming the noise is randomly distributed throughout the image. Simple highlighting may be performed by background removal, where the background is defined as anything which is of no interest. If the features of interest allow image capture before and after some event, then pixel by pixel subtraction of the two will yield a good picture of what has changed; the problem being then to align the two images when the camera or object may have moved in the intervening period of time. Histogram manipulations provide simple image improvement operations, either by grey-level shifting or equalisation. An image histogram is easily produced by recording the number of pixels at a particular grey level. If this shows a bias towards the lower intensity grey levels, then some transformation to achieve a more equitable sharing of pixels among the grey levels would enhance or alter the appearance of the image. Such transformations will simply enhance or suppress contrast, and stretch or compress grey levels, without any alteration in the structural information present in the image. It is convenient to consider the

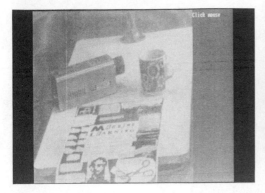

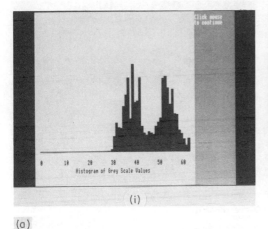
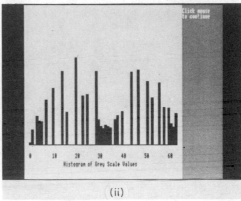

(i)   (ii)

(a)

Figure 7.7(a) Histogram equalisation. (i) Original image and its histogram; (ii) image after histogram equalisation has been performed, and the resulting histogram

grey-level histogram as a continuous probability density function, which is transformed to generate a uniform grey level distribution. The theory underpinning many of these ideas relies on continuous functions, and it is only a final step to move the results into the discrete domain (Gonzalez and Wintz, 1987).

Another class of algorithms are designed to perform pixel transformation, whose final value is calculated as a function of a group of pixel values in some specified spatial location in the original image. The many filtering algorithms for smoothing (low pass) and edge enhancement (high pass) are firmly in this category. This introduces the basic principle of 'windowing operations' in which a two-dimensional mask, or window, defining the neighbourhood of interest is moved across the image, taking each pixel in turn as the centre, and at each position the transformed value of the pixel of interest is calculated.

 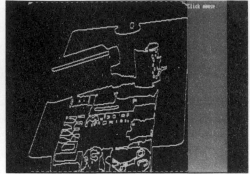

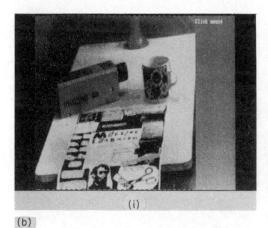 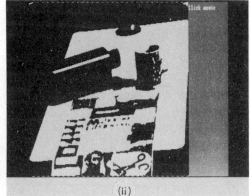

(i)   (ii)

(b)

Figure 7.7(b) Edge detection. (i) Sobell operator (top) applied to original image (bottom); (ii) Sobell operator (top) applied to a thresholded version of the original image (bottom)

These moving window operations are synonymous with digital convolution, in which the output at any given pixel point is given by the sum of the input values around the point, each multiplied by the corresponding term of the window array. The interrelationship between image manipulation in the spatial domain and the frequency domain, leads to the conclusion that convolution in the spatial domain is equivalent to multiplication in the frequency domain. This provides an alternative way of convolving two signals by taking their transforms, multiplying the transforms point by point, and taking the inverse transform of the result. Many filtering algorithms have been developed, but their results are sometimes disappointing when viewed subjectively. As mentioned above, the Fourier transform is widely adopted for filtering and image processing operations, and its application has been

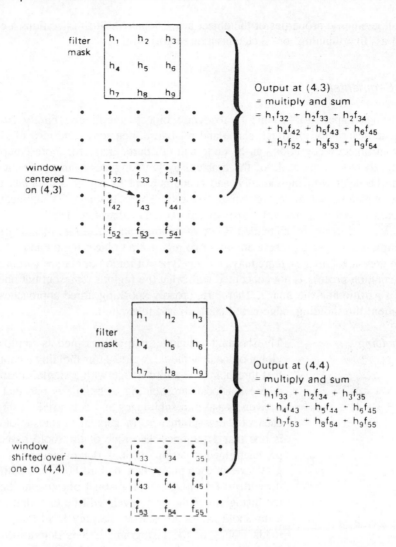

Figure 7.8 Digital convolution operations

facilitated by the development of the Discrete Fourier Transform (DFT) and Fast Fourier Transform (FFT), and both can be applied to two-dimensional digital images.

It is often the case that preprocessing operations require the implementation of a data transformation, able to 'normalise' certain specific properties within the image. Position normalisation requires the translation of image coordinates, so as to realign the extremities of the object of interest within the overall image matrix. Rotation normalisation allows for a rotation through some specified angle about the centre of gravity of the object, and again requires only a recalculation of the coordinates of the pixels defining the object. Where a grey-level image has been 'thresholded' to produce a binary image, it is sometimes useful to apply a transformation that reduces all lines to a single pixel thickness, while preserving the overall geometric properties of the object being depicted. Such operations are generally identified as 'line thinning' or 'skeletonisation' algorithms.

7.3.1.4 Segmentation

Human beings appear to be able to recognise objects intuitively, purely from a very approximate description of rough shape and relative dimensions. The idea of information reduction in object recognition can be quite helpful, particularly for more complex vision processing, to reduce the scale of the problem by breaking down an image into subareas which can be dealt with individually, and hence attempt to reduce a complex processing task to a sequential set of less complex subtasks. Thus, in many CV applications, it is appropriate to work with general shapes of objects rather than fine detail.

The process of image segmentation is one of dividing an image into meaningful regions. In the simplest case (binary) there are only two regions; an object region and a background region. In grey-level images, there may be many types of region or classes within the image; the segmentation process is not concerned with what the regions represent but only with the process of partitioning the image. There are, broadly speaking, three approaches to image segmentation: thresholding, edge-based methods and relaxation.

Thresholding   Thresholding techniques can be classified as employing either global or local methods. A global thresholding technique is one that thresholds the entire image with a single threshold value, whereas a local thresholding technique is one that partitions a given image into subimages and determines a threshold for each of these subimages. In general, a thresholding method is one that determines the value of threshold based on some predetermined criterion. If it is determined solely from the grey level of each pixel, then the thresholding method is point-dependent. For images with distinct objects and background, the histogram of the grey levels will be bimodal; in this case a threshold can be chosen as the grey level that corresponds to the valley of the histogram. If it is determined from the local property in the neighbourhood of each pixel, e.g. the local grey-level distribution, then the thresholding methods is region-dependent. Such a method is that of histogram transformation, where commonly a new histogram is obtained by weighting the pixels of the image according to the local property of the pixels.

Edge-based Segmentation These methods involve a number of steps. First, the image intensities are either smoothed or approximated locally by a smooth analytical function. Then the candidate edge elements, 'edgels', are detected, using a local edge operator. The output of local edge operators contain redundant edge data (due to noise or adjacent lines); these are removed before the result is thresholded into a binary image consisting of edgels and non-edgels. Finally, the individual edgels are grouped together in order to form one-dimensional features.

Smoothing reduces the effect of noise on the detection of intensity changes, and it sets the resolution or scale at which intensity changes are detected. Typically, smoothing filters perform some form of moving window operation in the spatial domain, that may be convolution or other local computation in the window. There are a variety of smoothing algorithms available including the mean filter, mode filter, and both first and second derivative Gaussian filters.

The classical approach to edge detection, makes use of digital versions of standard finite difference operators, as in the first-order gradient operator (e.g. Roberts, Sobell) or in the second-order Laplacian operator. The difference operation accentuates intensity changes, and transforms the image into a representation, from which properties of these changes can be extracted more easily. A significant intensity change gives rise to a peak in the first derivative or a zero crossing in the second derivative of the smoothed intensities. These peaks, or zero crossings, can be detected easily, and properties such as the position, sharpness and height of the peaks infer the location, sharpness and contrast of the intensity changes in the image. Directional or compass gradient operators, attempt to linearly convolve the image with a set of masks representing ideal step edges in various directions. The surface fitting class of operators (Hueckel) approximate the intensity profile by an analytical function which is then used to compute the derivatives or the best fit with an ideal edge.

Morphological edge operators view the image from the perspective of set theory (see Chapter 9) and geometry, and analyses them in terms of shape and size by the use of selected elemental patterns. A simple method of performing grey-scale edge detection in a morphology-based vision system is to take the difference between an image and its erosion by a small structuring element. The difference image is the image of edge strength, and an appropriate thresholding algorithm may then be applied to obtain a binary edge image.

Differing size masks can be utilised in edge detection. The basic idea is that the smaller windows can detect finer changes in intensity profile and are not affected by gradual changes in intensity, whilst the larger windows ignore fine details and therefore detect coarser changes in intensity profile. Important physical changes in the scene take place at different scales. At all resolutions, some of the detected features may not correspond to real physical changes in the scene. One multiple scale edge detector (Canny) first detects features at a set of discrete scales. The finest scale description is then used to predict the result of the next larger scale, assuming that the filter used to derive the larger scale description performs additional smoothing of the image.

Relaxation Techniques

These techniques attempt to assign pixels to segments, ensuring that neighbouring pixels are assigned in a compatible way. The method uses an initial set of probabilities that each pixel belongs to each possible class, it then uses an iterative technique to update the probabilities. The initial assignments can be based on probabilistic or fuzzy classification (see Chapter 9), and many algorithms exist. The advantage of the relaxation technique is the probability values of all orientations, for each adjacent point can make a contribution to the label assigned to the given point, and not just the best choice for adjacent elements.

7.3.1.5 Feature Extraction and Classification

Many of the foregoing operations which produce transformed image data will necessitate alternative image representations, in order to reduce the amount of storage necessary within the vision system. In many cases, it is useful to represent the object by applying a 'chain coding' technique. This employs the principle of finding an arbitrary edge point, and subsequently storing only the directional changes required to follow the edge, until the starting point is eventually 'recovered'. The outline of an object can then be chain-coded as a sequence of quantised directional transitions, to trace around the edge of the object. One advantage, which results from this approach, is the ease with which some of the basic physical properties of the object can be obtained. Basic parameters, such as area, perimeter, minimum enclosing rectangle, centre of area, are readily found. Furthermore, the shape factor, which provides a measure that is invariant to both size and orientation, can be readily derived. These parameters can facilitate simple classifications, or pattern matching, in applications where the problem can be sufficiently constrained.

Perhaps the simplest approach is that of template matching, which compares an edge-processed image with stored representations of various possible objects, and ascertains which is the best match. There are many variations on the basic template matching approach, such as those which utilise correlation and convolution techniques, and the selection of an optimum approach is closely related to the degree of constraint that can be imposed upon the problem within the industrial environment.

The basic Hough transform approach enables identification of straight line segments in an image. Here colinear points within an image can be detected by transforming their coordinates in image space (x,y) to coordinates in Hough space (r, ϕ). If image points are colinear then their Hough transforms intersect at a single point in Hough space. The transform can be particularly useful if segments of a straight line are missing, such as when overlapping objects occur or when degraded images result, but limitations can arise where edges of separate objects may be identified as one. It can be invariant to rotation and scaling, and can be generalised to recognise objects that are not readily expressed as a single function of several parameters (Ballard and Browen, 1982; Duda and Hart, 1973).

Within the visual inspection category, where the matching of an image with a predefined standard is performed, the derived description of the image should contain all the shape and size information originally contained within the stored image. In addition, the description should be invariant to both position and orientation of the object; the measurement of these parameters being an important function of the visual system for subsequent manipulation. Ideally, the vision system should be able to accommodate a number of objects randomly positioned in the field of view. However, within the controlled environment of most industrial inspection tasks, it is relatively easy to ensure mechanical assortment to ensure single object presentations to the data acquisition unit.

A number of basic parameters may be derived from an arbitrary shape, to provide valuable classification and position information; these include perimeter, minimum enclosing rectangle, centre of area, and holes (number, size, position) (Figure 7.9). Measurements of area and perimeter provide simple classification criteria, which are both position and orientation invariant. The dimensionless shape factor, (perimeter^2/area), has been used as a parameter in object recognition. The centre of area is a point that may be readily determined for any object, independent of orientation, and is thus of considerable importance for recognition and location purposes. It provides the origin for the radius vector, defined as the line from the centre of area to a point on the edge of an object; these radii vectors of maximum and minimum length are potentially useful parameters for determining both identification and orientation. Holes are common features of engineering components, and the number present in a part is a further suitable parameter for identification.

7.3.2 Visual Inspection: Measurement and Inspection Applications

7.3.2.1 Dimensional Inspection

Within this category, two types of objects may be identified. The first are those applications in which objects are inspected for relatively large defects, such as missing locks in doors. The second are those objects which require close dimensional tolerance checks.

Visual techniques for dimensional inspection have been applied widely at General Motors and Ford (Wilson, 1982; German, 1985). A prototype system, INSPECTOR (Perkins, 1983), was developed for the inspection of automobile parts; each part model being formed in an interactive session prior to inspection. The model contained a set of identifying points, in order to locate the parts, a set of critical areas for inspection and a set of inspection tests. The regions for inspection were obtained by comparing 'known good parts' against 'bad parts' with a single specified defect. The tests were employed to discriminate between good and bad parts and included an examination of the difference, either in intensity, or in 'edgeness' of superimposed images.

One of the first installations was KEYSIGHT (Rossol, 1981), a CV system to inspect valve spring assemblies on engine heads for the presence of valve spring cap keys. The vision system has to inspect a complete engine head in a time less than or equal to the cycle-time of the conveyor system. KEYSIGHT stores eight pictures as the engine

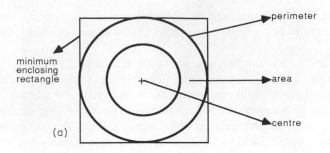

Figure 7.9(a) Parameters used in binary classification

head passes the camera, the system has eight seconds to inspect an engine head, or one second for inspecting each assembly. Since the variation in assembly location is greater than the half-width of the keys, KEYSIGHT must first find the exact location of the assembly before it can inspect for missing keys. Significant variation in the reflectances of different valve spring assemblies further complicates the task. High reliability, over thousands of cycles, is typically required in the production system.

A linear array camera, synchronised with the conveyor belt, digitises a picture of spatial resolution 64 × 64 points, with a grey-level resolution of 4 bit. Although valve spring assembly pictures show considerable variation from engine head to engine head, a strong circular pattern consistently appears under all conditions. Thus, a convolution method, based on circular symmetry, is used to find the centre of the assembly. Inspection then takes place. The program checks for the keys, by examining the intensity profile along the circumference of a circle through the keys. A threshold is calculated, using the average intensity of points along the circle. A valve spring assembly passes inspection, if the percentage of the circle that is dark lies in a specified range.

Motor car brake discs can be subjected to visual measurement to check that both sides of the disc has correct surface finish and porosity, Figure 7.10 (Batchelor *et al*., 1985). Oblique illumination is used to give a relief pattern which directly represents the surface structure. Three TV cameras are used to examine different facets of the disc. The disc is rotated in about five seconds, to allow its complete surface to be inspected. By performing a density selection process on the video signals from the cameras, it is possible to predict the surface finish and the porosity to an accuracy of 0.001 in in both cases.

Form measurement, from the profile of glass bottles or jars, has utilised a dark-ground lighting technique, as it produces better contrast pictures (Batchelor *et al*., 1985). The image processing necessitates an edge detector, followed by simple thresholding and noise removal, before finding the outer edge, Figure 7.11.

The deformation in large-scale car body panels has been detected using a single robot mounted camera (Schmidberger and Ahlers, 1984). The camera was moved to the point of interest, then the image was enhanced and converted to binary form. The resulting contour was compared to a reference to detect cracks, folds, missing holes and other defects.

For dimensional inspection, both two-dimensional linescan cameras and stereo pairs can often be employed to great effect. A high resolution linescan sensor allows close tolerance

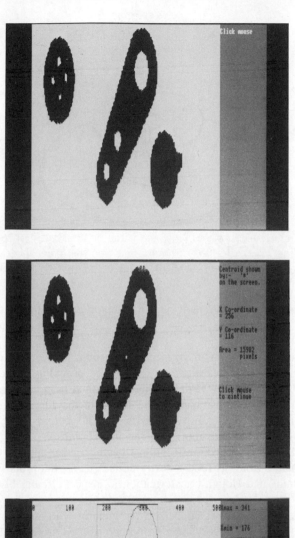

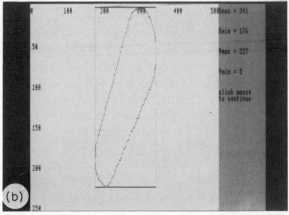

Figure 7.9(b) Binary image (top), processed to determine object area and centroid (centre), and minimise enclosing rectangle (bottom)

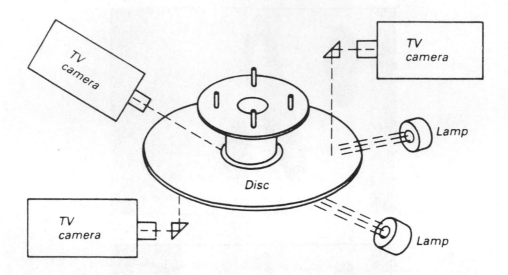

Figure 7.10 Brake drum measurement system. (Reproduced by permission of IFS Ltd.)

checks to be made, provided that movement of the object under the camera is inherent within the application. A stereo camera system, while providing more image data, may provide unacceptable accuracy by virtue of digitisation and triangulation errors.

Many visual inspection applications can be found in the literature. Examples include the inspection of machined and fabricated parts, and inspection using X-rays and ultra-sound (as well as the visible spectrum) (Kruger and Thompson, 1981; Chin and Harlow, 1983; Wallace, 1988). This inspection technique has also been applied to non-engineering industries. One example from the food industry utilises sequential processing stages of edge detection, line thinning and Hough transformation to inspect circular biscuits (Figure 7.12) (Davies, 1984).

The food-processing industry is concerned particularly with small products on rapidly moving conveyors. Moreover, a very large proportion of food products are circular; a factor which makes the generation of special inspection algorithms worthwhile. In addition, these products are often relatively thin, so the depth dimension need not complicate the inspection problem unduly. Grey-scale image capture is achieved via a line-scan camera, typically $256 \times 256 \times 8$ bit. Subsequent object location is performed by thresholding at a suitable value, followed by application of a 3×3 Sobell operator to produce estimates of edge orientation within 1°. Once the edge points of the circular object have been located, the generalised Hough transform technique is used to locate the centre of the object. The edge orientation is used to deduce the direction and possible location of the centre; the centres being located to within one pixel. This approach is capable of detecting and identifying a variety of faults, including broken and overlapping biscuits, products having only one coating of chocolate, and items having unwanted edge protruberances.

Visual inspection of instrumentation and control equipment is also well documented, (Wilder, 1983). A keyboard inspection system has been designed, which checks each

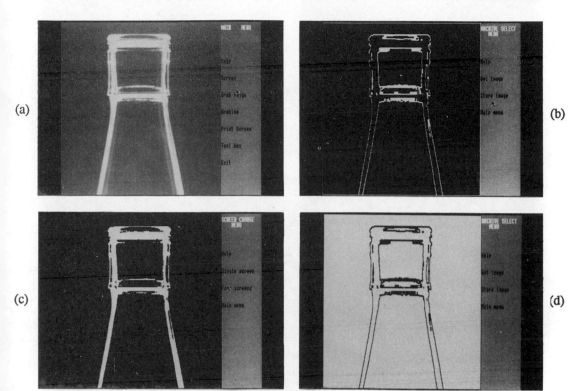

Figure 7.11 Finding the profile of a glass bottle. (a) Original image, (b) threshold applied to (a), (c) edge detected output, (d) negated version of (c)

keyboard location to ensure correct, properly oriented keys, and checks that the character depiction is not significantly degraded (Figure 7.13). The system may inspect assembled keyboards *in situ*, or inspect individual keys prior to assembly by a 'robotic station'. Where complete assemblies are inspected, it is necessary to provide an x–y table so that sufficient detail can be discriminated on each individual key. The tests included colour reflection, character size and width measurements, and two-dimensional correlation against predefined templates. Thus, a carefully controlled lighting system and accurate key alignment are prerequisite.

An on-line vision control system, designed to integrate into a flexible manufacturing system, is the AUTOVIEW system, which has an image processing unit at its centre (Atkinson, 1983). It is capable of examining real-time camera images of up to 256 light intensities. It has a range of over 200 commands, although it may only use a small number for any particular industrial application. The scope and versatility of the command set ensures that most problems can be solved quickly. AUTOVIEW performs four basic sequential operations; image acquisition, enhancement of features, extraction of features and making decisions about those features, see Figure 7.14.

Under normal operation, the system captures an image on an assembly line in about 0.1 s. It then performs a comparison of the assembly image unit with the previous image, as

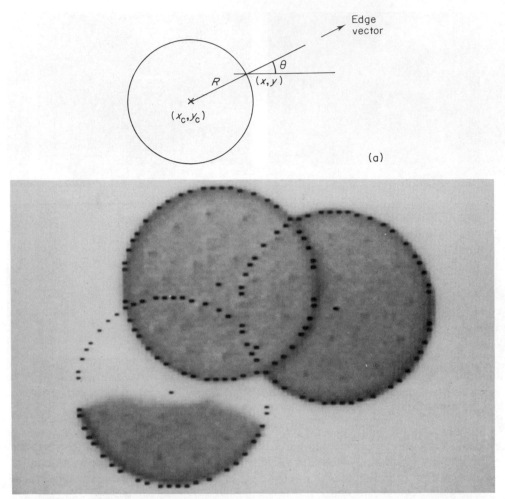

(a)

Figure 7.12 Inspection of biscuits. (a) Geometry of calculation, (b) superimposed edges of overlapping and broken biscuits. (Reproduced by permission of IFS Ltd.)

captured in the last successful cycle. It examines the key areas for any differences between the two images. If no differences are found, the assembly is passed, but if the differences exceed the acceptable level, then the assembly is rejected. The advantage of this method, is that the master image used for comparison is updated each time an assembly is accepted, thus gradual changes in ambient light conditions or position will not affect the process.

7.3.2.2 Surface Inspection

Automated visual inspection techniques have been applied to the surface inspection of a variety of materials including steel, wood and textile, although the majority of applications involve metallic materials at various stages of surface finish. The task of surface inspection may overlap with that of vision-controlled assembly, as surface defects may be removed

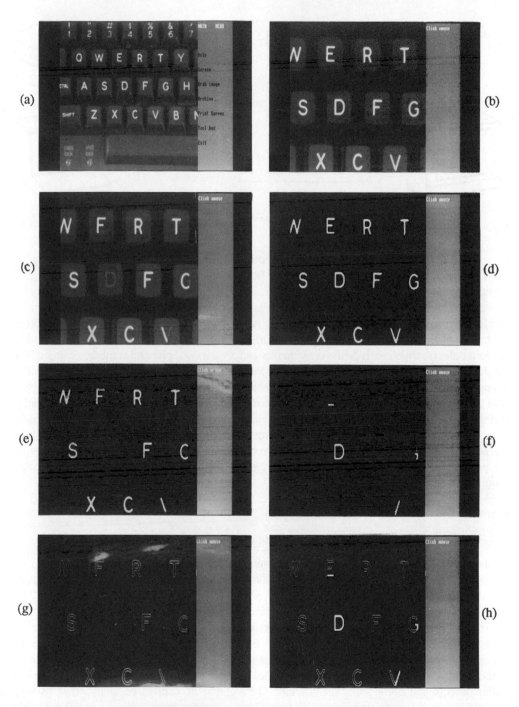

Figure 7.13 Inspection of keyboards. (a) General view of keyboard, (b) original image of good layout, (c) original image of faulty keyboard, (d) threshold version of (b), (e) threshold version of (c), (f) binary difference between (d) and (e), (g) edge detected version of (e), (h) binary addition of (f) and (g)

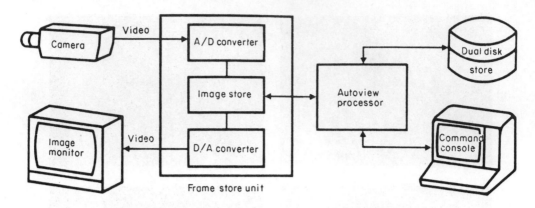

Figure 7.14 The AUTOVIEW system

by a surface finishing tool attached to a robotic manipulator. Defects in castings may be raised from the surfaces in swarf remains or burrs, or indented into the surface as pits and scratches.

Surface finishing operations on metal castings are not easy to automate, due to the irregular and random nature of the defects, and usually necessitates the use of some form of adaptive control mechanism. One such inspection system (Graham *et al.*, 1983) identifies and locates random defects, which include die craze marks and a variety of pits and scratches using a low resolution camera (Figure 7.15). The identification of a defect is dependent on correlating intensity variations within the image with local variations in reflectivity and topography. This is complicated by the fact that the part eometries are not easily illuminated uniformly, and the reflectivity of the whole surface area changes considerably as the abrasion proceeds. Thus, extreme conditions of shadow and direct specular reflection may occur which will mask the presence of defects. The use of a computer model of the component permits the intelligent selection of viewing geometry, which helps to overcome problems in image acquisition and analysis.

7.3.2.3 Electronic Circuit and Layout Inspection

Printed circuit board (pcb) inspection represents a significant application area for visual techniques. Generally such inspection is two-dimensional, and involves high precision optics and scanning systems.

Systems have been designed to inspect printed circuit boards (pcbs) before and after the insertion of electronic components, in order to detect both manufacturing and assembly faults. Typical pcb faults include missing tracks and pads, broken tracks and small indentations within tracks, and reduced substrate or track width caused by imperfections in the etching process. Typical assembly faults include drilling defects, incorrect insertion, solder defects and component substitution. The volume of data to be analysed is usually reduced by a windowing technique, whose position is determined accurately on the scanning system.

Generally, two approaches can be used. The first is a single camera detection method, in

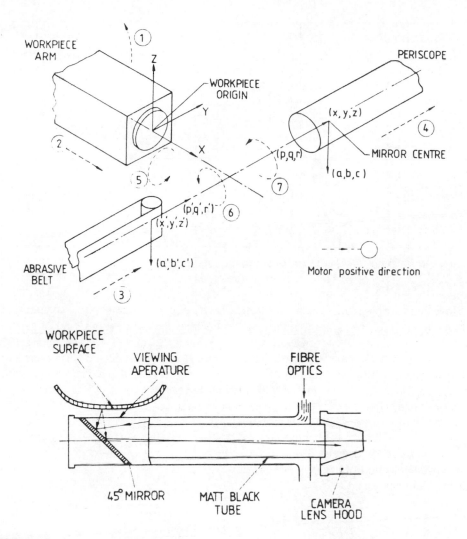

Figure 7.15 Image acquisition system

which the inspected pattern has to conform to specified rules. The second is a comparison method, in which the inspected pattern is compared with either a design pattern or a reference pattern. In the inspection system depicted in Figure 7.16(a) (Hara *et al*., 1983), a comparison technique is employed. The pcb is illuminated both vertically and diagonally by a system of lenses and mirrors, in order to produce near perfect images for the task in hand. Corresponding portions of the circuit pattern are projected onto a pair of linear image sensors, and the signals fed to the defect recognition circuit. The recognition hardware performs boundary line extraction and extraction of fine patterns, as shown in Figure 7.16(b). When the corresponding points on the reference pattern and board pattern exhibits the same features, the pattern is determined as having no defects (see Figure 7.16).

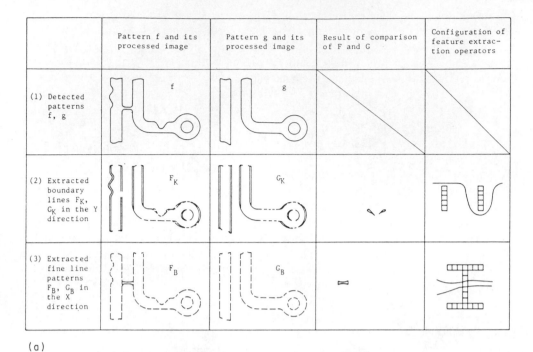

(a)

(b)

Figure 7.16 Inspection of printed circuit boards. (a) Schematic structure of the inspection system, (b) Extraction of features and comparison of the extracted feature patterns (from Hara *et al.*, 1983). (Reproduced by permission of IEEE © 1983 IEEE.)

7.4 ROBOTIC GUIDANCE

7.4.1 Three-dimensional Representations: Model-based Recognition

Most of the current generation of factory robots operate within a highly constrained environment, and therefore the initial object location prior to any inspection or assembly operation is relatively straightforward. In the future, it is probable that more flexible

CV systems will be developed with abilities to operate reliably and repeatably within a completely unstructured environment, and locate objects within three-dimensional (three-dimensional) space using appropriate three-dimensional imaging techniques. In addition, many of the low-level processing operations will have to be implemented in hardware in order to produce an acceptable speed of operation for the robot effector mechanism.

7.4.1.1 Image Representation

For a three-dimensional problem, the recognition process is one of relating representations of objects in an image with representations derived from models in memory. It is relatively easy to describe a range of two-dimensional image features, which result from projections of three-dimensional objects onto the two-dimensional image plane. Many of the above visual inspection problems are in themselves essentially two-dimensional in nature, and in other cases, the depth information available from a three-dimensional scene can be used to describe a range of salient two-dimensional features. Such a process can be executed using a top–down or *knowledge-driven* approach.

However, the inverse problem presents extreme difficulty; there is no clear way of transforming a two-dimensional image to produce a three-dimensional object model. The depth information has been lost, and any three-dimensional reconstruction is problematical. Current approaches to recapture the depth information, attempt to extract shape from the image through the use of stereo, motion, shading, texture, etc., and many of these issues are currently the focus of vision research within the A1 community. Marr's theory of vision provides a connection between the top–down knowledge-based path and the bottom–up perceptual path (Marr, 1982). To facilitate this connection, multiple data representations have been suggested (see Figure 7.17).

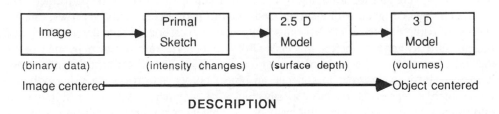

Figure 7.17 Representations required for image understanding

Here the term 'primal sketch' refers to intensity changes in the image which identifies the edges of objects. The 2.5 dimensional sketch describes depth structure and other characteristics of visible surface elements, both viewer centred. The three-dimensional model describes volumetric shape primitives of varying size and position, defined using an object centred coordinate system. Within the industrial domain, CV can be regarded as the act of matching perceptual primitives to an existing model however crude that model may be (Murray, 1988).

Regardless of how the models are organised and what particular model representations are chosen, the recognition process is essentially one of matching image data and stored model, either by some optimisation technique or by graph traversal. The three-dimensional nature of the problem means that a dramatic reduction in the size of the search space is necessary for industrial CV applications.

Three approaches to model-based recognition are readily identified (Grimson and Lozano-Perez, 1984):

(1) Two-dimensional models are matched with two-dimensional image data. Systems that utilise this approach assume that the third dimension is relatively small in relation to the other two. In other words, two-dimensional modelling is most applicable to objects that appear flat and are viewed in a plane that is parallel to the image plane of the camera. (Bolles and Cain, 1982)

(2) Three-dimensional models are matched with three-dimensional image data. This is most appropriate when an arbitrary viewing angle is desired, and the objects are more complex in nature than those assumed in (1).

(3) Hybrid systems, in which three-dimensional objects are represented as a number of two-dimensional models, or three-dimensional models are matched with two-dimensional image data. Difficulties associated with the stereo correspondence problem are alleviated, but exact object location can be problematical.

7.4.1.2 Two-dimensional Modelling and Matching

HYPER (HYpothesis Predicted and Evaluated Recursively) (Ayache and Faugeras, 1986) uses two-dimensional models of industrial parts to match against two-dimensional image data (although the authors state that the extension to three-dimensional analysis could be implemented without too many difficulties). The system has been coupled to a robot arm, in order that picking and repositioning of overlapped industrial parts can be performed. Thus the system was designed with a strong robustness to partial occlusions, shadows, touching and overlapping objects. HYPER begins analysis by generating a scene description (which is also used to generate object models). Object boundaries are detected by one of two methods, selected on the basis of lighting conditions. A list of connected border points is then determined, in order to generate an object border approximation using polygons. Using such a representation has the advantages of simplicity, with the result that fast implementation using commercially available equipment is possible, and local description. This results in different parts of the object being described independently of each other—crucial for identifying partially occluded objects.

Model-to-scene matching is initiated by considering a number of salient or privileged model features, selected on the bases of line length. Objects are identified and located by initially generating hypotheses. By matching the privileged model segments to the scene description, a few hundred hypotheses are usually generated and ranked on the basis of a local criterion of merit. The best of these hypotheses are used to predict object locations, and appropriate model transforms are calculated. Additional matches between the scene description and the transformed model may then be determined; the transform being iteratively refined. The matching process terminates either when a sufficient number of the high ranking hypotheses have been evaluated, or when a high degree of match between lengths of the identified scene segments, and the relative lengths of model segment lengths, has been obtained. The best hypothesis is finally re-examined before being accepted or rejected.

7.4.1.3 Three-dimensional Modelling and Matching

An example of a recently developed system that matches three-dimensional models against three-dimensional image data, is known as TINA (Porrill *et al.*, 1987). This is a three-dimensional vision system designed for an industrial pick-and-place robot. It is designed to support model-based recognition, and location of objects in a cluttered working environment, enabling a 'grasp plan' for the object to be computed.

Object location begins by recovering a depth map from two cameras positioned to one side of the working area. A Canny edge detector is used on both images before the edge-based depth map is recovered, by using the PMF stereo algorithm (Pollard *et al.*, 1985). The edge-based depth map is then used to recover a description of the scene in terms of straight lines and circular arcs. Instead of discarding any matches between scene description, they are used to form a composite two-dimensional and three-dimensional scene description, thereby forming a more complete description than using the stereo data alone.

Object models are matched to the scene description, by initially choosing a 'focus feature' from the model, and selecting a number of neighbouring salient features; the saliency of a feature being based on line length. From these matches, various model transforms are determined before the best match is selected.

In addition to object identification and location, TINA is also able to produce object models, by modelling objects in terms of straight lines and circular arcs, from a sequence of views. As the view is updated, features from the previous view are matched to ones from the current view. Novel features from the current view are added to the model. Only features that have appeared in more than one view are retained in the final model.

7.4.1.4 Three-dimensional Modelling and Matching to Two-dimensional Image Data

ACRONYM (Brooks, 1983) was designed to interpret two-dimensional scenes using three-dimensional models, and was applied to aerial image scenes of airports. Models of commercial aircraft were defined by using Binford's generalised cones (Figure 7.18) (Binford, 1971). A generalised cone, sometimes known as a generalised cylinder, is defined by sweeping a two-dimensional cross-section along a one-dimensional line or spline, according to some sweeping rule. The sweeping rule defines how the cross-section is changed as it is swept along the spine.

Object identification was performed in ACRONYM, by matching object models to special low level shape primitives from the two-dimensional image data (Figures 7.19 and 7.20). An edge detector is passed over the image before a line-fitting operation is performed on the resulting edges. These lines are then used to form ellipses and 'ribbon' shape primitives. Ellipses are useful for representing the ends of generalised cones, while ribbons are essentially two-dimensional strip-like projections of generalised cones (restricted to straight spines and linear sweeping rules in the ACRONYM implementation).

Matching is performed by predicting possible object appearances in the image and computing associated constraints on the model. These constraints are then used for object

Human three-dimensional model

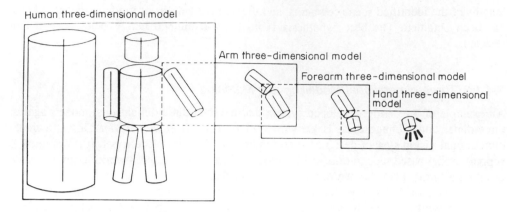

Figure 7.18 A hierarchical representation scheme based on generalised cylinders

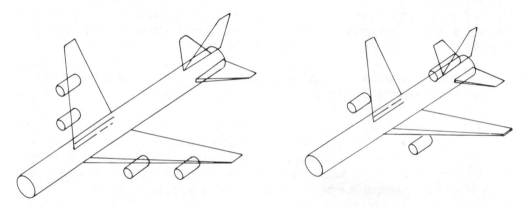

Figure 7.19 The ACRONYM system. Use of generalised cylinders as a primitive to describe components of models (from Hara *et al.*, 1983). (Reproduced by permission of IEEE © 1983 IEEE)

prediction and interpretation of the shape primitives. As objects are modelled in terms of object class, the most general set of constraints concerning the class of object (wide-bodied aircraft in the case of airport scenes) are initially applied when carrying out initial prediction and interpretation. Interpretation is achieved by using both geometric and symbolic reasoning. A consistent or partial match with a model initiates a subclass check (e.g. 'Boeing 747' or 'L-1011'), by applying the extra subclass constraints. Throughout the prediction and interpretation process, ACRONYM does not utilise any special knowledge of airport scenes; all its reasoning is based on geometric reasoning.

A recent approach to three-dimensional object recognition which does not attempt to reconstruct depth information (2.5-dimensional sketch), but instead uses the image properties directly (edges and corners), has been proposed by Lowe (1987). This alternative approach focusses on the process of perceptual organisation, which detects groupings and structures in the image that are likely to be invariant over a wide range of viewpoints. Although the appearance of a three-dimensional object can change markedly when viewed from different angles and positions, certain key attributes such as corners and edges will remain

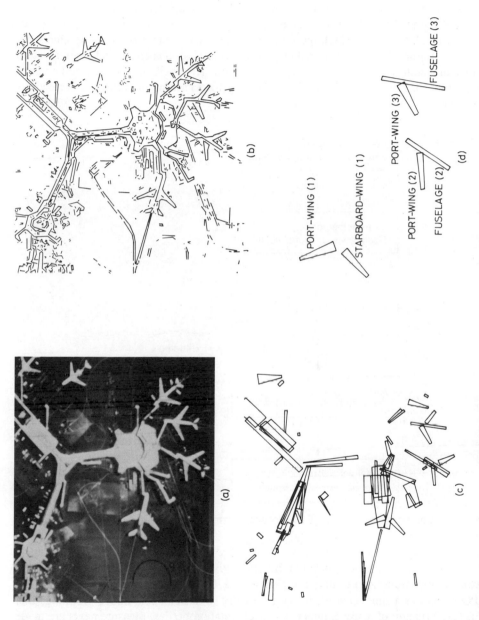

Figure 7.20 The ACRONYM image interpretation. (a) The original image, (b) the edge map of (a), (c) fitted ribbons and ellipses, and (d) identified aircraft parts (from Hara *et al.*, 1983). (Reproduced by permission of IEEE © 1983 IEEE.)

dominant. The role of perceptual organisation is to detect the most reliable occurrence of these attributes, as instances of proximity, parallelism and connectivity, in order that they can be matched to the corresponding three-dimensional models stored within the system database.

This concept has been incorporated into the SCERPO (Spatial Correspondence Evidential Reasoning Perceptual Organisation) vision system (Lowe, 1987) which recognises known three-dimensional objects in single grey-scale images. The main system components are identified in Figure 7.21, which includes low-level edge detection and segmentation operations, preceding the perceptual grouping operation. The figure also illustrates the match and verification processes which are used iteratively to determine the probable correctness of any match.

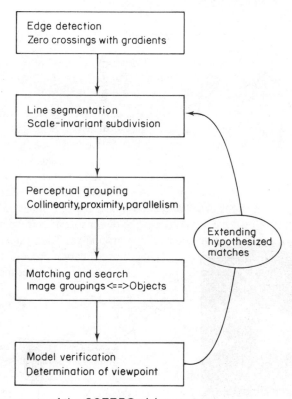

Figure 7.21   Components of the SCERPO vision system

The process of perceptual organisation in human vision results in spontaneous formation and detection of groupings, illustrated in Figure 7.22. This process is incorporated into the SCERPO system as a functional mathematical theory to detect stable image groupings that reflect actual structure of a scene, rather than accidental properties. Measurements are made to calculate the probability that instances of proximity, parallelism and colinearity arise from object features in the image, and not by accident from randomly distributed line elements. Proximity calculations are based on the expected number of end points within a radius of a given occurrences of parallelism and colinearity.

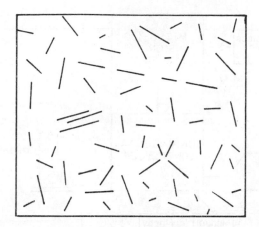

Figure 7.22 Perceptual organisation. This figure illustrates the human ability to identify non-random groupings, resulting from parallelism, colinearity and endpoint proximity (connectivity)

The matching process consists of comparing each of the above perceptual groupings detected within the image with each of the three-dimensional object model structures that could give rise to such groupings. To reduce the size of the search space, more complex structures can be defined, by combining the initial groupings into larger features. SCERPO uses the complexity of groupings as an estimate for the importance of features to be matched.

The final stage is a verification procedure, which utilises a viewpoint consistency constraint whereby the locations of all projected model features must be consistent with the projection from a single viewpoint. This checks the direct spatial correspondence between the three-dimensional model and the matched image groupings. The system utilised both line-to-line and point-to-point correspondence. Once the verification process has been performed, the initial match estimates can be used to predict the locations of other model features within the image and hence extend the matching procedure.

7.4.2 Visual Guidance Applications

The CV hierarchy, referred to earlier in this chapter, has already demonstrated that modular interdependence can be exploited to achieve optimum solutions. A grey-scale image necessitates more complex processing than binary images, and consequently binary image processing has proved more cost effective, and has therefore become more widely used within industry, where the environment can often be appropriately constrained. However, the greater detail inherent within a grey-scale image may be required for certain inspection operations. Here it is possible to utilise the modular interdependence and use binary images for the initial part location and orientation, followed by grey-scale processing for the more detailed analysis.

7.4.2.1 Manufacturing

Many systems perform the large-scale handling of manufactured items on the basis of a 'look then move' principle, starting with binary images or part silhouettes. The PUMA/VS100

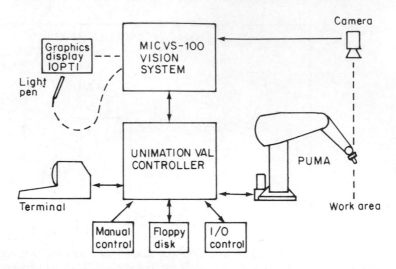

Figure 7.23 The PUMA/VS-100 robot system (from Pugh, 1983). (Reproduced by permission of IFS Ltd.)

robot vision system (Carlisle *et al*., 1983), integrates a multidegree of freedom manipulator arm, a binary image-processing system, and a computer communication interface (see Figure 7.23).

In this system, the vision module locates objects and communicates their position to the robot for manipulation. The location is specified as the two-dimensional centroid of the silhouette. The vision system characterises the objects on the basis of such features as area, perimeter, minimum and maximum radii, and number of holes. To recognise objects, the system has to be trained, using sample objects, which are captured at differing positions and orientation, in order that the object's visual features can be determined and stored.

The system incorporates a light pen controlled menu for ease of operation, direct connection to any suitable camera, flexible interfacing for remote control, and the system has been designed to cope with a rugged industrial environment. It has the ability to acquire randomly located workpieces and hence reduces jigging, fixturing and part presentation requirements. The system is most effective when parts can be identified by their silhouettes, and when extreme accuracy is not required.

The most well-known vision system is CONSIGHT (Holland *et al*., 1986; Baumann and Wilmshurst, 1983), developed and installed at General Motors. This system is capable of determining the position and orientation of parts randomly placed on a moving conveyor belt, and subsequently controlling a robot arm to track the parts and transfer them to a predetermined location (see Figure 7.24).

CONSIGHT uses a structured lighting approach, in which two linear light sources project sheets of light at 15° on either side of a linear diode array camera, which is mounted some two metres above the light housing. When a part passes through the line of light, the beams from the two sources are deflected out of the camera's field of view. The scanning action of the camera is able to determine the locations of discontinuities (edges or holes) as the part moves through the field of view. In this way, data is accumulated from which the vision system software constructs an image, calculates appropriate parameters (area, number of

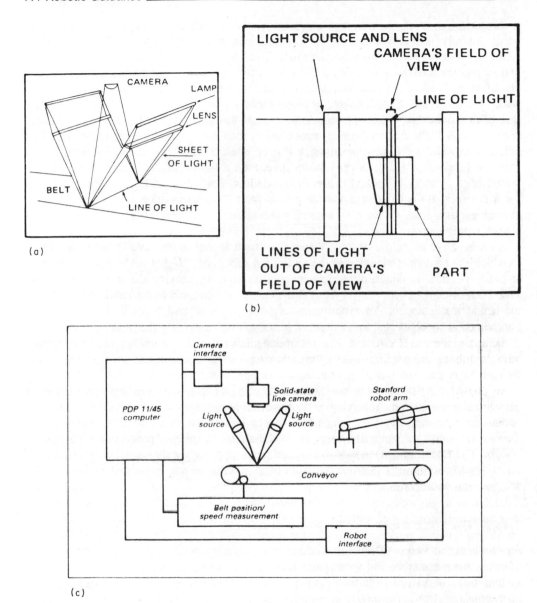

Figure 7.24 The CONSIGHT System. (a) Illumination system, (b) camera view and (c) hardware schematic (from Pugh, 1983). (Reproduced by permission of IFS Ltd.)

holes etc), and compares these with internal library of descriptions. When a match is found, the system calculates the angular position of the part on the conveyor belt and locates the pick-up point for use by the robot.

The above examples utilise binary image processing. Grey-level processing is more complex, although it can resolve ambiguities of orientation. One application of grey-scale analysis is that of recognising stable states of engine connecting rods (Faedo *et al.*, 1983).

A three-stage analysis is performed sequentially, and at each stage unique recognition may occur. Initially, a small window (16 × 16 pixels) within the total image (256 × 256 pixels) locate a small portion of the workpiece which uniquely defines the two stable faces. The first step in the analysis is an evaluation of the two-dimensional grey-scale histogram which identifies the background grey level (highest histogram peak), and object grey level (smaller peak). It also reveals a small deviation in grey levels, from a derived mean value, if the flat face of the connecting rod is uppermost and a larger deviation in grey levels if the boss face is uppermost. If the first analysis is inconclusive, then a second similar analysis is carried out, this time in one dimension along a line of interest within the located object. If this grey-level deviation again proves inconclusive, then a third analysis is performed. Here a Sobell edge operator is applied to the small window used previously. An edge is detected for a flat face, no edge is detected for a boss face. This three-stage analysis reduces the overall test time since the Sobell windowing operation is only used if the first two grey-level analyses prove inconclusive.

An alternative to the use of fixed camera positions is that of the 'eye-in-hand' approach, in which the camera is mounted at the end of the robot arm. Higher resolution is obtained, and it is possible to obtain different viewpoints simply by moving the manipulator hand. The robot-to-camera geometry is fixed if the camera is mounted in the robot that performs the actual manipulation. An experimental system (Agrawal and Epstein, 1983) utilises a photodiode array connected to coherent fibre-optic bundles to carry the light from the object to the array interface electronics. This separation allows a small scanner head to be integrated with the robot's gripper mechanism, thus allowing it to access positions which are normally difficult to reach. The scanning head and fibre optics are lightweight, and the fibre-optic cable provides good electrical noise immunity. Use of different interconnection patterns between the sensing and receiving end of the fibres provides the opportunity for simple geometric preprocessing. This can in turn reduce the amount of subsequent processing in the vision system, and again illustrates the modular interdependence properties of the overall system. The main disadvantage of the 'eye-in-hand' approach is the complexity of analysis necessary to recognise objects on the basis of three-dimensional models at variable scales and viewing coordinates.

7.4.2.2 Electronic Assembly

Assembly is one step in the manufacturing process which typically includes the presentation of parts, their assembly and subsequent joining or bonding together. Thus, any assembly system must successfully interface to the parts storage facility and to the subsequent joining operations. System performance is evaluated in terms of its speed, throughput, reliability and resulting product yield, as well as its impact on other aspects of the manufacturing process. Automated assembly of electronic components into printed circuit boards usually results in a large number of different components being handled, discrete and integrated, encapsulated in a variety of packages. The positioning of such components has to be precise (within 0.01 in), in order to achieve insertion of the component leads into holes in the printed circuit board. The requirement for high yield and reliability are paramount, owing to the high investment necessary to produce the board itself, and the inability to carry out full functional testing until the soldering is complete. The cost of reworking a fully assembled printed circuit board can be prohibitive.

A flexibility electronic assembly station has been developed by Westinghouse (Sanderson and Perry, 1983), in which the assembly is completed in accordance with 'mil-spec' requirements. The vision sensors within the system utilise binary processing to determine the location and orientation of principal axis of components, and grey-scale processing to estimate orientations of multilead components. The grey-level algorithms used are all model-based, since electronic components conform to precisely known shapes and sizes. However, it is more difficult to characterise the surface properties of components which have a wide variety of colour, texture, specularity and marking. The vision system is fully integrated with other sensor systems: the robot control system and overall supervisory computer which communicates between other areas of the manufacturing plant and other electronic assembly stations.

A system for unpacking and mounting television deflection units, using visually controlled robots, have been developed by Philips (Saraga *et al.*, 1983); the deflection units being loosely constrained in a large cardboard box. To reduce the vision preprocessing time, the scene constraints in this case have dictated the use of a television camera equipped with parallel projection optics. Such optics produce an image size independent of object distance, whilst providing immunity to ambient illumination, ease of calibration, and elimination of parallel errors. A number of sequential processing steps are performed to extract specific area and edge features, which are then used to determine the position and orientation of the deflection units within the box.

The second stage of assembly, involves placing the deflection units correctly into position on the neck of the tube. To facilitate this, three cameras are used, one below the tube conveyor belt. The upward-facing camera, equipped with parallel projection optics, locates the position of the tube on the conveyor belt. The neck of the tube is located by the other two cameras, using techniques of image thresholding, at a level which gives the correct neck width in the resulting binary image. The correspondence between these two binary images gives the exact location of the neck top. The vision system is fully integrated with the robot control system, to produce a flexible assembly unit.

7.4.2.3 Robot Bin-picking

The bin-picking problem has been described as one of the classic CV problems (Kelley *et al.*, 1983a and b; Bolles, 1986). The problem is for a robot to pick the top, or most easily accessible component, from a completely disordered pile within a bin, and to present the component to the next process that requires it. Locating the top-most component immediately suggests a three-dimensional element in the analysis. The main difficulty in this problem is that parts touch and overlap each other, and therefore cannot be 'seen' as individual objects by the vision system. In addition, there is generally poor contrast in illumination between the parts and their background, thus making it difficult to follow the outline of any of the components. From a systemic viewpoint it could be argued that, within a fully automated manufacturing plant, such random storage of components should not be allowed. However, this problem arises throughout industry. From the modular interdependence viewpoint, the more degrees of freedom that can be constrained, the easier it should be to apply standard vision analysis techniques. A fully constrained problem has precisely defined component positions and orientations, but unfortunately provides little or no flexibility for the operators.

The 'i-bot' approach (Ray and Wilder, 1983) to the bin-picking problem, uses binary processing for shrinking operations, in order to identify regions corresponding to smooth surfaces. It then uses grey-scale processing to determine edge data for identification of opposing parallel edges, where a gripper could be placed. Another approach (Ledoux and Bogaert, 1984) is to separate the object, localisation and recognition steps. The object is first located by processing the image to identify suitable edges (opposing faces) which are suitable for gripping, to remove the part from the bin. The second operation is then carried out, with this component in isolation from the bin on a separate surface, where it is viewed by a camera under controlled lighting conditions.

7.5 SUMMARY

This chapter has introduced the topic, and problems associated with CV. It has discussed the issues important and relevant in industry, and it has cited some significant industrial systems in use today. It is certainly true that the whole area of image acquisition, processing and scene understanding will play an increasingly important role in the future, and in all the technical processes involved, use will be made of embedded knowledge. This will take the form of knowledge to assist in the identification and extraction of significant image features, intelligent image classification, and finally the actual scene interpretation.

The problems faced by developers of AI-based vision systems are many, but some of them deserve a mention:

- Speed of operation. This is a problem not confined to vision of course, but it is perhaps more of a critical component in industrial vision systems.

- Type of knowledge and how it must be represented. In a system employing knowledge at various levels, i.e. image processing, right through to image understanding, relevant knowledge may include physical (geometric, etc.), logical and heuristic factors. As before, the representation problem is a general one in AI, and Chapters 2 and 8 of this book include a comprehensive treatment of the more important issues.

- The mixing of technology. Realistic systems will inevitably be hybrid in nature, employing a variety of technology best suited to a particular task (or subtask). AI represents an interesting challenge, but may not be appropriate at all levels within an 'integrated vision system'.

- Economic viability. Industry will only utilise computer vision equipment when it can perform a given task more accurately, reliably and effectively than an employee. For this reason many of the current research advances will take some time to filter down into the real world of engineering.

All these factors represent crucial decisions. However, advances in both hardware and software continue push AI techniques into the forefront of the computer vision arena, where it surely has a place.

BIBLIOGRAPHY

Agrawal, A. and Epstein, M. (1983) Robot Eye-in-Hand using Fibre Optics, *Proc. Third Int. Conf. on Robot Vision and Sensory Controls*, IFS Publications, Cambridge, MA, pp 257–62.
Atkinson, B. M. (1983) Some Applications of On-line Vision Sensing in Industry, *Proc.*

Third Int. Conf. on Robot Vision and Sensory Controls, IFS Publications, Cambridge, MA, pp 503–8.

Ayache, N. and Faugeras, O. D. (1986) HYPER: A New Approach for the Recognition and Positioning of Two-Dimensional Objects, *IEEE Trans.* **PAMI-18**.

Ballard, D. H. and Browen, C. M. (1982) *Computer Vision*, Prentice Hall, New York.

Batchelor, B. G., Hill, D.A. and Hodgson, D.C. (1985) *Automated Visual Inspection*, IFS Publications.

Baumann, R. D. and Wilmshurst, D. A. (1983) Vision System Sorts Castrings at General Motors Canada, In: *Robot Vision*, Ed. A. Pugh), IFS Publications, pp 256–66.

Binford, T. (1971) Visual Perception by Computer, *Proc. IEE Conf. on Systems & Control, Miami, FL. Dec.*, IEEE, New York.

Bolles, R. C. (1986) Three Dimensional Locating of Industrial Parts, In: *Robot Vision*, Vol 1, Ed. A. Pugh, IFS Publications, pp 245–53.

Bolles, R. C. and Cain, R. A. (1982) Recognising and Locating Partially Visible Objects: The Local-Feature-Focus Method, *International Journal of Robotics Research*, **1**, 57–82.

Brooks, R. A. (1983) Model Based Three-Dimensional Interpretations of Two-Dimensional Images, *IEEE Trans.*, **PAMI-5**, 140–150.

Carlisle, B. et al. (1983) The PUMA/VS100 Robot Vision System, In: *Robot Vision* Ed. A. Pugh, IFS Publications, pp 313–23.

Chin, R. T. and Harlow, C. A. (1983) Automated Visual Inspection: A Survey, *IEEE Trans.*, **PAMI-4**, 557–573.

Davies, E. R. (1984) Design of Cost-Effective Systems for the Inspection of Certain Food Products during Manufacture, *Proc. Fourth Int. Conf. on Robot Vision and Sensory Controls*, IFS Publications, London, pp 437–46.

Duda, R. O. and Hart, P. E. (1973) *Pattern Classification and Scene Analysis*, Wiley, Chichester.

Faedo, W., Foulloy, L. and Kelley, R.B. (1983) Three Stage Asgorithm for Recognising Stable States and Reorienting a Workpiece, *Proc. Third Int. Conf. on Robot Vision and Sensory Controls*, IFS Publications, Cambridge, MA, pp 461–69.

Fountain, T. (1987) *Processor Arrays*, Academic Press, New York.

Geisselmann, H., Ossenberg, K., Niepold, R. and Tropf, H. (1987) Vision Systems in Industry: Application Examples, *Robotics*, Vol. 2, Elsevier, Amsterdam.

German, D. (1985) Vision Aided Inspection Techniques for the Automotive Industry, *Proc. Soc. Manuf. Eng. Conf. on Vision*, Detroit, MI, Section 6, pp 1–7.

Gonzalez, R. C. and Wintz, P. (1987) *Digital Image Processing*, 2nd edn, Addison Wesley, Reading, MA.

Graham, D., Jenkins, S.A. and Woodwark, J. R. (1983) Model Driven Vision to Control a Surface Finishing Robot, *Proc. Third Int. Conf. on Robot Vision and Sensory Controls*, IFS Publications, Cambridge, MA, pp 433–40.

Grimson, W. E. L. and Lozano-Perez, T. (1984) Model Based Recognition and Localization from Sparse Range or Tactile Data, *International Journal of Robotics Research*, **3**, 3–35.

Hara, Y., Akiyama, N. and Karasaki, K. (1983) Automatic Inspection System for Printed Circuit Boards, *IEEE Trans.*, **PAMI-5**, 623–630.

Holland, S. W., Rossel, L. and Ward, M. R . (1986) Consight 1: A Vision Controlled Robot System for Transferring Parts from Belt Conveyors, In: *Robot Sensors*, Vol.1, Ed. A. Pugh, IFS Publications, pp 213–28.

Hollinghum, J. (1984) *Machine Vision*, IFS Publications.

Horn, B. K. P. (1979) Artificial Intelligence and the Science of Image Understanding, In: *Computer Vision and Sensor-Based Robots*, Ed. G. Dodd and L. Rossel, Plenum Press, New York.

Hunt, V. D. (1985) *Smart Robots*, Chapman & Hall, London.

Kelley, R. B., Birk, J., Dessimoz, J., Martins, H. and Tella, R. (1983a) A Robot System which Aquires Cylindrical Workpieces from Bins, In: *Robot Vision*, Ed. A. Pugh, IFS Publications, pp 225–44.

Kelley, R. B., Birk, J., Dessimoz, J., Martins, H. and Tella, R. (1983b) Forging: Feasible Robotic Techniques, In: *Robot Vision*, Ed. A. Pugh, IFS Publications, pp 286–94.

Kittler, J. and Duff, M. J. B. (1985) *Image Processing System Architectures*, RSP.

Kruger, R. P. and Thompson, W. B. (1981) A Technical and Economic Assessment of Computer Vision for Industrial Inspection and Robotic Control, *Proc. IEEE*, **69**, 1524–1538.

Ledoux, O. and Bogaert, M. (1984) Progmatic Approach to the Bin Picking Problem, *Proc. Fourth Int. Conf. on Robot Vision and Sensory Controls*, IFS Publications, London, pp 313–23.

Loughlin, C. (1987) Automatix Provides the Real Success for Austin Rover, *The Industrial Robot*, **14** 145–148.

Lowe, D. G. (1987) Three-Dimensional Object Recognition from Simple Two-Dimensional Images, *Artificial Intelligence*, **31**, 355–395.

Marr, D. (1982) *Vision*, Freeman, New York.

Murray, D. W. (1988) Strategies for Object Recognition, *GEC Journal of Research*, **16**, 80–95.

Nitzan, D. et al. (1979) Machine Intelligence Research Applied to Industrial Automation, SRI Project Reports 1979–82.

Perkins, W. A. (1983) Inspector: A Computer Vision System Which Learns to Inspect Parts, *IEEE Trans.*, **PAMI-5**, 584–592.

Pollard, S. B., Mayhew, J. E. W. and Frisby, J .P. (1985) PMF: A Stereo Correspondence Algorithm using a Disparity Gradient, *Perception*, **14**, 449–470.

Porrill, J., Pollard, S.B., Pridmore, T.P., Bowen, J.B., Mayhew, J.W. and Frisby, J.P. (1987) TINA: A 3D Vision System for Pick and Place, *Proc. Third AVC Cambridge UK, Alvey Vision Club*, pp 65–72.

Ray, R. and Wilder, J. (1983) Robotic Acquisition of Jumvbled Parts from Bins by Visual and Tactile Guidance, *Proc. Third Int. Conf. on Robot Vision and Sensory Controls*, pp 441–451.

Rossol, L. (1981) Vision and Adaptive Robots in General Motors, *Proc. First Int. Conf. on Robot Vision and Sensory Controls*, IFS Publications, Stratford-upon-Avon, pp 227–228.

Sanderson, A. C. and Perry, G. (1983) Sensor Based Robotic Assembly Syustems: Research and Applications in Electronic Manufacturing, *Proc. IEEE*, **71**, 541–548.

Saraga, P., Newcomb, C.V., Lloyd, P.R., Humphreys, D.R. and Burnett, D.J. (1983) Unpacking and Mounting TV Deflection Units using Visually Controlled Robots, *Proc. Third Int. Conf. on Robot Vision and Sensory Controls*, IFS Publications, Cambridge, MA, pp 541–548.

Schmidberger, E.J. and Ahlers, R.J. (1984) Quality Control with a Robot Guided Electro-optical Sensor, *Proc. Fourth Int. Conf. on Robot Vision and Sensory Controls*, IFS Publications, London, pp 27–36.

Sheldon, K. (1987) Probing Space by Camera, *Byte* **12**, 143–148.

Wallace, A. M. (1988) Industrial Applications of Computer Vision since 1982, *Proc. IEE, Pt-E*, **135**(3).

White, K. (1988) Automated Inspection Comes Into View, *Industrial Computing*,(Sept), pp 29–32.

Wilder, J. A. (1983) A Machine Vision System for Inspection of Keyboards, *Signal Press*, **5**, 413–442.

Wilson, W. S. (1982) The Role of Vision in a Dimensional Control Strategy, *Proc. Soc. Manuf. Eng. Conf. on Vision*, Detroit, MI, pp 160–67.

Wright, P.K. and Bourne, D.A. (1988) *Manufacturing Intelligence*, Addison Wesley, Reading, MA.

R. Thomas
Department of Electrical and Electronic Engineering
Brighton Polytechnic
Lewes Road
Brighton
East Sussex BN2 4GJ

8 Developments in Artificial Intelligence Reasoning

A.B. Hunter

Imperial College, UK

8.1 INTRODUCTION

Much of the attention of artificial intelligence (AI) has been at the symbolic level. This is the level at which the representation and manipulation of information or knowledge is in the form of tangible concepts such as dog, car, house, red, loud, happy, etc. Symbolic AI requires the manipulation of such symbols in order to satisfy goals of the system. From this, AI reasoning can be interpreted to mean the formalisms for representing and manipulating symbolic level knowledge. So for example it would be a straightforward exercise to implement an AI system to represent the following knowledge:

> All dogs are mammals.
> Fido is a dog.

and to reason with this knowledge so that for example we could derive:

> Fido is a mammal.

Production rule systems, such as OOPS5, or logic programming systems, such as Prolog, would be obvious starting points for implementing such reasoning. Chapters 2 and 9 contain a much fuller exposition of basic symbolic AI reasoning systems.

However an increasing number of those involved in AI are aware of the shortcomings of current AI reasoning technology. As outlined in earlier chapters, many approaches to reasoning rely on techniques such as semantic networks, frames and first-order predicate logic. Useful as these approaches may be, they are often found to be insufficiently sophisticated to cope with real-world problems.

Much current research is directed at addressing this lack of sophistication by analysing real-world problems in far greater detail and using these analyses to develop more appropriate formalisms. Of particular interest are logic-based formalisms that are built on

Artificial Intelligence in Engineering. Edited by Graham Winstanley
ⓒ1991 by John Wiley & Sons Ltd.

top of first-order logic and those that extend and/or deviate from first-order logic. Chapter 9 contains an introduction to first-order predicate logic (FOPL). A fuller introduction to FOPL can be found in Hodges (1977).

As discussed in Haack (1978), FOPL can be extended by adding extra vocabulary to the logic and allowing new theorems that involve the extra vocabulary. In this way, what can be said with FOPL can be said with the extended logic, but additionally extended logic can express concepts that cannot be said with FOPL. A logic can be deviant by allowing a different set of theorems from the same vocabulary as FOPL. This effectively means that some theorems in the deviant logic are false, whereas in FOPL they are true.

This chapter is biased towards the development of logic for addressing the shortcomings in current AI reasoning systems. However an argument will not be made for the logic approach here. The interested reader should refer to McCarthy and Hayes (1969).

For engineering AI, awareness of this expanding field of AI reasoning will provide not only the capability to harness more appropriate technology, but will also provide a deeper insight into the nature of the real-world situations in which they apply.

Producing a chapter of this sort is problematical. Firstly, it is difficult to delineate the boundaries of the subject; a far wider range of issues could be considered if space permitted. Secondly, it is difficult to decide what work to include; much of the work in this chapter is still in the AI research laboratories. Finally, in order to structure the chapter, arbitrary headings and subheadings have been chosen; yet there is much interconnectedness, and hence some of the ways that various sections are presented may not suit all tastes.

In considering interrelatedness, reasoning about space and time can be regarded as state-based and some of the formalisations reflect this. A state-based approach is reasoning in terms of states of the world (or situation) and the relationship between these states. For certain kinds of temporal logic, this relationship between states can be regarded as an ordering of worlds over time such that at a particular world, other worlds are either in the past or the future. By contrast, in reasoning about space, it is possible to consider states being related by particular events such that the event of an item changing location in a world creates a new world.

State-based reasoning is also significant in considering causality and action. The state before and after a particular action occurring can be represented and reasoned with. Causality can also be dealt with in terms of temporal reasoning. So for example, if A causes B, then B is true after A is true.

Uncertainty reasoning and non-monotonicity are interrelated issues. Uncertainty requires the representation and manipulation of uncertain information. Non-monotonic reasoning is the capability to make a deduction on the basis of a certain set of information, and to be able to retract that deduction on the basis of further information. So for example (taken from Nute (1988)), if a match is struck, it is assumed that it will light. However if it is discovered that it is a wet match, it is assumed that it will not light.

Uncertainty and non-monotonicity are aspects of all kinds of reasoning. For example, certain kinds of temporal reasoning can be considered to be non-monotonic. In work on temporal prediction (Shoham, 1987), an example of a snooker ball rolling across a snooker table is used. If the ball is rolling in a certain direction at time t, that at the next point in time t', the ball will still be rolling in that direction. This assumption can be defeated if something happens to stop the ball rolling in that direction at time t'. Hence this is an example of non-monotonic temporal reasoning. Similarly uncertainty and non-monotonic reasoning are important in other forms of state-based reasoning.

Why adopt formalisms to reason about time, space, etc? And why strive for some mathematical foundations for such formalisations? The answers lie in the benefits that can be accrued from such formalisations. Adopting such formalisations facilitates the construction of intelligent systems. The formalisations provide a level of abstraction that makes the representation more natural. This leads to benefits in the engineering processes and also supports enhanced user interaction by way of richer explanation facilities. Furthermore, mathematically based formalisations open the door to the potential of verification and validation of AI systems.

In each section of this chapter, some key approaches are briefly presented and some of the controversial issues are highlighted. Some of the work is of significant academic interest, even though the application to practical engineering AI problems is not entirely obvious. Other approaches show more obvious applicability.

Five main sections cover the areas of reasoning with uncertainty, non-monotonicity, time, space, and causality and actions. Much of the work is part of on-going research and its impact on real-world applications has yet to be seen. Yet it is unlikely that the development of enhanced AI technology is possible without some recourse to the considerations outlined in this chapter.

8.2 UNCERTAINTY REASONING

Uncertainty reasoning involves the representation and manipulation of uncertain information. In a rule-based system, rules act on assertions to derive conclusions, so for uncertainty reasoning, a rule-based system may involve using rules that are not absolutely correct. In other words, the rules may provide incorrect answers, to some degree or with some frequency. They may be approximations. The data used as assertions may also be subject to uncertainty. It may be ambiguous, partially incorrect or imprecise.

In most real-world applications of AI, capability to reason with uncertainty is required. There are now a number of AI techniques available. In mathematical terms these range from the continuous to the discrete. Both poles of the spectrum have their proponents, though pragmatism often dictates a pluralist approach.

The choice of methodology for uncertainty reasoning is complicated by the frequent requirement of an AI application to be able to 'explain' its reasoning. To a user, the machine explanation of a hypothesis generated by a complex statistical calculation might be less readily acceptable than one generated by qualitative knowledge.

Uncertainty reasoning should also be considered in the wider context of the end use of the reasoning from the system. For example, the degree of reliance a user may put on the advice proffered by a system may have important consequences in applications such as medicine and defence. In this way, considerations of accuracy, updating and completeness are important.

In order to set the scene for uncertainty reasoning, some natural reasoning scenarios are presented:

> In the Air Force, if an unidentified aircraft is seen on the radar in national airspace, it is assumed that it is provocative, and a fighter is scrambled. In this scenario, if uncertainty reasoning is not used by the Air Force, i.e. if they waited until they had identified the aircraft before they scrambled the fighter, it might be too late to provide suitable defence.

In the medical treatment of a growth, a clinician may be uncertain as to whether the growth is benign or malignant. However, if the clinician waited until she was certain as to the nature of the growth, it may be too late to save the patient. Therefore the clinician has to reason with uncertainty to determine what action to take.

A mountaineer may be unsure as to which is the safest route up an unclimbed peak. However in order to get to the top, a decision has to be made on the choice of route.

Before a horse race, a gambler does not know which horse is going to win, but information like the current odds, previous form and ground conditions can be used to make a prediction.

When an insurance company offers to insure a car at a premium, it does not know how many payments it is going to have to make. The company reasons with uncertainty, to predict how much it will pay out, and then charges the premiums at a rate to cover for this.

Attempts to classify uncertainty reasoning techniques categorically is arguably not possible. However, it is worthwhile to consider briefly some aspects of information that is subject to uncertainty. This is not a comprehensive or precise delineation of types of uncertainty, just an indication of the spectrum.

- Stochastic processes; information that cannot be described with certainty, though can be modelled usefully by use of probability theory.
- Complex, random-like, processes; information that appears to behave stochastically, but is actually a complex, non-stochastic process. These may be usefully modelled by probability theory.
- Information that is precise, but the truth value is unknown; some or all of it may be false.
- Ambiguous information; information that has more than one meaning.
- Absent information; often in natural reasoning, information is absent, and yet decisions have to be made.
- Non-strict information; some rules are not universally true—they may be either true most of the time or true for most cases. In other words, expressing some notion of modality.

There are a number of uncertainty reasoning systems that have been developed to encapsulate aspects of uncertainty in knowledge representation and manipulation. The majority of these are quantitative systems. Quantitative, or continuous, uncertainty reasoning systems have mainly been based on the probabilistic/Bayesian approach, or certainty factors. Discrete, or qualitative, methods encompass logics and related systems.

Pollock (1983) distinguishes between statistical and logical probability. Statistical probability is the kind of probability that is learnt by discovering relative frequencies, counting cases (using for example, cards, dice, etc). These are open to extensive mathematical manipulation and, in the case of an uncertainty reasoning system, a numerical/probabilistic method of representation and manipulation of statistical probability is preferable.

By contrast, logical probability of a proposition is intended to be a measure of the proportion of 'possible worlds' in which it is true. Logical probability assigns *a priori* probabilities to propositions unconditionally (i.e. with no knowledge to the contrary). Assignment and manipulation of probabilities is not as simple as in statistical probability, and in many cases it is preferable to use a non-numerical approach. For example, 'I will drive my car away from the car park where I left it; unless in the unlikely event that it has been stolen'. It is possible to estimate the probability of the car being stolen in a given geographical area and therefore it is possible to manipulate the information probabilisticly. However for many applications of such knowledge it is simpler to use a non-numerical method.

In some uncertainty reasoning applications, such as with statistical probability, probabilistic/numerical reasoning is entirely appropriate, Spiegelhalter (1984) amongst others has demonstrated this successfully. However in many cases, including logical probability, such an approach should be avoided. There has been much debate on the subject in the literature (for example Rich, 1983). Four arguments against the quantitative approach are presented here:

(1) As in much of AI, it is important to capture as many relevant generalisations about the world as possible. This is to enable manipulation of increasingly larger problems, and also to maintain a reasonable user interface. Probabilistic and numerical methods tend to require focussing at too detailed a level, when compared with other approaches which can take a more useful generalised overview.

(2) Many aspects of non-monotonic reasoning, such as default reasoning, do not fit the mould of a probabilistic/numerical approach. For example, it would not be possible to provide a numerical value to the default 'birds fly'.

(3) It is common for there to be a lack of quantitative knowledge with which to determine a probabilistic/numerical representation. This lack of knowledge can be due to insufficient data and/or lack of confidence by say a consulting expert to provide an accurate figure. This leads to a system with approximate/imprecise figures. This problem is exacerbated with relationships of more than two variables.

(4) Probabilistic/numerical systems are difficult to maintain and expand, because of the interrelated nature of the numerical factors.

From these points it does not appears as though quantitative reasoning encapsulates all kinds of uncertainty reasoning satisfactorily. Though in practice, the problem of developing suitable and meaningful interfaces to quantitative systems is perhaps the biggest barrier to their exploitation. Interpretation of a sophisticated and esoteric numerical model is difficult.

However, the result of all this argument is that there is an increasing consensus of opinion that neither a purely probabilistic nor a purely discrete methodology is appropriate to the exclusion of the other (Fox, 1986; Nutter, 1987; Pollock, 1987). The overriding conclusion is that no one methodology or technique is capable of representing all kinds of uncertainty in a knowledge-based system satisfactorily, and that a selection has to be made from the range of techniques spanning a spectrum from quantitative to qualitative. For choosing a formalism for uncertainty reasoning, there is a requirement for research into criteria to categorise applications according to appropriate formalisations.

Given the coverage of quantitative techniques for uncertainty reasoning, presented in Chapters 2 and 9, this chapter will be biased towards qualitative techniques and, in particular, logic. Many logics considered for uncertainty reasoning can be regarded as extensions to classical (first-order predicate logic).

A currently important research area for uncertainty reasoning is that of non-monotonicity.

A number of logics and systems have been developed for this. Other logics of interest include multivalued logics and conditional logics. Furthermore, formalisation of uncertainty reasoning in terms of logic programming, such as Prolog with negation-as-failure, has contributed to a qualitative knowledge approach. Representation of qualitative knowledge has also been explored in other symbolic computational formalisations, such as semantic networks, conceptual graphs and frames.

This section will look at a popular approach to uncertainty reasoning, that of fuzzy logic, and provide a critique of it. This is followed by describing an approach to formalising a language for uncertainty reasoning in Prolog. Finally, a brief introduction to modal logic, which was originally developed for formalising the concepts of necessity and possibility. The importance of modal logic, however, now goes well beyond the consideration of these concepts and is again considered in this chapter under temporal reasoning and action reasoning.

Systems combining continuous and discrete approaches include incidence calculus (Bundy 1985, 1986) which attempts to introduce truth functionality to probabilistic reasoning. This has been developed into a propositional incidence logic. Qualitative knowledge has also been used to augment expert systems that use a numerical method for the main form of uncertainty reasoning. In particular, the *ad hoc* use of default knowledge is common.

Further research in AI reasoning could be enhanced by a deeper understanding of uncertainty reasoning in natural cognition. Uncertainty, and its implications in epistemology, has long been important to philosophy. This has included the meaning and types of uncertainty, knowledge, belief and truth. Some of this work has important ramifications in AI.

8.2.1 Fuzzy and Many-valued Logic

Fuzzy logic provides a popular mechanism for uncertainty reasoning in a number of wide-ranging expert system applications. An introduction to fuzzy reasoning can be found in Chapter 5. The theory of fuzzy sets provides foundations for fuzzy reasoning. Zadeh (1965), who developed the theory, provides the following definition of a fuzzy set:

> A fuzzy set is a class of objects with a continuum of grades of membership. Such a set is characterised by a membership (characteristic) function which assigns to each object a grade of membership ranging between zero and one.

Taking the following definition from Zadeh (1965), let X be a space of points (objects), with a generic element of X denoted by x:

> A fuzzy set (class) A in X is characterised by a membership (characteristic) function $f_A(x)$ which associates with each point in X a real number in the interval [0,1] with the value of $f_A(x)$ at x representing the "grade of membership" of x in A. Thus, the nearer the value of $f_A(x)$ to unity, the higher the grade of membership of x in A.

Fuzzy set theory is reducable to set theory when the membership function can take on only two values, *viz*. 0 and 1.

Zadeh (1975) goes on to introduce the idea of fuzzy truth; so instead of just the truth values true or false, a range of truth values are used. For example:

very true
true
more or less true
rather true
not very true
not very false

⋮

false
very false

To incorporate this spectrum of truth values into a fuzzy reasoning system, the function f_{true} is augmented with a series of functions for the range of truth functions:

$$f_{\text{false}} = 1 - f_{\text{true}}$$
$$f_{\text{rather true}} = (f_{\text{true}})^2$$
$$f_{\text{very true}} = (f_{\text{true}})^{1/2}$$

There seems to be a strong argument against the utility of a range of truth values (Haack, 1978). The definition of intermediate truth values between the poles of 'absolutely true' and 'absolutely false' is subjective. Furthermore, the semantics of the terms to describe the various points on the truth value spectrum are dependent on the context. For example in natural reasoning, given an input of evidence and an output in terms of a truth value spectrum, the output would shift according to the implications of the reliance on the evidence. So if the output was used to make a life-critical decision the truth value used would be further towards false, than if the output was used for a non-life-critical decision. In this way confusion exists as to the interpretation to be applied to the meanings of the truth values.

Fuzzy reasoning has been incorporated into a meaning representation language called PROF (Zadeh, 1981). PROF represents the meaning of natural language sentences by use of datatypes. These datatypes convey the structure of objects, and the various attributes of these datatypes that pertain to the objects, have values that are described in terms of fuzzy theory.

The development of fuzzy logic has resulted in two stages of fuzzification. The first is the abandonment of two-valued logic for many-valued logic, by allowing object language predicates degrees of membership to sets. The second is the introduction of a number of fuzzy truth values to augment the truth values true and false. Though now described as a logic, fuzzy reasoning has developed with this second stage of fuzzification to the situation where not only does it deal with imprecise terms in arguments, but the whole system is also imprecise. The set of truth values of fuzzy logic is not closed under the operations of negation, conjunction, disjunction and implication. So taking two or more formulae that have truth values does not necessarily result in the formation of a formula with a truth value. Quoting from Zadeh (1975):

> (Fuzzy logic has) fuzzy truth values ... imprecise truth tables ... and ... rules of inference whose validity is approximate rather than exact.

Haack (1978) argues against fuzzy logic by claiming that the precision of more traditional logics is too important to be abandoned lightly. Furthermore the costs imposed by a fuzzy approach include increasing complexity and lack of manipulation capability. The imprecision built into the system by Zadeh brings into question the characterisation of the system as a logic.

An approach that is related to fuzzy logic is that of many-valued logic. Most many-valued logics are deviant, i.e. they share the same vocabulary as classical logic, but lack some of the axioms of classical logic such as the law of excluded middle, 'p v ¬ p'. In classical logic the law of excluded middle is fundamental: for any proposition, the proposition is either true or false. Some many-valued logics are also extensions of classical logic, i.e. they add new vocabulary.

Classical logic is two-valued, i.e., any statement can be valued as either 1 or 0 . 1 can be considered as true and 0 can be considered as false. Many-valued logics are n-valued, where $n > 2$.

Many-valued logics seem to lack the advantages of both a classical/extended classical logic approach and a probabilistic approach. It is difficult to see how to make sense of the truth values in the range between true and false. What do these truth values relate to? To some extent there is an overlap of notions between a range of truth values in a many-valued logic and a fuzzy logic. A many-valued logic moves away from the discrete sense of certainty that comes with a classical logic and moves towards a continuous range of certainty. Unfortunately the continuous representation of certainty does not benefit from the advantages that, say, a probabilistic approach does benefit. For these reasons many-valued logics are unlikely to be of use in practical AI situations; however, given the level of interest in fuzzy logic in AI, many systems developers do regard the approach favourably.

8.2.2 Qualitative Representation of Uncertainty in Prolog

So far in this chapter arguments have been made against a number of quantitative approaches. In the following two parts of this section we will shift the attention to qualitative techniques.

In reasoning about uncertainty, most logic approaches operate at the meta-level; uncertainty being indicated by some special addition to the language, and manipulated at the meta-level. However there is potential in representing a useful range of aspects of uncertainty at the object level. The argument for this approach is that the uncertainty can then be reasoned about at the object level in the same way any other information can be reasoned about. So for example it is argued that uncertainty about an object should reasoned about in the same way as its colour, size, age, etc.

In an attempt to reduce the need for precise quantification, Fox (1986), has explored the realm of qualitative knowledge, in an analysis of knowledge and belief. This has focussed on the use of terms such as 'possible', 'probable', 'plausible', etc., which are usually considered in AI as statistical representations in that they correspond to points, or regions, on some probability distribution. However this usage is contrary to the common usage of these terms, where they are used to make a different sort of distinction, but nonetheless valid, to that of the statistical usage.

To illustrate this, an analogous situation occurs in the representation and manipulation of space and time. Space can be represented by means of Cartesian coordinates, but it is often more appropriate to represent and manipulate information about places using terms like 'here', 'there', 'everywhere', etc. Similarly time can be represented in terms of a chronometer, but it is often appropriate to consider time by using terms such as 'now', 'after', 'before', 'never', etc. These terms can be ambiguous, but it is often the case that there is no need to resort to a quantitative representation. Furthermore without using a numerical representation, whose semantics are often uncertain anyway, logic and logic programming can be utilised more effectively.

In Fox's approach, the qualitative terms 'possible', 'plausible', 'probable' and 'certain' are defined in logic programming terms, together with related terms, including negative terms ('not possible', 'not probable'), contrast terms ('impossible', 'implausible'), and complex terms ('inconsistent', 'ambiguous'). Negative terms are usually the simple complement of the primary terms (e.g. 'X is not possible' implies 'not (X is possible)') whereas contrast terms are usually different ('not (X is improbable)' does *not imply* '(X is probable)'. Note in the following example 'NOT' is negation-as-failure, as in PROLOG (see Chapter 9):

P is possible if NOT (P is NOT possible)

P is impossible if C is a_necessary_condition_for P
         & C is NOT possible

P is NOT possible if P is impossible

P is impossible if NOT (P is impossible)

P is plausible if
         A is argument for P
         & NOT (P is exclusive of Q
         & Q is more plausible than P)

P is implausible if
         P is exclusive of Q
         & Q is more plausible than P

P is NOT plausible if
         NOT (P is plausible)

P is NOT plausible if
         P is implausible

P is NOT implausible if
         NOT (P is implausible)

P is probable if
         E is evidence for P
         & NOT (P is exclusive of Q
         & Q is more probable than P)

P is improbable if

> P is exclusive of Q
> & Q is more probable than P
>
> P is NOT probable if
> NOT (P is probable)
>
> P is NOT probable if
> P is improbable
>
> P is NOT improbable if
> NOT (P is improbable)
>
> P is certain if
> C is sufficient condition for P & C is certain
>
> P is uncertain if
> NOT (P is certain)

Reasoning about uncertainty at the object level allows for flexibility, and simplicity. However it can be argued that uncertainty is actually a meta-statement about the information/knowledge. In natural reasoning we may consider uncertainty in the object language, but often it is implicit in the reasoning.

Fox's approach appears attractive both for knowledge engineering and for end use. Further consideration, together with feedback from trials should provide some useful formalisations to complement quantitative methods. Trials for this, and in fact all the approaches presented in this chapter, need to identify the suitability of the formalisation in user terms. This requires analysis of the efficiency of interaction, the minimisation of ambiguity, and the accuracy of the computed solution to that required by the application.

8.2.3 Modal Logic

In order to round off our consideration of qualitative reasoning techniques, a brief exposition of modal logic should prove useful. Part of the motivation behind modal logic is to extend classical logic in a way that would more naturally allow the representation of arguments involving the concepts of necessity and possibility. In order to facilitate a more natural representation, two operators are introduced into the language. These operators, termed modal operators, precede a logic statement, or formula, to represent 'it is possible that' and 'it is necessary that'. The syntax of these operators is the same as that of negation. Possibility, denoted $\diamond$, can be defined in terms of necessity, denoted $\square$ and negation, denoted $\neg$, as follows where A is a formula:

$$\diamond A = \neg \square \neg A$$

This can be interpreted as meaning A is possible if and only if it is not the case that A is necessarily false.

A language of modal propositional logic, denoted here as L_{pm}, has atomic propositions, the truth functional connectives from classical logic, *viz*. ∨, ∧, →, etc., and the modal operators, ◇ and □. The formulae of L_{pm} are built up using the constructions from classical propositional logic, together with the use of the modal operators that can precede a formula. A formula with a modal operator is itself a formula. Thus modal formulae can then be built up into more complex formulae.

There are a range of modal logics that have been developed to manipulate L_{pm}. For the logics discussed here, all use the machinery of the underlying classical logic. However extra machinery is required to manipulate the modal operators. The smallest normal system, called K (for Kripke) is based on classical natural deduction rules together with the definition for ◇, as above, and the following inference rule RK:

$$\frac{A_1 \wedge \ldots \wedge A_n \to A}{(\Box A_1 \wedge \ldots \wedge A_n) \to \Box A}$$

Given these operators and the resulting language, what is the meaning of them? To answer this kind of question, recourse is often made to the notion of 'possible worlds'. Possible worlds, or Kripke semantics (after Saul Kripke, who made the significant contribution to logic of considering modal logics in terms of possible worlds) is a way of evaluating the truth of a modal sentence in terms of not just one world as in classical logic, but in a series of worlds. Accessibility relations are defined to describe the relationship between worlds. In other words accessibility relations describe which worlds are accessible from a particular world. Different definitions of the accessibility relation provide different logics. Accessibility relations can be introduced into a logic by use of axiom schemas. These schemas can be introduced in various combinations to produce the different logics. The following are the schemas T, 4 and B:

- T, which allows the accessibility relation the property of reflexivity, can be characterised by the schema:

$$\Box A \to A$$

So if $\Box A$ is read as necessary A, then it would seem reasonable to that if A is necessary in a world, then it is true in that world.

- 4 which allows the accessibility relation the property of transitivity can be characterised by:

$$\Box A \to \Box \Box A$$

So if □ is read as necessary , then if A is necessary in a world, it is necessary in all accessible worlds.

- B, which allows the accessibility relation the property of symmetry, can be characterised by:

$$A \to \Box \Diamond A$$

A necessary truth is true in every accessible world, whereas a possible truth is true in at least one accessible world. Therefore to provide a formal meaning or semantics to the notion of possible worlds and modal operators, an appropriate structure can be built that comprises a non-empty set of possible worlds and a formalisation of which atomic propositional letters are to be identified as being true in each world in the structure. One of the possible worlds will be the actual world: This is the world in which the formulae that are not modal are true.

A structure can be defined that can then be used to assist in ascertaining the truth of a formula in a world. Atomic formulae, the propositional letters, are specified as true at certain worlds in the structure. More complex formulae that are built up using truth functional operators can be considered as follows where 'iff' represents 'if and only if':

$A \land B$ is true at world w iff A is true at w & B is true at w
$A \lor B$ is true at world w iff A is true at w or B is true at w
$\neg A$ is true at world w iff A is not true at w

and more complex formulae built up using modal operators can be considered as follows:

$\Box A$ is true at w iff   for every accessible world w_i, A is true at w_i
$\Diamond A$ is true at w iff   for some accessible world w_i, A is true at w_i

A formula is true in a structure if and only if it is true in the actual world of that structure. This kind of structure, together with the above rules allows the computation of the truth value of any formula in the language L_{pm}.

So far a propositional language for modal language has been presented, together with some consideration of syntax and semantics. The modal logics presented here can be straightforwardly extended to the first-order case. For further information the reader is recommended to read Hughes and Cresswell (1972) or Chellas (1980).

How is this system of interest to AI and, more particularly, reasoning? In the first instance it provides a handle on the notions of possibility and necessity in knowledge and reasoning. Furthermore, it may provide a useful formalism for certain types of uncertainty reasoning. Finally, the work in modal logic has provided a number of interrelated systems whose foundations have been extensively explored. The mechanisms of these systems, and the apparatus developed to analyse them, have been adopted and adapted for representation, manipulation and analysis of other kinds of reasoning. For example, the work in modal logic has led to approaches in temporal, causal and action reasoning.

8.2.4 Discussion

In this section an argument has been made for the importance of a qualitative representation of uncertainty. A qualitative representation enhances both knowledge engineering and the end-user activities. However, to reiterate the conclusion drawn earlier, it is likely that an eclectic approach to the formalisation of uncertainty reasoning is required to address real-world problems.

As well as further research into developing qualitative forms of uncertainty reasoning, there is a requirement for the definition of criteria to match real-world applications to available uncertainty reasoning methods.

Of the two logic approaches discussed for uncertainty reasoning, we have not yet given consideration to the wider issues of non-monotonicity. Uncertainty and non-monotonicity are interrelated issues, and future work in cognitive science and AI is likely to deal with them as so.

Other work that has not been covered in this section but may be of interest includes probabilistic logic (Nilsson, 1986), and conditional logic (Lewis, 1973).

8.3 DEFAULT AND NON-MONOTONIC REASONING

Non-monotonic reasoning is an important aspect of natural reasoning. Non-monotonic reasoning occurs when if a decision is drawn on the basis of a set of facts, that decision may be withdrawn upon gaining a further assertion.

So what is the motivation behind non-monotonic reasoning? In McCarthy (1986), seven different uses of non-monotonic reasoning are discussed and the literature argues the case for others. Non-monotonic reasoning is important in dialogue, in database storage, as a rule of conjecture, and in the related issue of facilitating logical probability.

Dialogue is undertaken using many operations that might be described as conventions, with all involved normally accepting these conventions. So during the course of a conversation many assumptions are made; so for example if certain information is not made explicit, it assumed, by default, to be untrue. The following example from McCarthy (1986) serves as an example:

> Suppose A tells B about a situation involving a bird. If the bird cannot fly, and this is relevant, then A must say so. Whereas if the bird can fly, there is no requirement to mention the fact. For example, if I hire you to build me a bird cage and you don't put a top on it, I can get out of paying for it even if you tell the judge that I never said my bird could fly. However, if I complain that you wasted money by putting a top on a cage I intended for a penguin, the judge will agree with you that if the bird couldn't fly I should have said so.

Such conventions in dialogues have important implications for engineering intelligent systems. Natural language interfaces and user modelling techniques will have to cope with such non-monotonicity. However, such considerations may extend beyond man–machine interfaces, for such conventions may allow for more efficient agent to agent interactions and hence may be exploited in machine–machine interfaces.

In database storage, economies can be made by stating that whatever is not found to be true in the database is assumed to be false. This is effectively a closed-world assumption and is discussed further below.

As a rule of conjecture, if we have knowledge of the form 'most As are Bs', then it is reasonable to make the conjecture that any given A is also a B. For example, most swans are white, and therefore for any given swan we can make the default conjecture that it is a white swan.

Default reasoning is an aspect of non-monotonic reasoning. For example, by default we assume that any particular bird can fly unless we know something to the contrary; so if we then learn that our bird is a penguin we would retract that default assumption. Various formalisations of non-monotonic reasoning are discussed below.

In a monotonic system, if we draw a theorem from a theory, and then add another fact to the theory, we are guaranteed that the original theorem will still hold. In the following inference rule, let Δ denote the original theory, $\vdash$ denote the consequence relation, B denote an additional fact, and C denote a theorem:

$$\frac{\Delta \vdash C}{\Delta \cup \{B\} \vdash C}$$

In a non-monotonic system, the above is not guaranteed. If we draw a theorem from the theory, and then add an additional fact, we are not guaranteed that the original theorem still holds. It may be possible that the original theorem may have to be withdrawn. FOPL has the property of monotonicity; it therefore cannot be be used in a straightforward way for non-monotonic reasoning.

Rarely does a system have at its disposal all the information that would be desirable. However to wait until all required information was assimilated would involve delay, even infinite delay. Obviously this is not satisfactory. To ameliorate, some non-monotonic reasoning mechanism must be resorted to. Non-monotonicity is required when only a partial knowledge of a situation is possible. Where everything pertinent to the investigation is known, monotonicity suffices (Turner, 1984). The increasing importance being attached to non-monotonic reasoning is manifest by the increasing number of attempts to formalise aspects of it, with many of the approaches requiring fundamental alterations to classical logic. For a more thorough review of non-monotonic reasoning see Reiter (1987) or Ginsberg (1987).

8.3.1 Default Reasoning

Early attempts at formalising non-monotonic reasoning came from the consistency-based approaches of default logic (Reiter, 1980) and non-monotonic logic (McDermott and Doyle, 1980). These formalise non-monotonic reasoning by the adoption of logical consistency. In other words 'in the absence of information to the contrary, assume A' as approximately 'if A can be consistently assumed, then assume it'.

Default logic is based on a system of first-order sentences (denoted here as W) together with a set of default rules (denoted here as D). Default rules are of the form (where a, b and g are first-order sentences, and M means 'consistent'):

$$\frac{a : Mb}{g}$$

The intended interpretation of this is 'if you believe a, and it is consistent to believe b, then believe g'. For example:

$$\frac{\text{bird}(X) : M \text{ fly}(X)}{\text{fly}(X)}$$

which says that if X is a bird and it is consistent to believe that X can fly, then infer X can fly.

In this logic a notion of an 'extension' is introduced. An extension (denoted here as E) is a set of sentences that 'extend' the sentences of W according to E. A default theory defines zero or more extensions, such that each has the following properties:

(1) All the default rules that can be applied, without introducing inconsistency, are applied.

(2) Each extension is the smallest that can satisfy (1).

An extension reflects a 'possible world' and as such is internally consistent. Multiple extensions are possible with this logic, and a reasoner would then have to select an extension. However, even though an extension is consistent, the union of two extensions may be inconsistent. Furthermore, with this logic there is a problem of selecting the intuitively appropriate extension.

8.3.2 Non-monotonic Logic

Non-monotonic logic (McDermott and Doyle, 1980) introduces an object-level modal operator M to first-order logic. So for example, if A, B and C are first-order sentences, then the following says 'if A is true, and it is consistent to believe B, then derive C':

$$A \& MB \Rightarrow C$$

The problem identified above for default logic, the choice of extension, is transcended here by use of an inferencing mechanism described in Hanks and McDermott (1987) as: 'a sentence δ can be inferred from a default theory just in case δ is in every extension of that theory'. However this is not necessarily the most intuitive or efficient way of representing and manipulating defaults. The calculation of all extensions maybe considered inefficient. The rejection of all extensions given an inconsistency, where there is no consideration, or labelling of relative merits, of arguments used to derive the various extensions, maybe counter-intuitive. Other problems with this approach include the lack of a first-order system.

8.3.3 Circumscription

Another significant approach is circumscription (McCarthy, 1980, 1986) which is based on deriving minimal models. There are several versions of circumscription though for the purposes of this section we will follow McCarthy (1986).

As an example of circumscription for default reasoning we could describe the object-level representation of a default rule to be as follows, where A, B, C are first-order sentences, ab is a predicate representing abnormality, and $\neg$ represents negation:

$$A \& \neg ab1 \Rightarrow B$$
$$C \Rightarrow ab1$$

From this if A is known, and C is not known, then B is believed. Whereas, if A and C are known, then B is not believed. As in the approach of McDermott, if multiple models are derived, only those sentences that are consistent in all the models are deduced.

The basis of circumscription is the idea of minimisation. This is an idea that has also been exploited in negation-as-failure in Prolog, where 'NOT X' is true if and only if 'X' is not true in the logic program. In circumscription particular formulae can be minimised, so with abnormality the idea is to limit the number of abnormal objects, accepting as abnormal only those things that are known to be abnormal.

Circumscription is one of the richer formalisms available for non-monotonic reasoning. Though, as discussed by Reiter (1987), it is not obvious in general how to instantiate the circumscription axiom in order to derive interesting consequences of the circumscribed theory. Furthermore, there appears to be problems in the computability of the axioms for circumscription (Ginsberg, 1987). However, much work is being undertaken to develop the utility of the logic and to relate it to other logics (for example Lifschitz, 1985, 1987).

8.3.4 The Closed World Assumption and Negation-as-failure

Logic is becoming increasingly important in AI as a programming language. For practical applications Prolog (standing for PROGrammable LOGic) has been developed (Kowalski, 1979; Clocksin and Mellish, 1980). Prolog has a control structure, negation-as-failure (NBF), that enables non-monotonic reasoning, and furthermore NBF can be justified by the closed world assumption.

To demonstrate the closed world assumption (CWA) and NBF, we will take the example of 'all birds fly except ostriches':

> $fly(X)$ if $bird(X)$ & NOT $ostrich(X)$
> $bird(jim)$, $bird(bill)$, $ostrich(bill)$

If we ask the question 'can jim fly?', we can either treat the subgoal 'NOT ostrich(jim)' in the object language by using CWA, or in the meta-language by using NBF.

Using CWA we would rewrite our clauses so that:

> $bird(X)$ iff X = jim or X = bill
> $ostrich(X)$ iff X = bill

In this way we know that the only ostrich is bill, and therefore jim is not an ostrich. So we can answer the original question 'can jim fly?' with yes.

Using NBF we assume $\neg p$ holds whenever we fail to prove p. Since in the original program we cannot show ostrich(jim), we assume $\neg ostrich(jim)$. Therefore again the answer to the original question is yes.

NBF is a rule of inference; $\neg p$ holds if all ways of proving p fail in a finite time. The stipulation 'fail in finite time' is necessary because we cannot guarantee in general that a proof procedure will not loop.

NBF is justified by CWA, i.e. it assumes that all unstated 'only-if' assumptions hold. It interprets implicit 'only-if' assumptions in the meta-language. It can be implemented easily and efficiently by a simple extension to a Horn clause automated theorem-prover. Remember, Prolog is based on a subset of classical first-order predicate logic called Horn clause logic. A more detailed consideration of Prolog can be found in Lloyd (1984).

The CWA (Reiter, 1978) and NBF (Clark, 1978) are often not included in surveys of non-monotonic reasoning. However CWA/NBF offer the advantages, over some of the other approaches, of being better understood and of working satisfactorily in simple computer implementations.

We can show CWA and NBF are non-monotonic by using the example from above:

fly(X) if bird(X) & NOT ostrich(X)

bird(jim)
bird(bill)
ostrich(bill)

As before if we ask the question 'can jim fly?', we can either treat the subgoal 'NOT ostrich(jim)' in the object language by using CWA, or in the meta-language by using NBF. So if we use NBF, we show 'fly(jim)' by satisfying 'bird(jim)' and by failing to satisfy 'ostrich(jim)'.

However, if we now add the assertion 'ostrich(jim)' to the database, we have to withdraw the original conclusion of 'fly(jim)', because the 'ostrich(jim)' would be satisfied, and hence 'NOT ostrich(jim)' would fail, and so failing the whole clause.

NBF is similar to default reasoning in that default reasoning can be regarded as:

infer p if fail to show $\neg p$

whereas NBF can be regarded as:

infer $\neg p$ if fail to show p

So NBF can to some extent be used for default reasoning. For example:

fly(X) if bird(X) & fail to show $\neg$fly(X)

can be approximated by:

fly(X) if bird(X) & fail to show abnormal_bird(X)

which in Prolog is:

fly(X) if bird(X) & NOT abnormal_bird(X)

abnormal_bird(X) if ostrich(X)

abnormal_bird(X) if penguin(X)

.

.

However NBF is not ideal. For simple implementations of default reasoning it is adequate, but for more complex examples it lacks structure and flexibility. The following more complex example creates problems during updating of the program. Taking the example presented above, suppose we learn that there is a new genetic mutant of ostrich that can fly, we would add a new rule to the database:

fly(X) if bird(X) & abnormal2(X)

abnormal2(X) if ostrich(X) & new_genetic_mutant(X)

The problem arises when we then discover that some genetic-mutants are also diseased and are therefore too ill to fly. How do we now add the exception to this rule without altering what has already been written? What we want is to be able to add new rules to the program without having to delete or alter what is already in the program.

Other problems with CWA include the lack of capability to represent certain forms of knowledge in a structured way. Furthermore, in a default reasoning program based on NBF, there is no straightforward mode of indication of whether a consequent of the program is derived by default, and hence potentially subject to retraction on further information.

8.3.5 Discussion

In AI, the need to reason with defaults is widely recognised. However, the solution for many is that defaults can be satisfactorily be addressed with an *ad hoc* mechanism. Though, the lack of regard for a fuller understanding of the foundations of non-monotonic reasoning will lead to problems when trying to engineer more sophisticated cognitive systems.

Interconnections between the various approaches are being mapped, and hence a deeper understanding of the property of non-monotonicity is being achieved. However, the uptake of such research may only widen if the work in non-monotonicty comes with a deeper understanding of the kinds of reasoning in which it is important. For example, the understanding of non-monotonic reasoning, and that of temporal reasoning, has been advanced with an integrated consideration of the temporal qualification problem and the extended prediction problem (Shoham, 1988).

8.4 TEMPORAL REASONING

In many cognitive tasks, reasoning has a temporal aspect. This allows qualification of inferencing in terms of either intervals of time or points of time. As Shoham (1988) puts it:

> It is hard to think of a research area within artificial intelligence that does not use reasoning about time in one way or another: medical systems try to determine the time at which the virus infected the blood system; circuit debugging programs must reason about the period over which the charge in the capacitor increased; ... robot programmers must make sure that the robot meets various deadlines when carrying out a set of tasks ...

In the work that has been presented in this text so far, the reasoning is atemporal. So for example, if a deduction is made from a logic program, it is a truth that exists unconstrained by time. Yet in many applications, certain deductions are only true for certain periods, or points in time. In this section we will look at attempts to extend first-order logic by the addition of modal operators into the logical system, by reified formalisations, i.e formalisations that express time explicitly in the object language, and by the development of a calculus built on top of Horn clause logic.

8.4.1 Tense Logic

Classical logic is atemporal. In other words it conveys no concept of time. To represent that a sentence was true in the past but is no longer true, or that a sentence is not true now, but is true at some point in the future requires some axiomatisation of time, together with explicit predicates for time in the application representation. So for example in order to represent the following:

it is always going to be the case in the future that sid loves mary

We can represent 'sid loves mary' as L(s,m), have n meaning now, and quantify over time:

$$\forall X (X > n \rightarrow L(s, m))$$

This then says that for all points in time greater than now, (i.e. in the future) , sid loves mary. However this also requires axioms for the linearity of time:

- No point in time is earlier than itself:

$$\forall X \neg (X < X)$$

- Given a point in time Y, and a point in time X that is less than Y, and a point in time Z that is greater than Y, then Z is greater than X:

$$\forall X \forall Y \forall Z ((X < Y \land Y < Z) \rightarrow X < Z)$$

- For all points in time, other points are either greater than, less than or equal to that point in time:

$$\forall X \forall Y (X < Y \lor X = Y \lor Y < X)$$

But note that if there is quantification over variables, other than for time, a typed logic would be required.

For more complex sentences, readability and manipulation become far greater problems. Tense and temporal logics are in part a solution to this. These logics add extra operators into the logic that indicate the time periods over which the sentences are interpreted as being true. There are a number of variations of these logics, and are based on extending the logic with the following operators, where A is a propositional sentence:

GA, meaning A is always going to be the case
HA, meaning A has always been the case
PA, meaning A has been the case in the past
FA, meaning A will be the case in the future

Furthermore the following definitions can be made:

$$GA = \neg F \neg A$$
$$HA = \neg P \neg A$$

Using these operators various logics can be developed along the lines of modal logic. So for example with the statement:

'sid will love mary for ever'

if we represent 'sid loves mary' as L(s,m) and hence prefixing this with G, then:

G L(s,m)

This is obviously simpler to represent, and is easier to manipulate. Furthermore it increases the lucidity of the knowledge.

So what are the non-classical logics composed of? They are based on classical first-order logic, extended with modal operators, and axioms to manipulate this extended language.

The minimal temporal logic is Kt. In Kt, the following axioms apply:

A where A is a tautology
$G(A \rightarrow B) \rightarrow (GA \rightarrow GB)$
$H(A \rightarrow B) \rightarrow (HA \rightarrow HB)$
$A \rightarrow HFA$
$A \rightarrow GPA$
GA, if A is an axiom
HA, if A is an axiom

together with modus ponens as a the rule of inference. Remember, *modus ponens* allows the detachment of a consequent of a rule if the antecedent is assumed. This can be represented as:

$$\frac{A.A \rightarrow B}{B}$$

In order to gain a deeper insight into this approach, the notions of temporal precedence, and of possible worlds, require consideration. Possible worlds allow the consideration of a series of temporal states of the world, and the temporal precedence relation, R, allows an ordering of these worlds. Note the temporal precedence relation is analogous to the accessibility relation in modal logic.

Using these we can decide on properties that a temporal precedence relation, R, may have:

- Transitivity

  if X is earlier than Y
  & Y is earlier than Z
  then X is earlier than Z.

- Irreflexivity

  no X is earlier than itself.

- Asymmetry

  if X is earlier than Y
  then Y cannot be earlier than X.

Furthermore we can define the temporal operators in terms of the temporal precedence relation, R:

> FA is true at time *a*, if *A* is true at some time *b* such that R(*a,b*).
> PA is true at time *a*, if *A* is true at some time *b* such that R(*b,a*).
> GA is true at time *a*, if *A* is true at every time *b* such that R(*a,b*).
> HA is true at time *a*, if *A* is true at every time *b* such that R(*b,a*).

The connectives can be used to determine the truth value of compound formulae in terms of their constituents:

> $A \wedge B$ is true at time *t* iff *A* is true at time *t* & *B* is true at time *t*
> $\neg A$ is true at time *t* iff *A* is false at time *t*
> FA is true at time *t* iff $\exists t'(R(t,t'))$ & *A* is true at time *t'*)
> PA is true at time *t* iff $\exists t'(R(t',t))$ & *A* is true at time *t'*)

The temporal precedence relation can be restricted so as to develop further extensions of the minimal temporal logic such as introducing branching time (see Figure 8.1), where for example FA can be interpreted as true on some branch in the future, PA as true on some branch in the past, GA as true on all branches in the future, and HA as true on all branches in the past.

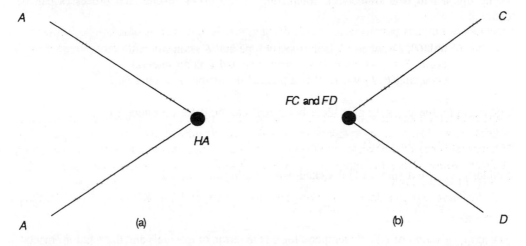

Figure 8.1 Branching time (a) left and (b) right

There are some versions of tense logic with explicit time (McArthur, 1976), though this is in one unit, so switching from reasoning in, say minutes, to reasoning in, say, years is difficult. However it must be noted that there are many applications in which specific time points are rare, and ordering is all important. Hence tense logics are found applicable to a range of applications.

In conclusion, tense logics are 'close' to natural language, and hence provide a more lucid representation, and information about the past and future can be easily recorded. However, updates do require a disproportionately complex modification of the knowledge-base, different occurrences of the same event type are difficult to distinguish, and explicit representation of time is not necessarily straightforward.

8.4.2 Allen's Temporal Logic

Allen's temporal logic (Allen, 1984, 1985) was one of the earlier attempts at developing a temporal formalism for AI. In this approach, the interval is the primitive temporal component. As discussed in Allen (1984), the system that is intended to deal with both static and dynamic aspects of the world:

> The static aspects are captured by properties (e.g., Cleo owning a car) that hold over stretches of time. The dynamic aspects are captured by occurrences (e.g., Cleo running a race) which describe change of form of resistance to change over stretches of time.

Furthermore, occurrences are subdivided into processes and events:

> Processes refer to some activity not involving a culmination in anticipated result, such as the process denoted by the sentence, "I am running". Events describe an activity that involves a product or outcome, such as the event denoted by the sentence "I walked to the store". A useful test for distinguishing between events and processes is that one can count the number of times an event occurs, but one can not count the number of times a process is occurring

Using this a temporal framework comprising of properties, events and processes can be defined:

> hold(P, I), where P is a property type and I is an interval
> occur(E, I), where E is an event type and I is an interval
> occurring(P, I) where P is a process occurring over interval I

For example, the property of a 'door being red' can hold for an interval $I1$:

> hold(door(red), $I1$)

Similarly, an event such as 'sid kicked harry' during interval $I2$:

> occur(kicked(sid, harry), $I2$)

The axiomatisation of Allen's temporal logic is in terms of intervals and the relation 'meets'. 'Meets' is defined as the situation where there is no time between two time intervals, i.e. one directly follows from the other. Using 'meets' all the following relationships between two time intervals can be described:

before(X, Y)	if	exists(Z) & meets(X, Z) & meets(Z, Y).
after(Y, X)	if	exists(Z) & meets(X, Z) & meets(Z, Y).
equal(X, Y)	if	exists(U),exists(V),meets(U,X),
		meets(X,V),meets(U,Y),meets((Y,X).
overlaps(X,Y)	if	exists(A) & exists(B)
		exists(C) & exists(D) & exists(E)
		meets(A,X) & meets(X, D) & meets(D, E)

$$\text{meets}(A, B) \ \& \ \text{meets}(B, Y) \ \& \ \text{meets}(Y, E)$$
$$\text{meets}(B, C) \ \& \ \text{meets}(C, D).$$

starts(X,Y) if   exists(A) & exists(B) & exists(C)
$$\text{meets}(A, X) \ \& \ \text{meets}(X, B) \ \& \ \text{meets}(B, C)$$
$$\text{meets}(A, Y) \ \& \ \text{meets}(Y, C).$$

finishes(X, Y) if   exists(A) & exists(B) & exists(C)
$$\text{meets}(A, B) \ \& \ \text{meets}(B, X) \ \& \ \text{meets}(X, C)$$
$$\text{meets}(A, Y) \ \& \ \text{meets}(Y, C).$$

during(X, Y) if   exists(A) & exists(B)
$$\text{exists}(C) \ \& \ \text{exists}(D)$$
$$\text{meets}(A,B) \ \& \ \text{meets}(B, X) \ \& \ \text{meets}(X, C)$$
$$\text{meets}(C, D) \ \& \ \text{meets}(A, Y) \ \& \ \text{meets}(Y, D)$$

Given these relationships and their inverses, axiomatisation of the uniqueness of 'meeting places', ordering, and of existence is required. Uniqueness of meeting places captures the intuition that intervals have a unique beginning position and a unique ending position. Therefore, in a situation where two intervals meet a third interval, then any interval that one meets, the other meets as also:

$$\forall A,B \ (\exists C \ \text{meets}(A,C) \ \& \ \text{meets}(B,C)) \Rightarrow$$
$$(\forall D \ \text{meets}(A,D) \Leftrightarrow \text{meets}(B,D))$$

The following axiom for ordering, states that given two 'places' where two intervals meet, then these are either equal, or one precedes the other:

$$\forall A,B,C,D \ \text{meets}(A,B) \ \& \ \text{meets}(C,D) \Rightarrow$$

(1) meets(A,D) XOR
(2)$\exists E$ meets(A,E) & meets(E,D) XOR
(3)$\exists F$ meets(C,F) & meets(F,B)

Axioms are also required to define existence. Firstly, for all intervals, there exists an interval that meets it, and an interval that it meets. An implication of this is that no infinite time interval is possible in this theory. Secondly, for the concatenation of two adjacent intervals, there exists an interval that is the union of the two intervals.

The following is a definition of occur, where E is an event:

$$\text{occur}(E, T) \ \& \ \text{in}(T', T) \Rightarrow \neg \ \text{occur}(E, T')$$

which effectively summarises the situation where occur is true only if the event happened over the time interval T, and there is no subinterval of T over which the event happened.

A process occurring over an interval, can be described as occurring over at least one subinterval of that interval. Hence:

$$\text{occurring}(P, T) \Rightarrow \exists T' \; \text{in}(T', T) \; \& \; \text{occurring}(P, T')$$

Shoham (1987) criticises the temporal framework developed here, see below. However, Allen's work has been adopted by a number of research and development projects, and is regarded as providing a valuable basis for developing temporal reasoning in AI systems.

8.4.3 McDermott's Temporal Logic

In McDermott's approach (1982), the temporal primitive is the time point. From this a temporal framework comprising of fact types and event types is defined:

$t(T, P)$ means that fact type P holds at time T

$tt(T1, T2, P)$ means that the fact type P holds from time point $T1$ to time point $T2$

$\text{occur}(T1, T2, E)$ means that the event type E occurred from time point $T1$ to time point $T2$

In this logic more than one thing can start happening at any one time. This is modelled by having many possible worlds. Furthermore in this logic, time is dense, i.e. there is a continuum of instances between any two instances.

The logic based on an infinite series of 'snapshots' of the universe. A state is an instantaneous snapshot of the universe. Each state is related to a 'date' which is the point in time which the universe is possibly in that state.

States are partially ordered by temporal ordering relation, $\leq$. Thus $a \leq b$ means that state a comes before or is identical to state b. States are arranged into chronicles, where a chronicle is total ordering of states that extends infinitely. A chronicle represents a possible way that events might go. There may be more than one chronicle, but chronicles only branch into the future, i.e, they are right-branching (see Figure 8.2).

Facts and events can be interpreted in terms of the states and chronicles. The truth value of a fact is a set of time points, i.e. the states in which the fact is true. In this way the truth value of a fact can change over time. This is related to the possible world semantics of modal logic, i.e. letting propositions denote sets of possible worlds. An event is described as a set of intervals. Intuitively those intervals over which the event happens once, with the first interval starting at the same time as the start of the event and the end of the interval ending at the same time as the end of the event.

Each interval consists of a beginning and an endpoint such that the beginning is constrained to precede or be coincident with the ending. Important intervals are referred to as tokens of a particular type, where a type is a formula such as '(operational-status, lathe 17, inservice)', '(operational-status, lathe 17, out-of-service)', and '(malfunction, assembly-unit 34)'.

An interval together with a token is a time token. Tokens corresponding to facts are referred to as persistences. Persistences are used to state that a fact will remain true for some period of time, without explicitly stating how long for. This allows a non-monotonic manipulation of facts over time, i.e. assume a fact is true if it becomes true and it is not known that it is no longer true.

McDermott argues against the representation of time in terms of events, since often it is difficult to find suitable events with which to reason. Instead he identifies an event as a

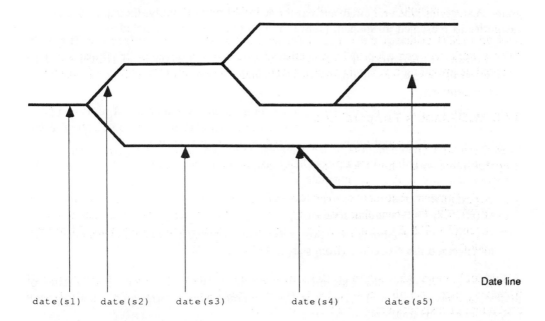

Figure 8.2 A tree of chronicles where s1, ..., s5 are states

set of intervals. The set of intervals being over which the event occurs and not extending beyond the end of the event.

8.4.4 Shoham's Temporal Logic

The motivation behind Shoham's work is to integrate and enhance some of the appealing characteristics of Allen's and McDermott's work. Shoham believes that both the previous logics lack sufficient semantics, i.e. there is an incomplete meaning behind the representation system. Furthermore, conceptually the assumptions and definitions behind both logics—Allen's trichotomy of properties/events/processes and McDermott's dichotomy of facts/events are unnecessary at some times and insufficient at others. Shoham's approach (1987) for the propositional case is a syntactic variant of McDermott's approach and for the first-order case is a generalisation of McDermott's approach. In this section we will consider the propositional case. For this, an assertion, denoted P, that is true over an interval is represented as follows:

$$\text{true}(T1, T2, P)$$

where true is not a predicate, only syntactic sugar, and $T1$ represents the start of the interval and $T2$ represents the end of the interval.

A precise syntax and semantics has been described for the system. In so doing a fixed categorisation of temporal propositions has not been attempted. Arguably this provides the opportunity to avoid distinctions when unnecessary, and also allows the description of finer grained categories when required, such as for description of causality. Examples of possible categories as presented in Shoham (1987) for a proposition type P include:

- P is downward-hereditary if whenever P holds over an interval it holds over all of its subintervals, possibly excluding the two endpoints. For example, 'The robot travelled less than two miles'.

- P is upward-hereditary if whenever P holds for all proper subintervals of some non-point interval (except possibly at its endpoints), it also holds over the non-point interval itself. For example, 'The robot travelled at a speed of two miles per hour'.

- P is liquid if P is both upward-hereditary and downward-hereditary. For example, 'The robot's arm was in the grasping state'.

- P is gestalt if P never holds over two intervals one of which contains the other. For example, 'Exactly six minutes passed'.

- P is solid if P never holds over two properly overlapping intervals. For example, 'The robot executed the procedure (from start to finish)'.

This is a non-exhaustive list. Categories from other systems can be compared in terms of these categories. For example, Allen's and McDermott's events correspond either to gestalt propositions, or to solid ones, or to both.

8.4.5 Event Calculus

Quite a different approach is a calculus based on events. Using Horn clause logic, together with negation-as-failure, event calculus is an approach to reasoning about time, and more particularly events (Kowalski and Sergot, 1986). This is with a view to developing a mechanism for updating of inferential databases. The calculus is intended to cope with updates that do not occur in the same order as in the real world.

This is a non-monotonic system, as default assumptions are made as to nature of the updates, and these updates can be amended in light of further information. The non-monotonic capability is mechanised by use of negation-as-failure. In many conventional databases, when new knowledge indicates that some existing knowledge is no longer true, the existing knowledge is deleted from the database. In this approach, new knowledge is used to delineate the time periods over which existing knowledge is true. So in the event calculus, updates are additive, they do not delete information.

A fundamental component of the calculus is that events are regarded as more primitive than time, and that events are represented explicitly in Horn clause logic, in order to reason about time. Time periods can be regarded as a function of an event and the relationship that started the event. So, for example, the following represents a time period, where R is a relationship, and E an event:

after(R, E).

Information about states can then be derived by when a relationship holds:

holds(R, after(R, E)).

This can be abbreviated to:

> holds(after(E, R)).

Similarly we can state this as:

> holds(before(R, E)).

An initial state can be given explicitly, or separate initialising events can result in an initial state. So for example:

> holds(after(E, R)) if initiates(E, R).

or:

> holds(before(E, R)) if terminates(E, R).

So using a general rule, as above, we can now state specific rules for different applications. For example, we can describe the event of hiring sid as chief-engineer being initiated as follows:

> initiates(e1, rank(sid, chief_engineer))   if act(e1, hire),
>   &   object(e1, sid),
>   &   destination(e1, chief_engineer).

Similarly, we could describe an event in terms of sid leaving:

> terminates(e2, rank(sid, chief_engineer))   if act(e2, leave),
>   &   object(e2, sid),
>   &   source(e2, chief_engineer).

The following general rules then allow us to derive knowledge about endpoints to events, where NOT refers to negation-as-failure, and $<$ is a temporal ordering relation such that '$a < b$' means event a occurs before event b:

> start(after(E, R), E).
> end(before(E, R), E).

> start(before(E^*, R), E) if after(E, R) = before(E^*, R).
> end(after(E, R) E^*) if after(E, R) = before(E^*, R).

> after(E, R) = before(E^*, R)   if holds(after(E, R))
>   &holds(before(E^*, R))
>   & $E < E^*$
>   and NOT broken(E, R, E^*).

> broken(E, R, E^*)   if holds(after($E'R'$))
>   & exclusive(R R')
>   & $E < E' < E^*$

$$broken(E, R, E*) \quad \text{if } holds(before(E'R'))$$
$$\&exclusive(R\ R')$$
$$\&\ E < E' < E*$$

Exclusive holds either when the relationships are identical, or when neither can hold simultaneously:

$$exclusive(R\ R)$$
$$exclusive(R\ R*) \text{ if } incompatible(R\ R*)$$

The predicate 'incompatible' is defined according to the application. So for the example above, we have:

$$incompatible(rank(X\ Y), rank(X, Z)) \text{ if NOT } Y = Z$$

General rules can then be developed that allow the determination of whether a relationship holds at a specific time instant:

$$holds\_at(R, T) \quad \text{if} \quad holds(after(E, R))$$
$$\& \quad T \text{ in } after(E, R)$$

$$holds\_at(R, T) \quad \text{if} \quad holds(before(E, R))$$
$$\& \quad T \text{ in } before(E, R)$$

$$T \text{ in } X \quad \text{if } start(X, E1) \ \& \ end(X, E2)$$
$$\& \ time(E1, T1) \ \& \ time(E2, T2)$$
$$\& \ T1 < T \ \& \ T < T2$$

$$T \text{ in } X \quad \text{if } start(X\ E) \ \& \ time(E, T*) \ \& \text{ NOT } end(X, E*)$$

The event calculus provides a framework for reasoning about events in time. This framework can be extended in various ways to account for different temporal reasoning situations. The non-monotonic aspect of the calculus is appealing for a range of applications, including temporal databases, and the relationship with Prolog provides a straightforward route to implementation.

8.4.6 Discussion

Formalisation of temporal reasoning is a controversial subject. There is debate as to the most appropriate temporal framework. In particular, the choice of having intervals or having points as the basic temporal element is open to debate. Allen takes the interval as the basic element whereas McDermott and Shoham take the point as the basic component. Kowalski avoids this problem by use of the event as the fundamental component. Depending on the kinds of application either Kowalski's or Shoham's work offers the most developed formalisations. However, there are still open questions in temporal reasoning, and given the increasing importance attached to temporality in AI, and also in information technology generally, much work will be directed towards developing these and other systems.

On the issue of using either points or intervals as the basic unit, it does not seem that the choice can be resolved at an abstract level but rather depends on the application. For example, some applications seem more naturally represented as points. For example, for reasoning about the Wall Street Dow Jones Closing Price, a point-based system would seem natural, with each point corresponding to a day. Whereas for a robotic system that monitors its environment and carries out tasks, it may be more appropriate to reason with intervals, since it may be difficult to model actions or events as instantaneous.

8.5 SPATIAL REASONING

Spatial reasoning is an important component of natural reasoning. It is important in perception and in higher level cognition. However for the purposes of AI, there has been surprisingly little work of a general nature that addresses spatial reasoning.

The most advanced and effective work has been in the area of machine vision. However, much of this work has been for frame-based technology (Walker *et al.*, 1988; Fisher, 1987), and this has not led towards a more general theory of space and the relationships between objects in space. Spatial reasoning is either based on very low level concepts, i.e. image-processing concepts, or on an application-specific representation of spatial knowledge.

But perhaps research in machine vision is the wrong place to look for a more general theory of spatial reasoning. Planning systems, as in robotics, is another area where reasoning about space is important. Some work of interest has been undertaken with the blocks-world domain. A good overview can be found in Rich (1983). However, again no general theory of spatial reasoning has come about.

In this section, two pieces of work are considered. The first system, situation calculus, is not a viable solution to the problem of spatial reasoning. However it does usefully highlight problems with developing a logic approach to spatial reasoning, in particular the problem of dealing with a global state. The second system, SPACES, does offer some useful ideas and there is potential for developing a logic for spatial reasoning along such lines.

8.5.1 Situation Calculus in the 'Blocks-world'

Situation calculus was originally developed as an attempt to formalise notions of situations and change of situations. In this sense it is related to the attempts to encapsulate notions of states and actions as in action and modal action logic, though the mechanisms are quite distinct.

As applied to the 'blocks-world' domain (Kowalski, 1979), there are interesting ramifications for the understanding of formalising spatial reasoning. In this formalisation of the blocks-world environment states are represented as the result of applying an action on the previous state. For example the result of applying action A to state S is represented as follows:

result(A, S)

An initial state can be explicitly expressed, denoted here as $\emptyset$. As well as representing states, statements about the location of items require representation, such as for the statement that X is on Y:

on(X, Y)

or the statement that there is nothing on top of X, i.e., X is clear:

clear(X)

Non-initial states are represented as a function of the previous state and the event that took it to the new state. For example, $S4$ maybe the result of moving block X from position Y to position Z in state $S3$:

$S4$ = result(move(X,Y,Z), $S3$)

Given these representations of states and statements, the relationship between a statement T holding in a particular state S is represented as follows:

holds(S, T)

So for example:

holds(on(a, b), result(move(a,c,b), $S0$))

which says that block a is on block b as a result of moving a from c to b during state $S0$. Similarly, for example:

holds(on(c, a), result(move(c, d, a), result(move(a,c,b), $S0$)))

which says that block c is on block a as a result of moving c from d to a during the state described by 'result(move(a,c,b), $S0$)', i.e. the previous state.

Describing states in terms of previous states allows the preservation of information about past states. Furthermore updates are given meaning in terms of event descriptions. However there is a problem of deleting old relations that no longer hold.

A solution to this last problem can be addressed by use of the frame axiom. This is effectively a formalisation of what holds in particular states holds in the subsequent state if it has not been terminated. Therefore by recursion, what holds in a particular state can be determined by analysis of previous states. For example:

holds(Statement, result(Event, State)) if holds(Statement, State)
& not terminates(Event, Statement)

Using this axiom would require some representation of what kind of statements each kind of event terminates.

It should be noted that situation calculus is not restricted to spatial reasoning but is intended as a formalism for reasoning about global states and can be applied to other state-based applications such as in temporal reasoning.

Unfortunately, the frame axiom is regarded as a computationally expensive mode of representing global states. Hence such a system may prove inefficient in all but small-scale exercises. A related approach that has been developed for temporal reasoning is the event calculus. In the event calculus, the representation of global states is dropped in favour of local states.

8.5.2 SPACES

SPACES is an attempt to develop a calculus for common-sense spatial knowledge (Green, 1986). This calculus is independent of a geometrical description of space. The calculus defines a set of primitive entities, attributes used to describe them more fully, and relationships which may exist between them. In Figure 8.3 there is a listing of the entities, together with attributes and relationships:

ENTITY	RELATIONSHIP	ATTRIBUTE
Space	Next_to Above Below Left_of Right_of	Dimensions Area Boundary Subspace
Boundary	Co-linear Parallel Right-angle	Length Orientation Point Sub-boundary
Point	Co-incident	

Figure 8.3 A listing of entities, attributes and relationships in SPACES

Each entity is defined in terms of a set of attributes. At the lowest level attributes are defined in terms of points:

- *Space* is an infinite set of points and the set of points is bounded by a closed set of boundaries.
- *Boundary* is a straight line which connects two endpoints.

Relationships such as closed, inside and outside are defined independently to avoid circular arguments. Other relationships and attributes are defined in terms of each other. For example:

A is next_to B
  if there is a boundary of A
  & a boundary of B
  & the two boundaries are co-linear
  & A is outside B

or

A is next_to B
  if B is next_to A

The calculus was developed for application in intelligent computer-aided design. However the kind of concepts considered here may be of interest in a wider range of application. It is likely that further work could result in a more sophisticated and far more applicable system.

8.5.3 Discussion

Out of the five sections presented in this chapter, spatial reasoning seems to be the least well developed. Is this because spatial reasoning is too context dependent, or is it because it is too difficult or is it of too little interest? As discussed in Section 8.6.3 in the 'Naive Physics Manifesto' (Hayes, 1979), common-sense reasoning will require consideration of concepts of space such as discussed in Green's spaces. However, it is likely for common-sense reasoning, a far more highly interconnected network of semantic relationships will be required.

 There are many current and potential applications that require spatial reasoning. For example there is much interest in automated car navigation systems. Such a system requires a considerable database on the layout of roads in a region. Similarly much effort has been directed at two- and three-dimensional vision understanding systems and robot navigation systems (see Chapter 7 of this book). In a similar way to temporal logics providing useful abstractions of temporal space, it seems likely that an appropriate calculus of Euclidean space will provide a system suitable for a range of applications. Such a calculus may optimise data storage and computation. It may also enhance man–machine interaction allowing more natural dialogue forms.

8.6 CAUSAL AND ACTION REASONING

Detailed consideration of any one of the concepts of causality, action and time often requires consideration of the other two. In AI the issue of causality is an important one. Many attempts at encapsulating the concept have involved a temporal logic approach. This includes work by Allen (1984) and McDermott (1982) who describe temporal ontologies that include events and processes. Shoham (1988) has extended this work to give a far clearer understanding of causality. Shoham has attempted to address the problems of temporal qualification, the problem of making predictions by making assumptions with regard to the past, and the extended prediction, the problem of the vast number of inferences required to make predictions in certain situations. The temporal qualification problem arises by not being sure of all the antecedents in the following kind of inference: 'if this is true now, then that is true later'. A solution to this problem requires some form of non-monotonic approach. The extended prediction problem occurs when the length of time over which a prediction refers is short, even instantaneous and hence requires a non-monotonic assumption that a prediction stands until there is reason to believe otherwise.

In AI relatively little consideration has been given to the area of action logics. Much of the motivation behind them comes from the field of formal specification in computer science, however there is potential in these formalisms for application to AI.

In this section a brief introduction to action logic is provided. This section also provides an overview of the wider problems of common-sense reasoning and of qualitative reasoning which was in part motivated by the problems of causality in common-sense reasoning.

8.6.1 Action and Dynamic Logics

One of the interesting aspects of the work in modal logics has been the adoption and adaptation of the formalisations, and analytic apparatus, for other notions. The idea of possible world is extended such that the notion of possible worlds correspond to computational notions of state. In this way modalities provide some relationships between states.

However it is likely that more than the existence of relationships between particular possible worlds is required. The cause of the relationship between particular possible worlds is required. In this way a family of modal operators is created. So instead of the modal operator $\Box$, we have $[X]$, where X can be instantiated by an action. Using this operator, an assertion can be qualified by an action. So for example in the following, after action A, b is true:

$$[A]b$$

Note that the reading of such a statement introduces a temporal bias, i.e. *after* action A, assertion b is true. However not all actions can operate in all circumstances. In the above statement as long as action A is conducted, assertion b is true. Yet, it may not always be possible for action A to operate in order to allow assertion b to be true. To account for this, a precondition can be applied to the statement. So for example if assertion a is true, then after action A, b is true:

$$a \rightarrow [A]b$$

This kind of system can be extended to the propositional and predicate cases.

If we consider such a system as a transition system:

where the transition system is a triple $\langle P, A, \rightarrow \rangle$ with
P as a set of possible worlds,
A as a set of actions
$\rightarrow$ is a family of binary relations on P, one for each member a of A, denoted $\rightarrow_A$

Then $\langle P, A, \rightarrow \rangle$ is finite branching if and only if for each $\rightarrow_A$ there is only a finite number of worlds to branch to:

$$\forall p \varepsilon P \{ q \,|\, p \rightarrow_A q \} \text{ is finite}$$

Much work in action logics has been directed towards program verification, concurrency analysis, etc. However action logic may be an appropriate formalism to describe other kinds of system. A number of results have been found for action logics, and these may have use in AI. For example, comparing actions may be required to identify sequences of actions that might be equivalent (Goldblatt, 1981).

8.6.2 Modal Action Logic

Modal action logic (MAL) was originally developed for software engineering (Goldsack, 1987; Maibaum, 1987). In essence, MAL contains a series of layers that allow the logic to be extended in various ways. The underlying logic is a many-sorted first-order predicate logic that is extended by agent/action modalities. Using this layer, structural rules can be described that relate actions to enabling and resulting contexts (states). A deontic layer then extends the logic to allow behavioural rules to be represented. These behavioural rules describe sequences of actions, combinations of actions, permission for actions and obliged actions. Finally a temporal layer allows the representation and manipulation of actions occurring in intervals of time, and conditions holding during periods.

This logic was originally developed for the formal requirements specification of real-time embedded systems (Jeremaes et al., 1986). However it is apparent that the logic may hold appeal to a wider range of activities. In particular the logic may be appropriate for AI reasoning.

For a structural rule, if C is an agent, A is an action and b is a formula which holds after C does A: this is expressed in the logic by the formula:

$$[C, A]b$$

b is the resulting context. In order to express some precondition, satisfied before an action occurs, i.e. an enabling context, the formula can be represented as:

$$a \Rightarrow (C, A)b$$

This is interpreted as if a holds, and C does A, then b will be the resulting state. Note a and b are MAL formulae.

Since MAL is intended to be used to describe systems, there is a requirement to describe the interaction or behaviour of actions. So for example, given a certain state there may be permission to carry out one or more actions, whereas, after an action has been completed some actions may no longer be permissible and/or further actions may be permissible. There may also be an obligation to carry out a particular action. Deontics are described as follows, where C is an agent and A is an action:

obl(C, A) is a formula which holds when an obligation exists for the agent C to perform the action A. This is a strong form of obligation; the next action by agent C must be A. It is blocked from performing other actions until the obligation is removed by performing the action, or until some other agent cancels the obligation.

> per(C, A) holds when C has permission to perform the action A
> $\neg$ per(C, A) holds if C is forbidden to perform A

So taking an example from Goldsack (1987):

> in_bed(john) $\Rightarrow$ [john, get_up] ($\neg$in_bed(john) & per(john, eat_breakfast))

which states that 'if john is in bed and john gets up, the resulting state is john is not in bed and it is permissible for john to eat breakfast'.

Modal action logic was developed to support the formal requirements specification of industrial real-time software. As such it is an appropriate formalism to represent a wide range of physical, and even conceptual, systems. For example, we could describe knowledge of the relationships between the throttle, gears, engine revolutions, fuel consumption, fuel quantity, etc. of a car. Some of the axioms are below, where 'succ' is the successor function or the 'add one function':

> gear(X) & engine_revs(R)
> $\Rightarrow$ [driver, change_gear_up] gear(succ(X)) & engine_revs(S) & $S \leq R$

This states that if the transmission is in gear X and the engine revs are R, then after the action change gear up, the gear is in the successor of X and the new engine revs, S, are less than R. Obviously if the transmission is in top gear, then the 'change_gear_up' is not possible, or forbidden, otherwise there is permission to conduct the action 'change_gear_up':

> gear(X) & $X = 4 \Rightarrow \neg$ per(driver, change_gear_up)
> gear(X) & $X \neq 4 \Rightarrow$ per(driver, change_gear_up)

If the engine revs increase so does the fuel consumption:

> fuel_consumption(J) & engine_revs(R) $\Rightarrow$
> [driver, accelerate] engine_revs(S) & $S > R$ & fuel_consumption(K) & $K > J$

Modal action logic provides a formalism that is appropriate for capturing real-world situations that involve actions and causality. There must be a large number of knowledge-based systems that have been restricted by traditional rule-based technology for engineering applications. Many of these would benefit from the relatively lucid representation, in terms of agents and actions, that this logic has to offer. Hence, building and updating would be assisted, and explanations would be more transparent.

8.6.3 Naive Physics

Naive physics was an attempt to stimulate an interest in formalising aspects of common-sense. Much of AI has been orientated towards either toy problems or towards developing an intelligent system, such as an expert system, for a narrow or restricted domain. The problem with developing a system that is designed for such a narrow domain is that, given a problem that is slightly outside the focus of the system, the system fails. The fall-off of understanding for the system is steep. This failing is often termed brittleness. By comparison, a human expert has a far shallower fall-off of understanding. Part of this shallower fall-off is due to knowledge on a wider range of related issues. However a significant contributory factor is the use of common-sense to solve problems.

The Naive Physics Manifesto (Hayes, 1979) was an attempt at addressing some of the problems of dealing with common-sense. In this, a proposal was made for constructing a formalism that encapsulates a 'sizable portion of common-sense knowledge about the everyday physical world: about objects, shape, space, movement, substances (solids and liquids), time, etc.'

In order to reason with common-sense about objects, it is argued that there should usually be rich linking between concepts. Such linking will reveal clusters of concepts, i.e. groups of concepts of particularly dense linking.

Of particular interest is the discussion of a (non-exhaustive) list of likely clusters and their concepts:

(1) Measuring scales and measuring spaces—including notions of approximation, nearness, 'typical' measures of various kinds, inequalities, etc.

(2) Shape, orientation and dimension—as discussed in the section above on spatial reasoning, this is a surprisingly undeveloped area. This proposal goes further in that consideration should include a wider range of notions, e.g. tall, wide, small, short, top, bottom, rim, lip, front, outline, back, etc.

(3) Inside and outside—an interesting motivation for this cluster is that 'containment can limit causality'. So for example, (door, portal, window, gate, way in, way out), (wall, boundary, container), (obstacle, barrier), (way past, way through).

(4) Energy and effort—including consideration of the distinction between things that 'happen naturally', e.g. things falling, and things that require some artificial intervention, e.g. throwing a rock in the air.

(5) Assemblies—including consideration of notions such as composed of, part of, member of, related to, etc.

(6) Support—including consideration of notions such as holding up, hanging from, attached from, floating, flying, etc.

(7) Substances and physical states—there are many different kinds of 'stuff', e.g. iron, water, wood, meat, stone, sand, etc, and these exist in different kinds of state, e.g. solid, liquid, powder, paste, etc. Furthermore, many kinds of stuff have a usual state, e.g. water as a liquid, sand as a powder, though these can change under various circumstances, e.g. wood can burn if it is hot. Sometimes, there are different terms for the same stuff in different states, e.g. water and ice, rock and sand, etc.

(8) Forces and movement—including notions of falling, pushing, pulling, rolling, trajectory, sliding, jumping, etc.

(9) Liquids—in the words of Hayes 'liquids pose special problems, since, unlike pieces of solid stuff, "pieces" of liquid are typically individuated not by being a particular piece

of liquid, but by being in a particular place (a lake), or in some special relationship to a solid object (inside a cup)'.

This work is not a solution to the problem of common-sense reasoning about the physical world. It is not even an attempt at a solution. It is more an attempt to initiate work directed towards such a solution. It is an interesting outline of the way such work may look, and has indeed helped to stimulate further work, including qualitative reasoning. There is, however, much opportunity for further work along the lines proposed, particularly in developing a logic-based formalism to represent naive physics.

8.6.4 Qualitative Reasoning

A non-logic approach to modelling causality that is receiving considerable interest in AI currently is qualitative reasoning. This is reasoning with a qualitative, as opposed to a quantitative model, of a system. In this section, consideration will be restricted to qualitative physics. As in traditional physics, qualitative physics is an attempt to model physical systems in the world and to provide mechanisms that predict and explain behaviour. However, qualitative physics diverges from traditional physics in that traditional physics uses models based on differential equations and provides analytical solutions, whereas qualitative physics provides results in the form of a 'common-sense description', by using qualitative differential equations that unlike ordinary differential equations are a function of a totally ordered set of discrete values.

Qualitative physics defines concepts in a formalisation that uses a symbolic representation. This representation avoids the complexity of the continuous real-valued variables and differential equations by only considering the underlying nature of the physical system. Useful inference can be drawn from a model that use only a small number values for qualitative variables. For the description of a quantity, the most important qualitative values are increase, decrease and no change; the corresponding derivatives being positive, negative and zero.

Qualitative physics is becoming an increasingly important research area in artificial intelligence. This is in part a response to the problems raised in the 'Naive Physics Manifesto' (Hayes, 1979). From the basic notion of qualitative physics there are a number of related theories (de Kleer, 1987).

By using a common-sense model of a physical system, it provides much essential and useful knowledge for many applications. For engineering applications it provides an attractive alternative to the difficult exercise of developing a traditional model. Interest in using such an approach in the engineering domain is increasing rapidly.

There are now a number of interesting experimental prototypes for the engineering domain, such as HELIX developed as a helicopter diagnostic system (Hamilton, 1988) and a qualitative analysis of MOS circuits (Williams, 1984). The core of HELIX is a qualitative model that represents a set of constraints that define normal behaviour of the engines, transmission, flight controls and rotors of the helicopter. HELIX assesses aircraft performance by determining whether observations (sensor readings and pilot control inputs) are consistent with the constraints of the model. If an inconsistency is detected, a process of systematic constraint suspension is used to test various failure hypotheses.

As expounded in de Kleer (1987), advantages of the qualitative approach include:

(1) it identifies all modes of system functioning;
(2) it functions with incomplete models;
(3) it functions with incomplete data;
(4) it is more efficient; and
(5) it provides explanations.

Currently the three most important ways of developing a qualitative description of a physical system are:

(1) Constraint-based—the physical system is described in terms of a set of variables and the constraints relating those variables. Furthermore landmark values, either numerical or symbolic, delimit the range of the variables.
(2) Process-based—the physical system is described in terms of the processes that are potentially present. A process is defined as something that causes changes in objects over time, e.g. heating, stretching, bending, boiling, flowing, etc.
(3) Component-based—the physical system is described in terms of the physical constituents.

A succinct overview of a protocol for using a qualitative simulation can be gained from (Kuipers, 1986):

> Qualitative simulation of a system starts with a description of the known structure of the system, and an initial state, and produces a directed graph consisting of the possible future states of the system and the "intermediate successor" relation between states. The possible behaviours of the system are the paths from the initial state through the graph.

In conclusion, proponents of qualitative reasoning believe that the approach is the most likely solution to the problem of common sense understanding. In engineering AI, it is likely that qualitative reasoning is likely to be crucial (Forbus, 1988). Reasoning in the engineering domain requires not only specific domain knowledge, but also a wider range of general knowledge. Traditional rule-based systems are narrow, and furthermore they are brittle. Qualitative seems to offer one of the more promising solutions to this.

8.6.5 Discussion

This section provides a view on a diverse range of topics. Actions logic, and more particularly temporally-based action logics, provide an intuitive representation for modelling a variety of dynamic systems. However, the 'Naive Physics Manifesto' provides a motivation for the development of more complex formalisms to capture a wider variety of notions of the real world.

Qualitative reasoning seems to offer a partial answer to some of the questions raised in the 'Naive Physics Manifesto'. However, other approaches do require development, and perhaps the interrelationships between quantitative reasoning and logic require exploration.

8.7 CONCLUSIONS

It is perhaps worth briefly considering the ramifications some of the work presented in this chapter will have on knowledge engineering.

Uncertainty reasoning is an important aspect of knowledge representation in a range of applications. A fuller understanding of the techniques available should be sought by anyone intending to engineer an intelligent system. Non-monotonic reasoning, apart from negation-as-failure, is unlikely to offer solutions to many problems at the current state of understanding. However, if AI is to achieve the aim of getting closer to formalising intelligence, non-monotonic reasoning capability will be critically important.

Now, temporal reasoning formalisms are being harnessed for AI systems under trial, and it is likely that in the not too distant future knowledge engineers will be using formalised temporal reasoning facilities in the same way production rules are straightforwardly used now.

As with non-monotonic reasoning, it is likely there will be some time before action, causality and qualitative reasoning become mainstream knowledge-engineering technology. However it seems difficult at this stage to envisage how AI can progress without exploiting such potential.

For engineering AI, this chapter is a snapshot of what is likely to be coming into the knowledge engineering toolkit in the not too distant future. However it is likely that much of what has actually been presented here will be superceded by more intuitive, effective and efficient formalisations. In fact in some of the work presented has already been updated, particularly in the area of temporal reasoning. However consideration of the underlying issues, as covered here, should pave the way to a deeper understanding of AI, and hence facilitate a more appropriate engineering solution.

In conclusion, it is worth reiterating the comment that the utility of the various approaches developed for AI reasoning should be examined in the context of knowledge engineering and of end use. There is a need to identify the suitability of the formalisation in user terms. This requires analysis of the efficiency of interaction, the minimisation of ambiguity, and the accuracy of the computed solution to that required by the application.

REFERENCES

Allen, J. F. (1984) Towards a General Theory of Action and Time, *Artificial Intelligence*, **23**, 123–154.

Allen, J. F. (1985) A Common-Sense Theory of Time, *Proc. Ninth Int. Conference on Artificial Intelligence*, Morgan Kaufmann Publishers Inc., Los Altos, CA.

Bundy, A. (1985) Incidence Calculus: A Mechanism for Probabilistic Reasoning, *Journal of Automated Reasoning* 1, 263–284.

Bundy, A. (1986) Correctness Criteria of Some Algorithms for Uncertainty Reasoning Using Calculus', *Journal of Automated Reasoning*, 2, 109–126.

Chellas, B. (1980) *Modal Logic: An Introduction*, Cambridge University Press, Cambridge.

Clark, K. (1978) Negation as Failure, In: *Logic and Databases*, Eds H. Galliere and J. Minker, Plenum Press, New York, pp 293–322.

Clocksin, W. F. and Mellish, C. S. (1980) *Programming in Prolog*, Springer-Verlag, Berlin.

de Kleer, J. (1987) Qualitative Physics, In: *Encyclopedia of Artificial Intelligence*, Ed. S.C. Shapiro, Wiley, New York, pp 807–814.

Fisher, R. B. (1987) Representing Three-Dimensional Structures for Visual Recognition, *Artificial Intelligence Review*, 1, 183–200.

Forbus, K. (1988) Intelligent Computer-Aided Engineering, *AI Magazine*, 9, (Fall), 23–36.

Fox, J. (1986) Knowledge, Decision Making and Uncertainty, In: *Artificial Intelligence and Statistics*, Ed. W. Gale, Addison-Wesley, Reading, MA, pp 57–76.

Ginsberg, M. (1987) *Readings in Non-monotonic Reasoning*, Morgan Kaufmann Publishers Inc., Los Altos, CA.

Goldblatt, R. (1981) Axiomatising the Logic of Computer Programming, *Lecture Notes in Computing Science*, Vol. 130, Springer Verlag, Berlin.

Goldsack, S. J. (1987) Specifying Requirements: An Introduction to the FOREST Approach, Department of Computing Research Report, Imperial College, London.

Green, S. (1986) *Proceedings of Expert Systems '86*, Cambridge University Press, Cambridge.

Haack, S. (1978) *Philosophy of Logics*, Cambridge University Press, Cambridge.

Hamilton, T. P. (1988) HELIX: A helicopter diagnostic system based on qualitative physics, *Artificial Intelligence in Engineering*, 3, 141–150.

Hanks, S. and McDermott, D. (1987) Non-Monotonic Logic and Temporal Projection, *Artificial Intelligence*, 33, 379–412.

Hayes, P. (1979) Naive Physics Manifesto, In: *Expert Systems in the Micro-electronic Age*, Ed. D. Michie, Edinburgh University Press, Edinburgh.

Hodges , W. (1977) *Logic*, Penguin, Harmondsworth, Middx.

Hughes, G. E. and Cresswell, M. J. (1972) *An Introduction to Modal Logic*, Methuen, London.

Jeremaes, P., Khosla, S. and Maibaum, T. S. E. (1986) A Modal (Action) Logic for Requirements Specification, In: *Software Engineering '86*, Eds P.J. Brown and D.J. Barnes, Peter Peregrinus, London.

Kowalski, R. A. (1979) *Logic for Problem Solving*, Elsevier-North Holland, Amsterdam.

Kowalski, R. A. and Sergot, M. (1986) A Logic-Based Calculus of Events, *New Generation Computing*, 4, 67–95.

Kuipers, B. (1986) Qualitative Simulation, *Artificial Intelligence* 29, 289–338.

Lewis, D. (1973) *Counterfactuals*, Harvard University Press, Cambridge MA.

Lifschitz, V. (1985) Closed World Assumption and Circumscription, *Artificial Intelligence*, 27, 229–235.

Lifschitz, V. (1987) Circumscription Theories: A Logic Based Framework for Knowledge-Representation (preliminary report), *Proc. 6th National Conf. on AI(AAAI87)*, Morgan Kaufmann Publishers Inc., Los Altos, CA, pp 364–368.

Lloyd, J. (1984) *Foundations of Logic Programming*, Springer-Verlag, Berlin.

McArthur, R. (1976) *Tense Logic*, Reidel, Dordrecht

McCarthy, J. (1980) Circumscription—A Form of Non-Monotonic Reasoning, *Artificial Intelligence*, 13, 27–39.

McCarthy, J. (1986) Applications of Circumscription to Formalizing Common-Sense Knowledge, *Artificial Intelligence*, 28, 89–116.

McCarthy, J. and Hayes, P. J. (1969) Some Philosophical Problems from the Standpoint of Artificial Intelligence, In: *Machine Intelligence* Vol. 4, Eds B. Meltzer and D. Mitchie, Edinburgh Univertsity Press, Edinburgh.

McDermott, D. and Doyle, J. (1980) Non-Monotonic Logic 1, *Artificial Intelligence*, **25**, 41–72.

McDermott, D. (1982) A Temporal Logic for Reasoning about Processes and Plans, *Cognitive Science*, **6**, 101–155.

Maibaum, T. S. E. (1987) A Logic for the Formal Requirements Specification of Real-Time Embedded Systems, Department of Computing Research Report, Imperial College, London.

Nilsson, N. J. (1986) Probabilistic Logic, *Artificial Intelligence*, **28**, 71–87.

Nute, D. (1988) Defeasible Reasoning and Decision Support Systems, *Decision Support Systems*, **4**, 97–110.

Nutter, J. T. (1987) Uncertainty and Probability, *Proc. Tenth Int. Conference on Artificial Intelligence, Milan*, Morgan Kaufmann Publishers Inc., Los Altos, CA.

Pollock, J. (1983) Epistemology and Probability, *Synthese*, **55**, 231–252.

Pollock, J. (1987) Defeasible Reasoning, *Cognitive Science*, **11**, 481–518.

Reiter, R. (1978) On Closed World Databases, In: *Logic and Databases*, Eds H. Galliere and J. Minker, Plenum Press, New York, pp 55–76.

Reiter, R. (1980) A Logic for Default Reasoning, *Artificial Intelligence*, **13**, 81–132.

Reiter, R. (1987) Non-Monotonic Reasoning, *Annual Review of Computer Science*

Rich, E. (1983) *Artificial intelligence*, McGraw Hill, New York.

Shoham, Y. (1987) Temporal Logics in AI: Semantical and Ontological Considerations, *Artificial Intelligence*, **33**, 89–104.

Shoham, Y. (1988) Chronological Ignorance: Experiments in Non-Monotonic Temporal Reasoning, *Artificial Intelligence*, **36**, 279–331.

Spiegelhalter, D. J. (1984) Statistical and Knowledge-Based Approaches to Clinical Decision-Support Systems, with an Application in Gastroenterology, *Journal of Royal Statistical Society*, Series A, **147**, 35–77.

Turner, R. (1984) *Logics for Artificial Intelligence*, Ellis Horwood, Chichester.

Walker, W. L., Herman, M. and Kanade, T. (1988) A Framework for Representing and Reasoning about Three-Dimensional Objects for Vision, *AI Magazine*, **9**(2), 47–58.

Williams, B. C. (1984) Qualitaive Analysis of MOS Circuits, *Artificial Intelligence*, **24**, 281–346.

Zadeh, L .A. (1965) Fuzzy Sets, *Information and Control*, **8**, 338–353.

Zadeh, L. A. (1975) Fuzzy Logic and Approximate Reasoning, *Synthese*, **30**, 407–428.

Zadeh, L. A. (1981) PROF—A Meaning Representation Language for Natural Language, In: *Fuzzy Reasoning and its Applications*, Eds E.H. Mamdani and B.R. Gaines, Academic Press, New York.

A.B. Hunter
Department of Computing
Imperial College of Science and Technology
South Kensington
London SW7 2BZ

9 Mathematical Foundations of Artificial Intelligence

D.K. Bose, and S.W. Ellacott

Brighton Polytechnic, UK

9.1 INTRODUCTION

Engineering AI applications try to construct an abstract representation of the data, physical laws and logistic constraints of the real-world problem under consideration. To do this a suitable abstract language is needed. It must be rich enough to model the wide variety of problems encountered and precise so that a rigorous model can be built. The language of mathematical analysis was found to be particularly suited to the classical problems of engineering, but in AI applications we have to model non-numeric data in the form of rules, and structural information often in the form of interrelationships. Discrete mathematics such as set theory, logic, etc. has been found to be a suitable language to describe these newer problems. We shall see how set theory helps in the construction of abstract data structures, how relations aid the study of problem structure and how predicate calculus is used to generate rules which embody knowledge about the problem. In essence, the aim of discrete mathematics is to help build an abstract 'mirror description' of the real world problem.

9.2 THEORY OF SETS

9.2.1 Collections of Objects

Given a collection of objects, we might be interested merely in what objects are in the collection, or in the order in which they occur and/or the frequency of each object (i.e. the number of times it occurs in the collection). If we are interested in neither the frequency nor the order, the collection is described by a set. Thus a *set* is a collection of *distinct* objects where the order is not important. If we are interested in the frequency of occurrence of objects, but not the order, the term *bag* is used to describe the collection. Conversely, if we are interested in order, but not frequency, we have an *n-tuple*. Finally, if both order and

Artificial Intelligence in Engineering. Edited by Graham Winstanley
ⓒ1991 by John Wiley & Sons Ltd.

frequency are important, we have a *list* or *sequence*. However, all these more complicated structures can be built up from the basic idea of a set.

9.2.2 Set Notation

The standard notation is to put the objects in the set between braces { }. Thus $\{a,b,c\}$ is a set which has a, b and c as its distinct objects. These objects are called the *elements* of the set. Two sets are equal if they have the same elements. Thus

$$\{a,b,c\} = \{c,a,b,\} = \{b,c,b,a,a\}.$$

Note that

$$\big\{\{a\}\big\} \neq \{a\}.$$

The set that contains no elements is called the empty set and is written { } or $\emptyset$.

Objects are present, or belong to a set ($\in$) or absent, or do not belong in it ($\notin$). Thus

$$a \in \{a,b,c\} \text{ but } d \notin \{a,b,c\}.$$

Sets can be defined by enumerating the elements as shown above, or by *set specifications*. The latter method defines a set which contains all elements satisfying some property. For example,

$$S = \{X \,|\, X \text{ is a pixel whose grey level } > 30\}.$$

Read this as 'The set of X, such that X is a pixel whose grey level is greater than 30'. Some authors use a colon ':' instead of the symbol '|'.

9.2.3 Set Manufacture

We now consider various operations which allow us to manufacture new sets from old ones.

T is said to be a *subset* of S if every element of T is an element of S. We can illustrate this using a *Venn diagram* (Figure 9.1).

We write $T \subset S$ ('T is contained in S') or $S \supset T$ ('S contains T'). A powerful constructor of new sets is the '*set of*' operation. This forms a set whose elements are subsets of S. We shall assume S is finite. Formally,

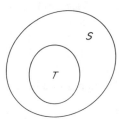

Figure 9.1 A subset

$$\text{set of } S = \{T \mid T \subset S\}.$$

For example, if Bool = {True, False}, then

$$\text{set of Bool} = \Big\{\{\,\}, \{\text{True}\}, \{\text{False}\}, \{\text{True, False}\}\Big\}.$$

Another name for set of S is the *power set* of S, written

$$P(S) \text{ or } 2^S.$$

2^S is an obvious notation because if S has n elements, the set of S has 2^n elements.

The *union* of two sets yields a set containing the elements of both sets. Thus the union of sets S and V, written $S \cup V$, is defined by

$$S \cup V = \{x \mid x \in S \text{ or } x \in V\}.$$

This can be seen diagrammatically in Figure 9.2.

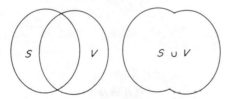

Figure 9.2 Union

For example if S is the set Bool above, i.e.

$$S = \{\text{True, False}\}$$

and

$$V = \{\text{Unknown, Meaningless}\},$$

then

$$S \cup V = \{\text{True, Meaningless, Unknown, False}\}.$$

Here are some simple facts about unions:

$$A \cup \{\,\} = A$$
$$A \cup B = B \cup A$$
$$A \cup (B \cup C) = (A \cup B) \cup C$$
$$A \cup A = A$$
$$A \subset B \text{ if and only if } A \cup B = B.$$

A dual notion to that of union is set *intersection*. The intersection of two sets S and V, written $S \cap V$, is the set which contains those elements common to both sets. Formally

$$S \cap V = \{x \mid x \in S \text{ and } x \in V\}.$$

The corresponding diagram is represented in Figure 9.3.

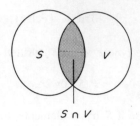

$$S \cap V$$

Figure 9.3 Intersection

For example if:

$$S = \{a, b, c, d\}$$

and

$$V = \{c, a, e, f\}$$

then

$$S \cap V = \{a, c\}.$$

Basic facts about intersections are:

$$A \cap \{\} = \{\}$$
$$A \cap B = B \cap A$$
$$A \cap (B \cap C) = (A \cap B) \cap C$$
$$A \cap A = A$$
$$A \cap B \text{ if and only if } A \cap B = A.$$

If $A \cap B = \{\}$ then A and B are said to be *disjoint* . Two useful facts relating $\cup$ and $\cap$ are the *distributive laws*:

$$A \cap (B \cup C) = (A \cap B) \cup (A \cap C)$$

and

$$A \cup (B \cap C) = (A \cup B) \cap (A \cup C).$$

Normally, one discusses subsets of some large set U called the *universal set*: for instance U might be the set of all real numbers, and we then talk about subsets of this. If we have a universal set U we have also the notion of the *complement* of a set.

The complement of S, written $\overline{S}$, is defined as:

$$\overline{S} = \{x \mid x \in U \text{ and } x \notin S\}.$$

This can also be written as $U - S$ where:

$$A - B = \{x \mid x \in A \text{ and } x \notin B\}.$$

These are illustrated in Figure 9.4.

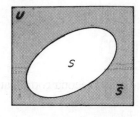 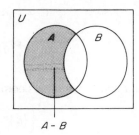

$A - B$

Figure 9.4 Complement and set subtraction

The duality of $\cup$ and $\cap$ is brought out by De Morgan's laws (see Denver, 1986):

$$\overline{A \cap B} = \overline{A} \cup \overline{B}$$

and

$$\overline{A \cup B} = \overline{A} \cap \overline{B}.$$

It is useful to have a set whose elements belong to one or other of two sets but not to both. This corresponds to the notion of an 'Exclusive OR' in logic. In set theory, this construct is known as the *symmetric difference*. Thus given sets A and B, the symmetric difference of A and B, written $A \triangle B$ is defined as

$$A \triangle B = \{x \mid x \in A \cup B \text{ and } x \notin A \cap B\}.$$

This completes our discussion of the elementary notions of sets.

9.2.4 Order

We now consider the connection between an ordered arrangement of objects, say, c, b, d, a, and a set of elements. (We shall closely follow the exposition in Halmos (1960).) First of all, we associate with each object a set whose elements are the objects that occur at or before that point. Thus c is associated with $\{c\}$, b with $\{c, b\}$, d with $\{c, b, d\}$ and the last object a with $\{c, b, d, a\}$. We now construct the set which has precisely these associated sets as elements, i.e. the set:

$$C = \Big\{\{d, a, b, c\}, \ \{b, c\}, \ \{c\}, \ \{d, c, b\}\Big\}.$$

Note that we have deliberately scrambled the order of the sets and the order of the elements of the set, since these are not important.

Conversely, given the set C, we can reconstruct the sequence c, b, d, a as follows:

> Take the intersection of all the sets that are the elements of C. This gives $\{c\}$. Thus c is the first item. Update C as $C - \{c\}$ and repeat. This gives $\{b, c\}$. Thus b is the second item, and so on.

As a particular case of the argument above, we have for a, b:

$$C = \left\{ \{a\}, \ \{a,b\} \right\}$$

and for b,a:

$$C = \left\{ \{b\}, \ \{a,b\} \right\}.$$

The *ordered pair* of a and b, with first coordinate a and second coordinate b, written (a,b), is the set

$$\left\{ \{a\}, \ \{a,b\} \right\}.$$

Note that $(a,b) = (x,y)$ if and only if $a = x$ and $b = y$. (Why?)

9.2.5 Cartesian Products

If A and B are sets, the set that contains all the ordered pairs (a,b) with $a \in A$ and $b \in B$ is called the *Cartesian Product* of A and B, and is written $A \times B$. Formally:

$$A \times B = \{(a,b) \mid a \in A \text{ and } b \in B\}.$$

Some basic properties are:

$$(A \cup B) \times C = (A \times C) \cup (B \times C)$$

and

$$(A \times B) \times C = (A \cap C) \cap (B \times C).$$

One can generalise the definition above to $A \times B \times C$ as being:

$$\{(a,b,c) \mid a \in A, \ b \in B, \ c \in C\}.$$

(a,b,c) can be thought of as the set:

$$\left\{ \{a\}, \ \{a,b\}, \ \{a,b,c\} \right\}.$$

Similarly:

$$A_1 \times A_2 \ \ldots \ \times A_n = \{(a_1, a_2, \ \ldots \ , a_n) \mid a_i \in A_i \text{ for each } i\}.$$

Note that this construction is appropriate for modelling records with fields A_1, A_2, ... , A_n.
As a simple example, let: A be the set of objects { man, mouse, horse } and B the set or attributes { 4legs, 2legs, furry }.
Thus:

$$A \times B = \{(\text{man,4legs}),(\text{man,2legs}), \ \ldots \ (\text{mouse,2legs}), \ \ldots \ (\text{horse,furry})\}.$$

Obviously, not all of these attributes are appropriate to all the objects: we wish to specify subsets of $A \times B$. This brings us to the subject of relations.

9.3 RELATIONS

Loosely speaking, a relation is a pairing of objects in one set A, to objects in another set B. For the example in the previous section, the pairing:

$$(\text{man,2legs}), (\text{mouse,4legs}), (\text{horse,4legs})$$

defines some sort of relationship which we might call 'legs'. More precisely, we can think of:

$$\text{legs} = \{(\text{man,2legs}), (\text{mouse,4legs}), (\text{horse,4legs})\} \subset A \times B$$

as a relation.

Another, more natural example of a relation as a set of pairings of objects, is when A is a set of men, B is a set of women and we consider the relation 'is married to'. This obviously again gives a subset of $A \times B$, namely those (a, b) such that a is married to b.

Formally, we say that R is a *relation* from A to B if $R \subset A \times B$. If $(a, b) \in R$, we say that a stands in the relation R to b, which we sometimes write in the shorthand notation aRb. This is actually the definition of a *binary relation*. We can generalise to higher order relations as subsets of $A_1 \times A_2 \times \ldots \times A_n$. Consider, for instance the ternary relations 'a and b are the parents of c' or (thinking of real numbers) 'x lies between y and z'. We can define the latter formally as:

$$\text{lies\_between} = \{(x, y, z) | y < x < z \text{ or } z < x < y\}.$$

Two particularly important types of relation are order relations and equivalence relations.

9.3.1 Order Relations

For a binary relation R to qualify as an order relation it should satisfy the same properties that $\leq$ has on numbers. We know

$a \leq a$.
If $a \leq b$ and $b \leq a$ then $a = b$.
If $a \leq b$ and $b \leq c$ then $a \leq c$.

Thus a natural definition of an *order relation* from A to A is a relation R, such that:

aRa (We say R is *reflexive*).
If aRb and bRa then $a = b$ (R is *antisymmetric*).
If aRb and bRc then aRc (R is *transitive*).

As a simple example, consider a set S and let A = set of S (see Section 9.2.3).
We may order the elements of A as: XRY if $X \subset Y$.

9.3.2 Equivalence Relations

Just as an order relation generalises the notion of $\leq$ on the set of real numbers IR, so an equivalence relation generalises the notion of $=$ on IR. R is an *equivalence relation* from A to A if and only if:

> aRa.
>
> If aRb then bRa (R is *symmetric*).
>
> If aRb and bRc then aRc.

For example, if A is a set of towns and aRb means that a and b are in the same county.

If R is an equivalence relation, the set of objects b such that aRb for any given object a is called the *equivalence class* of a. Obviously any object a is related to every object in its equivalence class, and to no object in any other equivalence class.

9.3.3 Converse Relations and 01-matrices

Given a relation R from A to B, we can define a converse relation from B to A by interchanging the order in each ordered pair.

Formally, the *converse* of R, denoted R^{-1}, is the relation from B to A defined by:

$$R^{-1} = \{(b,a) \mid (a,b) \in R\}.$$

If the sets A and B are finite, we can represent a binary relation R from A to B by means of a 01-matrix as shown in Figure 9.5 where $r_{ij} = 1$ if $(a_i, b_j) \in R$ else $r_{ij} = 0$. (What is the matrix representation of R^{-1}?)

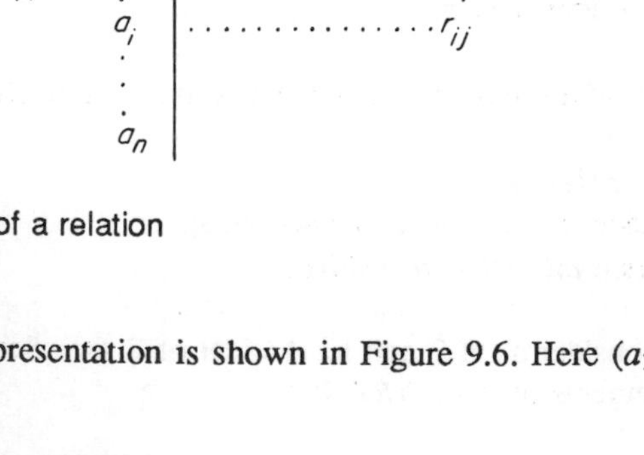

$$
\begin{array}{c|ccccccc}
 & & & & B & & & \\
R & b_1 & b_2 & \cdots\cdots & b_j & \cdots\cdots & b_m \\
\hline
a_1 & & & & \cdot & & \\
a_2 & & & & \cdot & & \\
\cdot & & & & \cdot & & \\
\cdot & & & & \cdot & & \\
a_i & \cdots\cdots\cdots\cdots\cdots & r_{ij} & & \\
\cdot & & & & & & \\
\cdot & & & & & & \\
a_n & & & & & &
\end{array}
$$

Figure 9.5   01 matrix of a relation

Another pictorial representation is shown in Figure 9.6. Here $(a_1, b_1) \in R$, $(a_1, b_i) \in R$ etc.

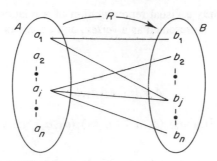

Figure 9.6 A relation

9.4 FUNCTIONS

If, in Figure 9.6, we have at most one arc emanating from any $a_i \in A$ to a $b_j \in B$, we say that the relation R is a partial function from A to B.

Formally, f is a *partial function* from A to B if, and only if, $f \subset A \times B$ and from each $a \in A$ there is at most one $b \in B$ with $(a,b) \in f$. We write $f : A \to B$, and $f(a) = b$. If, in addition, for *every* $a \in A$ there is a $b \in B$ such that $(a,b) \in f$, f is said to be a *function*, and in this case we call A and B respectively the *domain* and *codomain* of f.

Note that if $f : A \to B$ is a partial function, we can make it into a function $: A \to B \cup \{undefined\}$ by defining $f(a) = undefined$ whenever there is no $b \in B$ with $(a,b) \in f$. This trick is useful when we wish to treat computer software as a function, as we may wish not to specify a value for all cases. Of'course, *undefined* does not really mean the function is not defined: it is defined to have this value! However this case is usually treated differently.

The idea of a function allows us to define what we mean by a list of elements A. Let *IN* denote the natural numbers, i.e.

$$IN = \{1, 2, 3, \ldots\}.$$

A *list of elements* of A is a function $: IN \to A$.

A *string over A* is a function $f : \{1, 2, 3, \ldots n\} \to A$. For example, let $A = \{a, b, c\}$ and (with $n = 4$) let:

$$f(1) = b, \qquad f(2) = b, \qquad f(3) = a, \qquad f(4) = a.$$

Then we have the string '*bbaa*'.

9.5 GRAPH THEORY

One of the most powerful tools for studying the structure of reasoning systems is the directed graph. Roughly speaking, they can be used to study the global structure from local information.

A *directed graph* is simply a visual representation of a relation R from a finite set A to itself. Each element of A is represented as a *vertex*, and an *arc* or *edge* is drawn from a_i to a_j if and only if $(a_i, a_j) \in R$.

For example if:

$$A = \{a_1, a_2, a_3, a_4\}$$

and

$$R = \{(a_1, a_2), (a_1, a_3), (a_2, a_3), (a_3, a_1), (a_4, a_2)\}$$

then the representation is as shown in Figure 9.7. Note the arrows in the diagram to indicate the direction of the arcs. The case when the arcs do not have direction yields the simpler structure called a *graph*, but applications in artificial intelligence are almost always for directed graphs. The name is sometimes abbreviated to *digraph*.

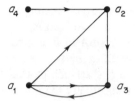

Figure 9.7 A directed graph

9.5.1 Graphs and Problem Representation

The study and possible solution of a given problem can be helped by starting off with a precise and suitably tailored representation of the problem. This normally takes the form of a mathematical model of the problem called a *state space description*. In many finite problems, graph theory provides the appropriate language for defining the model. We shall illustrate this by some examples, starting with two simple puzzles (cf. Deo, 1974, Doughety and Giardina, 1988).

9.5.1.1 The Decanting Problem

You are given three vessels A, B and C of capacities 8, 5 and 3 pints, respectively. A is filled while B and C are empty. Divide the liquid in A into two equal quantities.

Let a, b and c be the amounts of liquid in A, B and C respectively. Then $a + b + c = 8$ at all times. Further, assuming we cannot judge a half full vessel (otherwise the problem is trivial!) at least one of the vessels is always empty or full. Hence least one of the following equations must hold:

$$a = 0, \qquad a = 8, \qquad b = 0, \qquad b = 5, \qquad c = 0, \qquad c = 3.$$

With these constraints there are 16 possible problem states:

$$(8,0,0)\ (7,1,0)\ (6,2,0)\ (5,3,0)$$
$$(4,4,0)\ (3,5,0)\ (7,0,1)\ (2,5,1)$$
$$(6,0,2)\ (1,5,2)\ (5,0,3)\ (4,1,3)$$
$$(3,2,3)\ (2,3,3)\ (1,4,3)\ (0,5,3)$$

Note that only $(4,4,0)$ has A divided equally: this is therefore our *target* or *goal* state. We now model the problem by letting each 'problem state' be represented by a vertex (a,b,c). If we can move from one state to another by a decanting, then we join the corresponding vertices by an arc with an arrow to show which was the original state, and which the state after decanting. When we do this for all the possible states and decantings, we end up with a directed graph, which is shown (with some of the arcs omitted) in Figure 9.8.

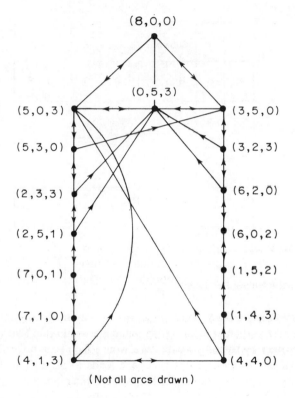

Figure 9.8 The decanting problem

The following sequences of decantings will divide A equally:

$$(8,0,0) \to (3,5,0) \to (3,2,3) \to (6,2,0) \to (6,0,2) \to (1,5,2) \to (1,4,3) \to (4,4,0)$$

or $(8,0,0) \to (5,0,3) \to (5,3,0) \to (2,3,3) \to (2,5,1) \to (7,0,1) \to (7,1,0)$
$$\to (4,1,3) \to (4,4,0).$$

These routes from one vertex to another are called *paths*. Since all the arcs in the second path can also be traced in reverse, we could also perform a circular 'walk' from (8,0,0) via (4,4,0) using these two paths. Note that we do not meet any vertex or edge twice. This is an example of a (directed) *circuit* or *cycle* in the graph.

9.5.1.2 Instant Insanity

Next we consider a simple example of the type of game often considered in AI research (see Shirai and Tsujii, 1982).

We are given four cubes. The six faces of every cube are variously coloured blue, green, red or white as shown (when unfolded) in Figure 9.9. Is it possible to stack the cubes, one on top of the other, to form a column such that no colour appears twice on any four sides of the column?

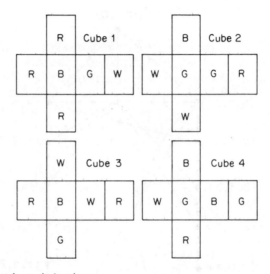

Figure 9.9   The 'Instant Insanity' cubes

We might first try to solve the problem in a similar way to the decanting problem. Each vertex could represent a particular stacking, and an edge would join two vertices if we could go from one to the other by rotating a cube. We would then search for the path to the goal state from the initial one. However, this approach leads to a graph with 331 776 vertices (the possible stackings), so this is not really helpful. On the other hand, the following graph representation (due to de Carteblanche in 1947) is perfect!

Let the four colours B,G,R and W be represented by four vertices: one for each colour. Pick a cube, say cube 1, and represent its three pairs of opposite faces by three edges drawn between the vertices with appropriate colours. Repeat this for the three other cubes. We then have the graph shown in Figure 9.10, in which the labels on the edges refer to the cube number.

Now consider the two opposite vertical sides of the desired stacking: say facing North–South. A *subgraph* (i.e. a collection of vertices and edges from the graph) with four edges will represent these eight faces. Each edge will have a different label 1, 2, 3 and 4. Each vertex of this subgraph will have *degree* two (i.e. two edges incident on it), since

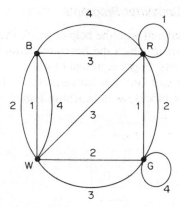

Figure 9.10 The 'Instant Insanity' state space

each colour appears exactly twice on the two vertical sides. The same argument applies to the East–West faces, and obviously the two subgraphs should not have any edges in common. Thus the cubes can be stacked as required if and only if there exist two edge-disjoint subgraphs each with four edges, each of the edges labelled differently and such that each vertex has degree two. For the set of cubes above we have the two subgraphs of Figure 9.11, which satisfy the conditions above. We can use these to construct the N–S and E–W faces of the required column (see Figure 9.12).

Is this the only solution?

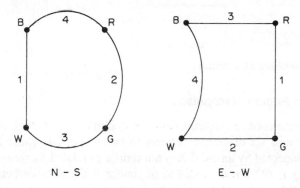

Figure 9.11 The 'Instant Insanity' solution graphs

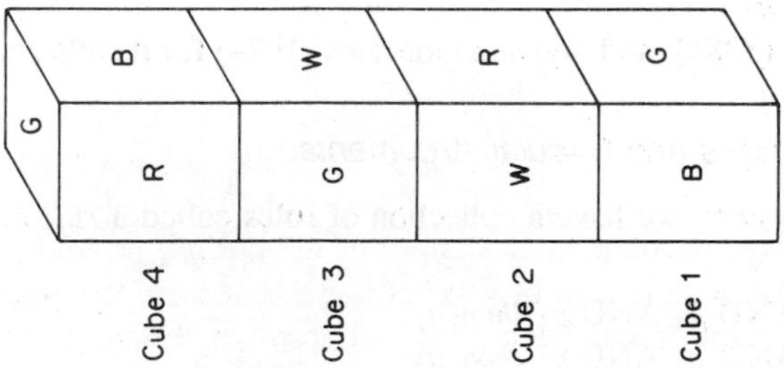

Figure 9.12 'Instant Insanity' solved!

9.5.1.3 Representation of Computer Programs

Computer programs can be analysed with the help of graph theory. The graph is essentially an abstraction of the flow chart in which the boxes are shrunk to vertices and the arrows become the edges. More precisely, edges represent blocks of statements each with one entry (the first instruction) and one exit (the last instruction). Each edge represents a possible transfer of control between the blocks, as indicated in Figure 9.13.

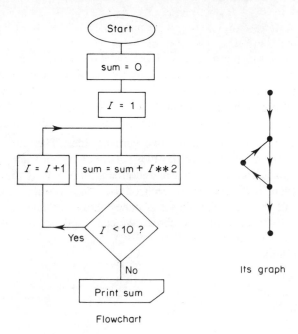

Figure 9.13 A flow chart as a graph

9.5.1.4 Graphs in Pattern Recognition

Graph theory has been used to construct pattern recognition algorithms. A *pattern graph* is first constructed from a set of sample patterns to be recognised. These form the vertices. Two vertices are connected by an arc if they are similar or related by some suitable measure.

For example, $x, y \in IR^N$ could be said to be similar if the distance between them is less than some specified threshold. One can then classify a given pattern vector z by seeking a subgraph of the pattern graph $\cup\{z\}$ which is a *clique*, i.e. every two vertices of this subgraph are connected by an arc.

See Johnsonbaugh (1984), and Tou and Gonzales (1974) for details and further examples.

9.5.1.5 Expert Systems and Circular Arguments

In a typical expert system, we have a collection of rules called a *rule base* of the form:

Rule1: *If* p_1 AND p_3 AND p_4 *then* q_1
Rule2: *If* q_1 AND p_4 AND p_2 *then* q_2
Rule3: *If* q_2 AND p_4 *then* p_1

Here the p_i and q_i are statements that are either true or false. (See Section 9 below for a more detailed description, and Chapters 2 and 3 of this book for a practical discussion on inference methodology.) This particular rule base can give rise to a circular argument. To see how this occurs, we note that expert systems usually work by a method called *backward chaining*. The user might specify that (s)he wishes to know if the proposition q_1 is true. A procedure called the *inferencing engine* examines the rule base to find a rule able to conclude q_1, and finds Rule1. It then proceeds recursively to try and conclude the truth of the so called *premises* of Rule1, namely p_1, p_3 and p_4. To conclude p_1, Rule3 is used (*'fired'*, in expert system terminology.) A premise of Rule3 is q_2, so Rule2 is fired. But a premise of Rule2 is q_1, the conclusion we are trying to verify!

Clearly a rule base of this type is undesirable, and many expert systems are equipped with routines to detect this situation. They do this by means of a graph theoretic representation of the rule base. The statements (the p_i and q_i) form the vertices of the graph, and we draw an arc from a vertex representing a conclusion of a rule, to the vertices that are its premises. Thus for the rule base above we arrive at the representation shown in Figure 9.14.

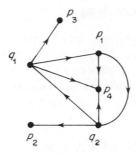

Figure 9.14 Graph of a cyclic rule base

It can be seen that there is a directed cycle $q_1 p_1 q_2 q_1$ corresponding to our circular argument. Thus we have reduced our problem of checking the rule base, to a standard graph theorectic problem of finding cycles. An algorithm for doing this is given in Section 9.5.2.6.

9.5.1.6 'is-a' Graphs and Non-monotonicity

Suppose we have the following statements:

> John is a cricketer and plays rugby. Cricketers are gentlemen while rugby players are rough. Also, gentlemen never shout, while rough people are loud.

We can represent this information by the graph shown in Figure 9.15. Paths in a graph such as Figure 9.15 correspond to lines of argument. The graph of Figure 9.15 represents an ambiguous situation, since John seems to be both loud and not loud. To resolve contradictions in such arguments, we require 'backtracking' and the rejection of previous tentative conclusions: this form of reasoning is called *non-monotonic logic*. This topic is of considerable importance in practical AI, and reference should be made to Chapter 8 of this book for further discussion.

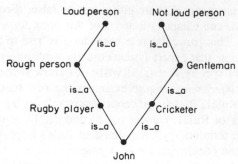

Figure 9.15 An 'is_a' graph

9.5.2 Fundamentals of Graph Theory

We begin by summarising and formalising the ideas about graphs introduced above:

> A *graph* $G = (V, E)$ consists of a set of objects $V = \{v_1, v_2, \ldots v_n\}$ called *vertices* and another set $E = \{e_1, e_2, \ldots e_n\}$ called *edges* or *arcs*. Each edge e_k is identified with a pair of vertices $v_i v_j$. If the order of v_i and v_j is significant, then we direct the edge with an arrow, otherwise the edge is undirected. Thus in the first case we have a *directed* graph, otherwise it is undirected and simply called a graph (see Figures 9.16 and 9.17).

The number of edges identified with a particular vertex v is called the *degree* of V.

9.5.2.1 New Graphs from Old

It is often convenient to consider a large graph as a combination of small ones and to derive its properties from those of the small ones. The *union* of graphs $G_1 = (V_1, E_1)$ and $G_2 = (V_2, E_2)$ is another graph G_3 written $G_1 \cup G_2$ whose vertex set $V_3 = V_1 \cup V_2$ and edge set $E_3 = E_1 \cup E_2$. The *ring sum* of G_1 and G_2, written $G_1 \oplus G_2$ is the graph whose vertex set is $V_1 \cup V_2$ and whose edges are those either in G_1 or G_2 but *not* in both. For an example of this, see Figure 9.18.

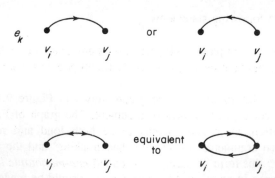

Figure 9.16 Edges and vertices

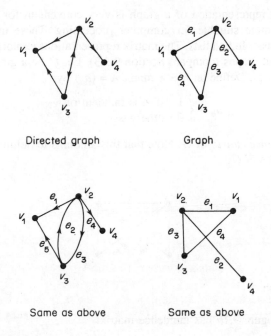

Figure 9.17 Types of graph

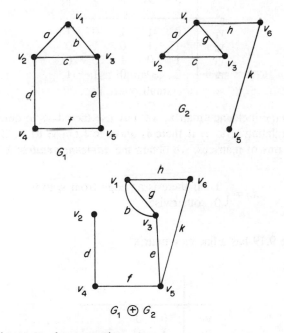

Figure 9.18 The ring sum of two graphs

9.5.2.2 *Computer Representations of Graphs*

Although a pictorial representation of a graph is very convenient for a visual study, other representations are more suitable for computer processing. These usually employ either matrix representations or linked lists. The matrix representations exploit obvious connections between relations and graphs (compare Section 9.3.3). Let G be a graph with n vertices, e edges and no self loops. Define an $n \times e$ matix $A = (a_{ij})$ by:

$$a_{ij} = \begin{cases} 1 & \text{if } e_j \text{ is incident on } v_1 \\ 0 & \text{otherwise} \end{cases}$$

A is called the *incidence matrix* of G. Note that this amounts to defining a relation between the vertices and edges of G.

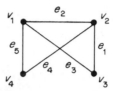

Figure 9.19 A graph

The example of Figure 9.19 has incidence matrix:

	e_1	e_2	e_3	e_4	e_5
v_1	0	1	1	0	1
v_2	1	1	0	1	0
v_3	1	0	1	0	0
v_4	0	0	0	1	1

As an alternative to the incidence matrix, we may use the edges to define a relation on the set of vertices, by relating v_i to v_j if there is an edge from v_i to v_j: if so v_j is said to be *adjacent* to v_i. In terms of matrices, we obtain the *adjacency matrix* $X = (x_{ij})$ by:

$$x_{ij} = \begin{cases} 1 & \text{if there is an edge from } v_i \text{ to } v_j \\ 0 & \text{otherwise.} \end{cases}$$

The graph of Figure 9.19 has adjacency matrix

	v_1	v_2	v_3	v_4
v_1	0	1	1	1
v_2	1	0	1	1
v_3	1	1	0	0
v_4	1	1	0	0

Note that, since the graph is not directed, the adjacency matrix is symmetric. With this representation, we do not need the restriction of no self loops: these simply appear as diagonal elements.

For many problems, the closely related *adjacency list* representation is preferable to either of the above. (We shall use it in path-finding algorithms.) In this case we associate a linked list L_v with each vertex v, consisting of the vertices adjacent to v (see Figure 9.20 for an example).

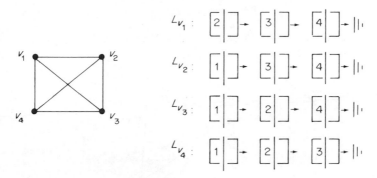

Figure 9.20 A graph and its adjacency lists

One way of implementing these lists is to use arrays $E(j,k)$, $1 \le k \le 2$, and $P(i)$, $1 \le i \le n$, where $E(j,1)$ stores a vertex number, $E(j,2)$ stores a link and $P(i)$ points to the start of the list for v_i in E. If v_i has no adjacent vertices we set $P(i) = 0$, otherwise $E(P(i),1)$ is adjacent to v_i.; similarly if $E(P(i),2) \ne 0$, then $E(E(P(i),2),1)$ is adjacent to v_i.

In Section 9.5.2.5 we shall introduce another efficient way of describing a graph called the 'fundamental cycle matrix'.

9.5.2.3 Paths and Components

One of the first things to find out about a graph is whether it is in 'one piece' or has various pieces (or components) as shown in Figure 9.21.

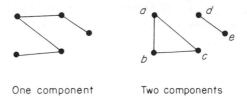

One component   Two components

Figure 9.21 Components of graphs

To make these ideas precise, we first of all define what is meant by a *path* from vertex v_0 to vertex v_r (not necessarily distinct). This is defined to be a sequence of alternating vertices and edges of the form:

$$\begin{array}{cccccc} e_1 & e_2 & e_3 & e_{r-1} & & e_r \\ v_0 & v_1 & v_2 & \cdots & v_{r-1} & v_r \end{array}$$

where e_j is an edge of the graph from v_{j-1} to v_j, and all the edges e_j $(j = 1, \dots, r)$ and vertices v_j $(j = 0, \dots, r - 1)$ are distinct. If $v_0 = v_r$ the path is called a *cycle* or *circuit*.

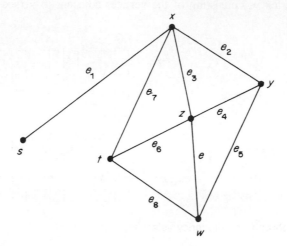

Figure 9.22 Paths in a graph

With reference to Figure 9.22,

$$\begin{array}{ccccc} e_1 & e_2 & e_5 & e_8 \\ s & x & y & w & t \end{array}$$

is a path from s to t.

$$\begin{array}{cccccc} e_1 & e_2 & e_4 & e_3 & e_7 \\ s & x & y & z & x & t \end{array}$$

is *not* a path, since it goes through x twice.

A graph is said to be *connected* (i.e. in one piece) if there is at least one path between every pair of vertices. It is easy to see that a disconnected graph consists of two or more connected graphs called *components*. Note that the collection of sets of vertices, representing the vertices of the various components, partitions the set of all vertices. Thus in the second graph of Figure 9.21, these subsets are $\{a, b, c\}$ and $\{d, e\}$.

9.5.2.4 Trees

Probably the most important type of graph is the *tree* (Figure 9.23). This is a connected graph without any circuits. It is not difficult to prove the following (see Deo (1974) for details):

A graph G, with n vertices, is a tree if any of the following conditions hold:

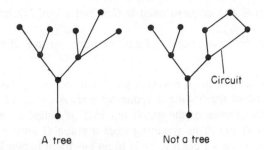

Figure 9.23 A tree or not a tree, that is the question!

(1) G is connected and has $n - 1$ edges.
(2) There is exactly one path between every pair of vertices in G.
(3) G is a minimally connected graph, i.e. removal of any edge disconnects the graph.

A *binary tree* is defined as a tree in which there is exactly one vertex of degree 2, and all the other vertices have degree 1 or 3. The vertex of degree 2 is called the *root* of the tree, and vertices of degree 1 are called *leaves*; for example, see Figure 9.24. Trees are used extensively in data organisation, variable length binary codes, etc.

Figure 9.24 A binary tree

Now suppose we have a connected graph G which is not a tree. We pose the question,

'Does there exist a subgraph of G which contains all the vertices of G, and is a tree?'

If it exists, it will look like a 'skeleton' of G, as shown in Figure 9.25. In fact, such a tree (called a *spanning tree* of G) always exists. To show this (and to find the tree), we proceed as follows:

Figure 9.25 Graph skeletons

(1) Locate a circuit in G (which must exist if G is not a tree). Delete an edge from it. G is still connected (why?).
(2) Check if there are any more circuits. If not, we have our tree. If there are , repeat from 1).

The number of spanning trees in an n-vertex graph can be as high as n^{n-2}. Later we shall give an efficient algorithm for finding a spanning tree. An edge of a spanning tree T is called a *branch*, while an edge of the graph not in T is called a *chord* (or *tie* or *link*)). Note that with respect to any of its spanning trees a graph G with n vertices and e edges has $n - 1$ branches and $e - n + 1$ chords. Thus in an electrical network with e elements and n junctions, the removal of $e - n + 1$ elements will remove all circuits in the network.

9.5.2.5 Fundamental Cycles

Let T be a spanning tree in a connected graph G. Adding a chord to T will create exactly one circuit. Such a circuit is called a *fundamental circuit* or *fundamental cycle*. Since each chord gives rise to exactly one fundamental cycle, there are $e - n + 1$ fundamental cycles in a connected graph with n vertices and e edges, as can be seen in Figure 9.26.

A set of fundamental cycles with respect to any spanning tree are the only 'independent' circuits in a graph. The rest of the circuits can be obtained as ring sums of these circuits (see Figure 9.27).

The ideas illustrated in Figures 9.26 and 9.27 can be expressed in terms of matrices and row operations. The *circuit matrix* $B = (b_{ij})$ of a graph G with q different circuits and e edges is a $q \times e$ matrix defined as

$$b_{ij} = \begin{cases} 1 & \text{if the } i\text{th circuit includes the } j\text{th edge} \\ 0 & \text{otherwise .} \end{cases}$$

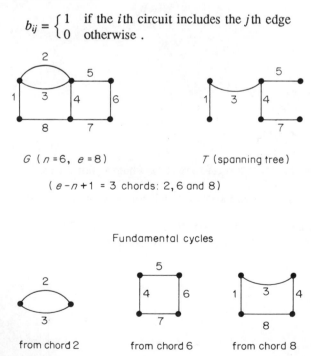

$G\ (n=6,\ e=8)$ T (spanning tree)

$(e-n+1 = 3$ chords: 2, 6 and 8)

Fundamental cycles

from chord 2   from chord 6   from chord 8

Figure 9.26   Spanning trees and fundamental cycles

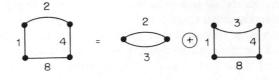

and

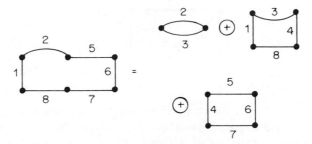

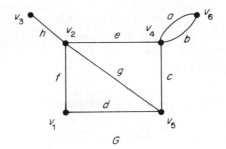

Figure 9.27 Cycles as ring sums of fundamental cycles

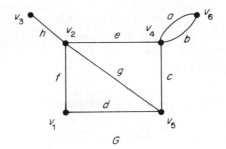

Figure 9.28 A graph G with four circuits

The graph shown in Figure 9.28 has four different circuits:

        (1.) $\{a,b\}$

        (2.) $\{c,e,g\}$

        (3.) $\{d,f,g\}$

and

        (4.) $\{c,d,f,e\}$.

Therefore the circuit matrix is:

$$
B = \begin{array}{c}
 \\ 1 \\ 2 \\ 3 \\ 4
\end{array}
\begin{array}{c}
\begin{array}{cccccccc} a & b & c & d & e & f & g & h \end{array} \\
\left(\begin{array}{cccccccc}
1 & 1 & 0 & 0 & 0 & 0 & 0 & 0 \\
0 & 0 & 1 & 0 & 1 & 0 & 1 & 0 \\
0 & 0 & 0 & 1 & 0 & 1 & 1 & 0 \\
0 & 0 & 1 & 1 & 1 & 1 & 0 & 0
\end{array}\right)
\end{array}.
$$

The *fundamental cycle matrix* B_f is a submatrix of the circuit matrix in which all rows correspond to a set of fundamental circuits with respect to a given spanning tree. Normally the columns of B_f are arranged so that all the $e - n + 1$ chords corresponding to the first $e - n + 1$ columns, and the rows are arranged so that the first row corresponds to the fundamental cycle made by the chord in the first column, the second row to the fundamental cycle made by the second, and so on.

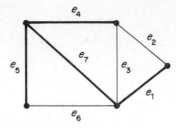

Figure 9.29 A graph and a spanning tree

Figure 9.29 provides an example. The fundamental cycle matrix is:

$$B_f = \begin{array}{c} \\ 2 \\ 3 \\ 6 \end{array} \begin{array}{ccc} e_2 & e_3 & e_6 \\ \left(\begin{array}{ccc} 1 & 0 & 0 \\ 0 & 1 & 0 \\ 0 & 0 & 1 \end{array} \right. \end{array} \begin{array}{c} : \\ : \\ : \end{array} \begin{array}{cccc} e_1 & e_4 & e_5 & e_7 \\ 1 & 1 & 0 & 1 \\ 0 & 1 & 0 & 1 \\ 0 & 0 & 1 & 1 \end{array} \left. \begin{array}{c} \\ \\ \end{array} \right) .$$

Note that B_f is an $(e - n + 1) \times e$ matrix and arranged as above it can be written:

$$B_f = \left(I_\mu : B_t \right)$$

where I_μ is the identity matrix of order $\mu = e - n + 1$, and B_t is the remaining $\mu \times (n - 1)$ submatrix.

If we regard the entries of the incidence matrix A, circuit matrix B and fundamental cycle matrix B_f as being added according to the rules:

$$1 + 0 = 0 + 1 = 1$$

and

$$0 + 0 = 1 + 1 = 0$$

then taking the ring sum of two circuits corresponds to adding the corresponding rows.
To illustrate this with the example above,

$$\begin{aligned} (\text{circuit 2}) \oplus (\text{circuit 6}) &= (1001101) + (0010011) \\ &= (1011110) \\ &= e_2 e_6 e_1 e_4 e_5 \\ &= e_1 e_2 e_4 e_5 e_6 . \end{aligned}$$

as shown in Figure 9.30.

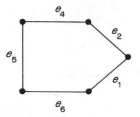

Figure 9.30 The ring sum of two circuits from Figure 9.29

One can also show that:

$$AB^T = BA^T = 0.$$

where ^{T} denotes the matrix transpose, and we assume that the columns are arranged using the same order of edges. Other important identities relating A, B, B_f, B_t etc. may be found in Seshu and Reed (1961).

9.5.2.6 An Algortihm for Spanning Trees and Fundamental Cycles

Given a graph theoretic representation of a problem (e.g. as in Section 9.5.1), we may wish to investigate all the vertices. In order to make sure we visit each precisely once, it is desirable to construct a spanning tree (otherwise our search strategy might literally cause us to go round in cycles). The following algorithm will construct the tree.

We assume that the graph G is represented in adjacency list form (see Section 9.5.2.2).

(1) Choose a vertex for the root. Label it 'active'.
(2) Find an edge incident with the active vertex.
(3) If the other end is a vertex already encountered, flag it as examined and ignore it (otherwise we would complete a cycle). Repeat this step until we find an adjacent vertex not yet visited.
   If one such is found, label this vertex active and repeat from (2)
   Else backtrack to a previous active vertex. Label this 'active' and repeat from (2).

The algorithm terminates when there are no more unused edges for the root. If G is connected, then the algorithm generates a spanning tree of G. If not, it generates a spanning tree for the component of G in which the root lies.

The algorithm is example of the general graph searching technique known as 'depth-first search'. As an example, consider the seven vertex graph shown in Figure 9.31 which has adjacency list:

$$
\begin{aligned}
&1: 2,5,6,7 \\
&2: 1,3,5 \\
&3: 2,4,5 \\
&4: 3,5,6 \\
&5: 1,2,3,4 \\
&6: 1,4,7 \\
&7: 1,6
\end{aligned}
$$

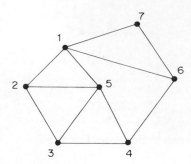

Figure 9.31 Graph to illustrate the algorithm for fundamental cycles

The algorithm proceeds as follows:
Choose vertex 1 as the root.

Edge	Action taken	Active vertex
1–2	Include in spanning tree	2
2–3	Include in spanning tree	3
3–4	Include in spanning tree	4
4–5	Include in spanning tree	5
5–1	Reject this edge	5
5–2	Reject this edge	5
5–3	Reject this edge	5
None left on 5	Backtrack to 4	4
4–6	Include in spanning tree	6
6–1	Reject this edge	6
6–7	Include in spanning tree	7
7–1	Reject this edge	7
None left on 7	Backtrack to 6	6
None left on 6	Backrack to 4	4
None left on 4	Backrack to 3	3
None left on 3	Backrack to 2	2
None left on 2	Backrack to 1	1
None left on 1	Terminate	

The spanning tree thus consists of the edges

$$1–2,\ 2–3,\ 3–4,\ 4–5,\ 4–6\ \text{and}\ 6–7,$$

as shown in Figure 9.32.

Whenever an edge e is examined and rejected in the algorithm above, the edge would make a fundamental circuit. By recording this circumstance, and retaining the list of predecessors used for the backtracking, we can also generate the fundamental cycles corresponding to the spanning tree. (To generate the actual cycles, we just follow the predecessor list until we reach the vertex at the other end of the edge forming the cycle.)

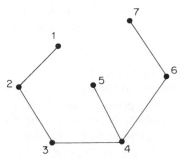

Figure 9.32 A spanning tree for the graph of Figure 9.31

9.5.2.7 Other Aspects of Graph Theory

There are many important areas we have not covered here. Amongst these should be mentioned *cut-nets* and *network flows*. Details may be found in Deo (1974) and Seshu and Reed (1961).

Some generalisations of graphs are also worthy of note. If the underlying relation is not binary, we can employ *hypergraphs* and *matroids*. If the structure develops with time (e.g. a state space description of logic circuits), *Petri nets* are appropriate. *Q Analysis* is used for investigating the actual structure of relations. For further information, see Johnsonbaugh (1984) and Atkin (1974).

9.6 THE MATHEMATICS OF REASONING

9.6.1 Formal Reasoning and Mathematical Logic

The term 'computer' suggests a machine for performing numerical calculations. However the basic electronic building blocks that make up a computer actually perform logical operations; the arithmetic operations with which we are familiar are defined in terms of these. The study of logic thus forms the basis of computer science itself, and the logician George Boole who formally stated the rules in the mathematical way that we use today is often regarded as the father of computation. Notwithstanding, we may make little use of logical operations in much engineering computation. However, in AI, the precise statement of rules for reasoning and inferencing is crucial.

The basic object that we discuss in formal symbolic logic is the *logical statement* or *predicate*. Such a statement must either be True (T) or False (F). We represent predicates by lower case letters.

Examples

(1) Let A and B be sets, and $\sim$ a relation between them. Then given $\alpha \in A$ and $\beta \in B$, p could be the statement '$\alpha \sim \beta$'. Since by definition the relation $\sim$ defines a subset of $A \times B$, either (α, β) is in this subset (so $p = $ T), or it is not (whence $p = $ F).

(2) In a non-mathematical context, which is more appropriate for AI, p could be a statement such as 'It is raining.'

Note, that unlike Example 1, even with such a simple statement about the 'real world' as Example 2, we cannot be as precise about the truth or otherwise of p. There are times when it is definitely raining, and times when it is definitely not. However there are other times (e.g. when there is a very heavy and damp mist, or when there is a very light drizzle), when one cannot really be sure. This lack of precision in natural language statements is a major problem in AI. The mathematical language of the statement of Example 1 avoids the problem of imprecision, but at the expense of the subtlety of natural language. One cannot write poetry in mathematical symbols! Various ways of getting round this problem have been proposed, and we shall discuss some in Section 9.6.3, but for the moment we can be certain about the value of the predicates we consider.

9.6.1.1 Production Rules and Implication

We will now proceed to develop some symbolism and formal rules for reasoning. The basic operation we require is the *implication*, or as it is commonly called in AI, the *production rule*. Consider the English statement:

$$\text{'If there is no rain then the crops will die'.} \tag{1}$$

Note that the 'if–then' construct is defining a relationship between the existence of rain, and the state of the crops. If we were building an agricultural expert system, we might write this rule in the form:

$$
\begin{aligned}
&\textit{If} \\
&\quad \text{rain} = \text{no} \\
&\textit{Then} \\
&\quad \text{future-of-crops} = \text{die}
\end{aligned}
\tag{2}
$$

If we let p denote the statement 'there is no rain' (or 'rain = none') and q the statement 'the crops will die' (or 'future_of_crops = die') then the rule is of the form:

$$\textit{If p then q} \tag{3}$$

We refer to p as a *premise* of the rule, and q as a *conclusion* or *consequence*. (In practice, the rules of an expert system will almost certainly have more than one premise, and may have more than one conclusion: we will see how these are combined later.) The 'if–then' construct as used in expert systems is convenient in that it is close to the natural language statement, but it is long-winded for mathematical manipulation. Instead, we write simply:

$$p \rightarrow q \tag{4}$$

(read this as 'p implies q'). To summarise, then, note that (1), (2), (3) and (4) are all ways of writing the same thing.

9.6.1.2 Logical Negation and the Contrapositive

Note that it follows from (1) that:

$$\text{'If the crops do not die, then it has rained'.} \tag{5}$$

This statement is called the *contrapositive* of (1). To write it symbolically, we introduce the logical operator $\neg$. The predicate $\neg p$ (read '*notp*') is simply the negation of p, i.e if p = True then $\neg p$ = False, and *vice versa*. Thus (5) may be written

$$\neg q \rightarrow \neg p \tag{6}$$

and the fact that a predicate implies its contrapositive can be written as

$$(p \rightarrow q) \rightarrow (\neg q \rightarrow \neg p) \tag{7}$$

Read this statement carefully, and make sure that you understand what it means. In fact the implication in the middle of (7) is reversible, i.e. the statement $(p \rightarrow q)$ and its contrapositve are exactly equivalent. To see this, note that $\neg\neg p = p$ and $\neg\neg q = q$. Thus if we substitute $\neg q$ for p and $\neg p$ for q in (7) we get

$$(\neg q \rightarrow \neg p) \rightarrow (p \rightarrow q) \tag{8}$$

as required.

9.6.1.3 Truth Tables

Since a predicate can only take two possible values, it is easy to prove statements about them by listing all the values in a table called a *truth table*. The truth table can also be used to give formal definitions for the operators. For instance, the truth table for $\neg$ is:

p	$\neg p$
T	F
F	T

By convention, we write the basic predicate or predicates in the left-hand columns, enumerating all the possible values, and then write the things to be defined or proved in columns to the right. Writing down the definitions of the operators in this way can throw up some surprises. The truth table for $\rightarrow$ is:

p	q	$p \rightarrow q$
T	T	T
T	F	F
F	T	T
F	F	T

If p is known to be true, and q is known to be false, then $(p \rightarrow q)$ must be false: this gives the second row. Conversely, if p is true and $(p \rightarrow q)$ is true then obviously q is true: hence the first row.

But what about the rows when the premise p is false? To get the fourth row, look at the implication (8). Since p and q are false, $\neg p$ and $\neg q$ are true. In view of the first row of the table, we must conclude that $(\neg q \rightarrow \neg p)$ is also true. Since we also know that (8) is true, we conclude, again from the first row of the table, that $(p \rightarrow q)$ is true, as required.

The most difficult case to decide is row three. To assign a value here, we need to think about the nature of proof or logical argument, and hence about the way in which predicates are combined together. Thus a justification of the third row is deferred to the next section.

9.6.1.4 The $\wedge$ and $\vee$ Operators

In a sensible expert system, our rules will have more than one premise, e.g.

> *If*
>> rain = no and
>> irrigation = no (9)
> *then*
>> future-of-crops = die

The corresponding 'and' operator of symbolic logic is $\wedge$, i.e.

$$(p \wedge q) \rightarrow r.$$

The conclusion r is true if and only if p and q are both true, i.e. the truth table for $\wedge$ is:

p	q	$p \wedge q$
T	T	T
T	F	F
F	T	F
F	F	F

Armed with this, we can see where the third row of the truth table for $\rightarrow$ comes from (see the previous section). In fact this concerns the nature of proof or logical argument. Consider the two statements $p \rightarrow q$ and $q \rightarrow r$. From this we want to draw the conclusion $p \rightarrow r$. It is not reasonable to insist that whether we can do this should depend on the *values* of p, q or r: the *process of deduction* should remain valid, even if the premise p is not. Symbolically, we want:

$$[(p \rightarrow q) \wedge (q \rightarrow r)] \rightarrow (p \rightarrow r)$$

to be true, *whatever* the value of p, q, r. (A statement that is always true is called a *tautology*.) Here is the complete truth table for this statement, the values being filled in from the tables for $\wedge$ and $\rightarrow$, except that we have put ? where we have to use row 3 of the table for $\rightarrow$. You should examine carefully how this table is constructed. Note that we have put all Ts in the last column, as this is what we are aiming to achieve.

p	q	r	$p \rightarrow q$	$q \rightarrow r$	$(p \rightarrow q) \wedge (q \rightarrow r)$	$p \rightarrow r$	$[(p \rightarrow q) \wedge (q \rightarrow r)]$ $\rightarrow (p \rightarrow r)$
T	T	T	T	T	T	T	T
T	T	F	T	F	F	F	T
T	F	T	F	?	F	T	T
T	F	F	F	T	F	F	T
F	T	T	?	T	?	?	T
F	T	F	?	F	F	T	T
F	F	T	T	?	?	?	T
F	F	F	T	T	T	T	T

The interesting row for our purposes is the third one. Here the predicate:

$$u = (p \rightarrow q) \wedge (q \rightarrow r) = \text{false}$$

and

$$v = (p \rightarrow r) = \text{true}.$$

The last column is then:

$$(u \rightarrow v) = \text{true}$$

and this will only work if the third row of the table for $\rightarrow$ is as we have given it! You might like to fill in the correct values for the ?s and convince yourself that the whole table is correct.

Now that we have the complete table for $\rightarrow$, we note an interesting feature. If $p \rightarrow q$ and p is false, then we can conclude *both* $q = \text{true}$ and $q = \text{false}$. This is the well known technique of 'proof by contradiction'. If a premise p leads to the truth of both q and $\neg q$, then we deduce that the premise itself is false.

Another way to write rule (9) is

> *If*
> > rain = yes or
> > irrigation = yes
>
> *then*   (10)
> > future-of-crops = live.

In symbolic logic, the 'or' operator is given the symbol $\vee$. Its truth table is:

p	q	$p \vee q$
T	T	T
T	F	T
F	T	T
F	F	F

Note that if p is 'There is no rain', and q is 'There is no irrigation', then comparing (9) and (10) suggests that

$$\neg(p \wedge q) = (\neg p \vee \neg q).$$

If, instead we let p be 'There is rain' and q 'There is irrigation', then we get:

$$(\neg p \wedge \neg q) = \neg(p \vee q).$$

These two identities are known as De Morgan's Laws (compare De Morgan's laws for sets in Section 9.2.3). As an exercise, provide them using truth tables.

In fact, once we have the $\vee$ and $\neg$ operations, it is possible to dispence with $\rightarrow$ altogether. Again, you should easily be able to prove using a truth table that:

$$(\neg p \vee q) = (p \rightarrow q).$$

Finally in this section, we mention (for completeness) two other operators, 'XOR' and 'NAND'. These are very useful in systems design, but are rarely used in inferencing systems. The $\vee$ operator is an 'inclusive or', i.e. $p \vee q$ is true if either p or q *or both* are true. In natural English, 'or' is ambiguous, for instance in the statement 'I will go swimming or I will go to the cinema'. Here the implication is that I will do exactly one of two things, but not both. It is this type of 'or' that we call 'XOR'. The 'NAND' operation is simply a combination of $\neg$ and $\wedge$. Its main use is that it is possible to express any other operation in terms of 'NAND', but we will omit a proof here. The truth tables are:

p	q	pXORq	p	q	pNANDq
T	T	F	T	T	F
T	F	T	T	F	T
F	T	T	F	T	T
F	F	F	F	F	T.

Up to now, we have been regarding our predicates as representing definite statements. However it is possible to introduce the concept of 'logical variables'. This is the idea behind the predicate calculus.

9.6.2 Predicate Calculus

The predicates we have considered up to now involve only single statements: however in many applications we may wish to make use of variables.

9.6.2.1 Variables and Quantifiers

Like variables in any other branch of mathematics, variables in predicate calculus exist in some predefined domain set. Consider for example the set $I\!R$ of real numbers. Typical predicates on $I\!R$ would be:

$$p(x) \text{ meaning } `x^2 + 4x + 4 = 0` \qquad \text{and}$$

$$q(x) \text{ meaning } `(x + 2)^2 = x^2 + 4x + 4`.$$

Each of these statements is either true or false. You will observe, however, that $p(x)$ is true only for a particular value of x, i.e. $x = -2$, whereas $q(x)$ is true for all $x \in IR$. A predicate which holds for all values of the variable is said to be *valid*: if it only holds for some values it is said to be *satisfiable*. To distinguish between these two types of predicate, we introduce the *quantifiers* $\exists$ ('there exists') and $\forall$ ('for all'). As the names suggest, $\exists$ means that a predicate evaluates to true *for at least one* x, whereas $\forall$ insists that it must hold for *every* x. Thus:

$$(\exists x p(x)) \text{ is true} \quad \text{and} \quad (\forall x q(x)) \text{ is true} \quad \text{but} \quad (\forall x p(x)) \text{ is false.}$$

To take an illustration which looks more like AI, consider our agricultural example. Suppose W denotes the set of values for 'weather' that we will consider, for example,

$$W = \{\text{sunny, cloudy-dry, fog, rain, snow}\}$$

We can define a predicate function.

$$d(\text{weather}) \text{ to mean 'The crops will die'.}$$

Since the set W is finite, we can can enumerate all the values of d: perhaps:

$$
\begin{aligned}
d(\text{sunny}) &= \text{true} \\
d(\text{cloudy-dry}) &= \text{true} \\
d(\text{fog}) &= \text{true} \\
d(\text{rain}) &= \text{false} \\
d(\text{snow}) &= \text{false}
\end{aligned}
$$

Thus

$$\exists \text{weather } d(\text{weather}) \text{ is true,}$$

but

$$\forall \text{weather } d(\text{weather}) \text{ is false.}$$

Notice that the quantifiers have the effect of 'removing' the variables and converting the predicate functgion back to an ordinary predicate.

Obviously we are not restricted to one variable here. If $x, y, \in IR$, and if:

$$r(x, y) \text{ means } `(x + y)^2 = x^2 + 2xy + y^2`$$

then

$$\forall x \forall y \; r(x, y) \text{ is true, or we may more briefly just write}$$
$$\forall x \forall y \; r(x, y), \text{ the 'is true' being understood.}$$

Composite predicates can of course be built up from simpler ones, just as in ordinary symbolic logic.

9.6.2.2 Relations and Clauses

Observe that predicates define relations, and vice versa. For instance, if x, y, $\in IR$, the binary relation:

$$x > y$$

or more precisely (regarding the relation as defining a subset of $IR \times IR$ as described in Section 9.3) the relation given by:

$$\{(x,y)|x > y\}.$$

is exactly equivalent to defining the predicate:

$$s(x,y) \text{ to mean`}x > y'$$

where we interpret the truth of s to specify whether or not the pair (x,y) satisfies the relation. The AI language Prolog depends on the specification of predicates in relational form. Further references to the Prolog programming language can be found in Chapters 2, 3 and 8 of this book.

9.6.3 Reasoning with Uncertainty

Up to now, we have have been assuming that for all predicates p, it is possible to be certain whether p is true or false. Of course, in real life this is often not the case, and practical intelligent system must be able to cope with uncertainty. There are two basic approaches to this problem. One is to work with tentative propositions that may be retracted at a later stage. This idea, known as 'non-monotonic logic', is explored in detail in Chapter 8 of this book. The alternative 'line of attack', which we pursue here, is to try to assign some numerical value to the truth of a predicate. Assuming that 0 represents false, and 1 represents true, values between 0 and 1 will represent levels of certainty. There are two commonly used systems for assigning these certainties; fuzzy logic and probability theory. We will discuss both of them.

9.6.3.1 Fuzzy Reasoning

We start by establishing a correspondence between predicate logic, set theory, and functions. For clarity, we will look at a specific case; our agricultural example. Recall that W denotes the set of values for 'weather' that we will consider:

$$W = \{\text{sunny, cloudy-dry, fog, rain, snow}\}$$

A subset of W is those types of weather involving precipitation, i.e.:

$$C = \{\text{rain, snow}\}.$$

Further subsets R and S refer just to rain, and to snow:

$$R = \{\text{rain}\}. \qquad S = \{\text{snow}\}.$$

Now think of 'weather' as being a variable that can take any value in W.
  The predicate

$$r = \text{'The weather is rain'}.$$

can thus be expressed as:

$$\text{weather} \in R.$$

Similarly,

$$S = \text{'The weather is snow'}$$

becomes:

$$\text{weather} \in S.$$

The composite predicate:

$$p = \text{'There is precipitation'} = r \vee s$$

in set theory terms becomes:

$$\text{weather} \in C = R \cup S. \tag{11}$$

Clearly there is an exact correspondence between the $\vee$ operation of predicate logic and the $\cup$ operation of set theory. Similarly there is a correspondence between the $\wedge$ operation and $\cap$.

$$t = \text{'There is rain and snow'} = r \wedge s \tag{12}$$

becomes:

$$\text{weather} \in P \cap S$$

(which in this case is empty, since in our model rain and snow are mutually exclusive). The $\neg$ operator corresponds to the set theoretic complement.
  Now for any subset $A \subset W$, consider a function $f_A : W \rightarrow \{0, 1\}$ defined by:

$$f_a(\text{weather}) = \begin{cases} 1 & \text{if weather} \in A \\ 0 & \text{if weather} \notin A \end{cases}.$$

This function is called the *characteristic function* or *indicator function* of the set A. The following rules about characteristic functions of sets A and B are obvious:

$$f_{A \cup B} = \text{maximum } [f_A, f_B]$$

and

$$f_{A \cap B} = \text{minimum } [f_A, f_B].$$

Combining these with the correspondences established above, we are now able to give yet another representation of the $\wedge$ and $\vee$ operations of symbolic logic. Recall that we agree that a certainty of 1 represents true and 0 represents false. We can now write the certainties of predicates (11) and (12) and negation as:

$$\text{certainty}(r \vee s) = \text{maximum}[f_R, f_S]$$
$$\text{certainty}(r \wedge s) = \text{minimum}[f_R, f_S]$$

and $\hspace{10cm}$ (13)

$$\text{certainty}(\neg r) = 1 - f_R$$

We are finally in a position to introduce our measure of uncertainty. Suppose that on a particular day, we are uncertain of the weather. Perhaps we are 75% certain that it is snow, 20% that it is rain (note that these do not add up to 100%). The certainty factors (CFs) for the predicates can then be expressed as:

$$r = \text{'The weather is rain'. } CF = 0.2$$

and

$$s = \text{'The weather is snow'. } CF = 0.75$$

We combine these certainties by the rules (13), thus:

$$p = r \vee s, \ CF = \max[0.75, \ 0.2] = 0.75,$$
$$t = r \wedge s, \ CF = \min[0.75, \ 0.2] = 0.2.$$

This type of reasoning is known as *fuzzy logic*. You will observe that we do not get a certainty of 0 for the predicate t, even though the sets R and S are mutually exclusive. In fact, if we increased the cf r to 0.25, so that the certainties add up to 1, the CF of t actually goes up! A major problem with fuzzy logic is that it does not behave consistently under $\neg$ (i.e. set theoretic complementation). De Morgan's laws do not work correctly. For this reason, various attempts to modify the rules (13) have been suggested.

Complementary to the idea of fuzzy logic, we can define a *fuzzy set*. This is a 'set' whose characteristic function can take values between 0 and 1. More precisely, we can think of the statement '$a \in A$' as a fuzzy predicate. Of course such as object is not really a set: it is simply a function:

$$f_A : A \rightarrow \{X \in I\!R \,|\, 0 \leq x \leq 1\}.$$

In spite of the theoretical problems, systems built on fuzzy logic do seem to work reasonably well in practice. In order to build such a system, we need to define the certainty behaviour for fuzzy implication $\rightarrow$. This is normally performed as follows. Suppose we have a rule:

If
> *a*

then
> *b*, *CF* = *x*,

where *x* is a real number, $0 < x \leq 1$. This is taken to mean that if we know *a* with certainty 1, we can conclude *b* with certainty *x*. If in fact we only know *a* with certainty *y*, then we conclude *b* with certainty *xy*. For instance if $x = 0.75$ and we have elsewhere concluded *a* with certainty 0.5, then we conclude *b* with certainty 0.375. (In practice *a* will be a composite predicate, built up using the rules (13).)

9.6.3.2 Probability Theory and the Laws of Probability

A much older and more firmly established approach to certainty is provided by probability theory. This avoids much of the arbitrariness and inconsistency of fuzzy logic, but at the expense of requiring much more precise information about the propositions under discussion. This information may not be available. However in certain applications, for instance in computer vision, where good statistical information is known or can be obtained, probability theory provides a vital tool.

We will continue to discuss the example of the previous section, and retain the same sets. In the context of probability theory, the set *W* is known as the *sample space*. The variable 'weather' is termed a *random variable*, and its possible values (the elements of W) are called *outcomes*. The various subsets of W are known as *events*.

If $a, b \in IR$, let $[a,b] = \{X \in IR \mid a \leq x \leq b\}$.

We define a function $P : W \rightarrow [0,1]$ with the property that *the sum of all the values of P*(weather) *is 1*. We refer to this number as the P *probability* of the outcome 'weather'. For instance we might have:

$$P(\text{sun}) = 0.2$$
$$P(\text{cloudy-dry}) = 0.4$$
$$P(\text{fog}) = 0.05$$
$$P(\text{rain}) = 0.3$$
$$P(\text{snow}) = 0.05$$

The probability represents the expected proportion of times that this weather will occur; the table above means that we expect sun one fifth of the time. If possible these probabilities should be assigned on the basis of statistical information, in this case presumably by consulting the weather records for the relevant area. If this detailed information is not available, we just have to guess suitable values.

We will also refer to the probability of an event (i.e. subset of W). The probability of an event containing one outcome is the same as the probability of that outcome, for example

$$P(R) = P(\{rain\}) = P(\text{rain}) = 0.3.$$

More generally, the probability of an event should be the aggregate probability of its outcomes occurring, e.g.

$$P(\{\text{sun, fog}\}) = P(\text{sun}) + P(\text{fog}) = 0.2 + 0.05 = 0.25.$$

For mutually disjoint events (i.e. sets whose intersection is empty). We want the probability of the union to be the sum of the individual probabilities, for example

$$P(C) = P(R \cup S) = P(R) + P(S) = 0.3 + 0.05 = 0.35.$$

For the general case of two events, A, B which are not disjoint, if we simply add the probabilities we will count those events in $A \cap B$ twice. Thus we obtain the rule:

$$P(A \cup B) = P(A) + P(B) - P(A \cap B).$$

Armed with these results, we can easily see that $P(W) = 1$ and that if $W = A \cup B$ with $A \cap B = \emptyset$, then $P(A) = 1 - P(B)$, avoiding the problems of complementation which occur with fuzzy logic.

Recalling the correspondences between $\wedge$ and $\cap$, and $\vee$ and $\cup$ discussed in the previous section, we can now write down the analogous formulae to (4.13) for a probabilistic reasoning system. (Here ':=' means 'is defined to be'.)

$$P(r \vee s) := P(R \cup S) = P(R) + P(S) - P(R \cap S)$$
$$P(r \wedge s) := P(R \cap S) \text{ and}$$
$$P(\neg r) := 1 - P(R)$$

For our example, since $R \cap S = \emptyset$,

$$P(\text{'rain and snow'}) = P(\emptyset) = 0$$

as we would reasonably expect.

However, a fundamental difficulty of probabilistic reasoning systems concerns the evaluation of $P(A \cap B)$ in the general case.

Two events A, B are said to be *statistically independent* if

$$P(A \cap B) = P(A)P(B).$$

Roughly speaking, A and B are independent if event A has no effect on B and *vice versa*. Consider, for instance, the experiment of tossing a coin twice. If H represents a head and T a tail, the sample space is:

$$\{ \text{ HH, HT, TH. TT } \}.$$

Assume each outcome has probability 0.25. Let

$$A = \text{'First toss a head'} = \{ \text{ HH, HT } \},$$

and

$$B = \text{'Second toss a head'} = \{ \text{ HH, TH } \}.$$

whence $P(A) = P(B) = 0.5$.

But also $A \cap B = \{HH\}$. Thus $P(A \cap B) = 0.25$.

Hence $P(A)P(B) = 0.5 \times 0.5 = 0.25 = P(A \cap B)$, as we expect, since the fact that the first toss is a head should have no effect on the second toss.

Take care not to confuse disjointness with independency. In this example, A and B are not disjoint: $A \cap B = \{HH\}$.

To summarise then, the basic laws for combining probabilities are:

$$P(A \cup B) = P(A) + P(B) - P(A \cap B)$$

$$P(A \cap B) = P(A)P(B) \text{ provided } A \text{ and } B \text{ are independent, or}$$
$$P(A \cap B) = 0 \text{ provided } A \text{ and } B \text{ are disjoint.}$$

Unfortunately, however, it is very difficult to be sure in practice whether our predicates correspond to independent events. Generally we just have to set up the system to make them as 'independent' as possible.

9.6.3.2.1 Conditional Probability and Bayes' Theorem. In order to make inferences in a probabilistic system, we have to introduce the idea of conditional probability. Consider again the experiment of tossing a coin twice. As we have seen, the probability of obtaining two heads is 0.25. This quantity is known as the *prior* probability. Suppose we then toss one coin. What is the probability now of two heads? Clearly this depends on the result of the first throw. If the first throw was a tail, the chance of two heads is now 0. On the other hand, if it was a head, then the chance of two heads is now 0.5, because we now only need to look at the two outcomes $\{HH, HT\}$. By performing the first toss, we have effectively restricted or redefined the sample space. In general, if the initial sample space is U, and if the event F is known to have happened, then the 'new' sample space is $U \cap F$, and all the probabilities of the outcomes in this set need to be divided $P(F)$ so that they sum to 1 again. However, if S is an event with prior probability $P(S)$, then the S is itself restricted. Now only $S \cap F$ can occur. It follows that the 'new' or *conditional* probability of S, which we denote by $P(S|F)$ (read 'the probability of S given F'), is given by:

$$P(S|F) = \frac{P(S \cap F)}{P(F)}. \tag{14}$$

Note that if S and F are statistically independent, then

$$P(S|F) = P(S)$$

as we would expect. A simple but fundamentally important consequence of (14) is Bayes' rule:

$$P(S|F) = P(F|S)\frac{P(S)}{P(F)}. \tag{15}$$

Rules in probabilistic inferencing systems amount to assignment of conditional probabilities. Suppose that in a medical system:

p is 'The patient has eaten sour fruit'

q is 'The patient feels sick'

and

r is 'The patient is pregnant'

A rule such as:

If

The patient has eaten sour fruit

then   (16)

The patient feels sick, probability 0.6

means that:

$$P(q|p) = 0.6.$$

However, what do we do with another rule:

If

The patient is pregnant

then   (17)

The patient feels sick, probability 0.3?

Suppose we know that in fact the patient is pregnant *and* has eaten sour fruit. It is far from clear that one can combine these numbers into a new estimate of the probability of nausea. However Bayesian inferencing systems get round this problem in a most ingenious fashion, effectively using the rules in reverse. Whereas a conventional inferencing system would use the rules (16) and (17) to conclude information about q from p and r. Bayesian systems conclude information about p and r from q. More precisely, suppose we assign prior probabilities of 0.03, 0.1 and (assuming the patient is female!) 0.05 to p, q and r respectively. If we now ascertain that the patient feels sick, we can conclude using (15) that:

$$P(p|q) = P(q|p)\frac{P(p)}{P(q)} = 0.6 \times 0.03/0.1 = 0.18$$

and

$$P(r|q) = P(q|r)\frac{P(r)}{P(q)} = 0.3 \times 0.05/0.1 = 0.15.$$

Thus, on the basis of the existence of sickness, we have increased our estimate of the likelihood of both indigestion and motherhood. The simplest form of Bayesian inferencing system will start by assigning prior probabilities to all the possible conclusions. It will then run through all the rules, determining the values of the premises (e.g. by asking the user if she feels sick), and update the conditional probabilities. Finally it will report either the most probable conclusion, or alternatively all conclusions above a certain threshold value. In practice, such a simple system is really only practicable if direct interaction with the user is not involved (e.g. signature recognition systems, where the premises such as angle of the letters can be verified automatically), since it requires reference to all the rules known, whether or not they are actually relevant to the problem in hand. A more sophisticated system would always try to pursue the most probable conclusion at any particular stage, and stop when some conclusion has reached a preset threshold.

9.6.3.3 The Pragmatic Compromise

In practice, no numerical way of estimating uncertainty is entirely satisfactory in all respects. Fuzzy logic methods have problems of consistency, and probabilistic methods suffer from the difficulty of supplying accurate probabilities, and determining the independency of rules. Practical certainty factors used in expert systems are often a kind of hybrid of fuzzy logic and probabilistic methods that can only really be justified on the grounds that they work reasonably well—sometimes!

BIBLIOGRAPHY

Atkin, R. (1974) *Mathematical Structures in Human Affairs*. Heineman, London.

Denvir, T. (1986) *Introduction to Discrete Mathematics for Software Engineering*, Macmillan, London.

Deo, N. (1974) *Graph Theory with Application to Engineering and Computer Science*, Prentice Hall, New York.

Dougherty, E.R. and Giardina, C.K. (1988) *Mathematical Methods for Artifical Intelligence and Autonomous Systems*, Prentice Hall, New York.

Halmos, P.R. (1960) *Naive Set Theory*, D. Van Nostrand, New York.

James, M. (1984) *Artificial Intelligence in Basic*, Butterworth, London.

Johnsonbaugh, R. (1984) *Discrete Mathematics*, Macmillan, London.

Seshu, S. and Reed, M.B. (1961) *Linear Graphs and Electrical Networks*, Addison-Wesley, Reading, MA.

Shirai, Y. and Tsujii, J. (1982) *Artificial Intelligence*, Wiley, Chichester.

Tou, J.T. and Gonzales, R.C. (1974) *Pattern Recognition Principles*, Addison-Wesley, Reading, MA.

Wilson, R.J. and Keineke, L.W. (1979) *Applications of Graph Theory*, Academic Press, New York.

D.K. Bose
Department of Mathematical Sciences
Brighton Polytechnic
Lewes Road
Brighton
East Sussex BN2 4GJ

S. W. Ellacott
Information Technology Research
  Institute
Brighton Polytechnic
Lewes Road
Brighton
East Sussex BN2 4AT

10 Computer Architectures for Artificial Intelligence

N.J. Avis

University of Sheffield, UK

10.1 INTRODUCTION

In this section an overview of the current and emerging computer architectures for executing Artificial Intelligence applications is presented. It is assumed that the reader has some knowledge of both the operation of the von Neumann computer and of programming languages.

The discussion is centred around the hardware platforms required in the development of large-scale AI applications. The performance of the development vehicle is extremely important if restrictions of development or design choices are to be avoided. The section looks at those performance aspects which are specific to AI processing, and not those which would benefit all types of computation in general.

The aim of this section is not to provide an exhaustive survey of currently available commercial systems, or of the huge number of research machines currently under development worldwide. Instead, the aim is to illustrate those factors which have influenced current designs and those which are likely to influence future systems, where possible using commercially available products.

Procedural languages have not, to date, been widely used for implementing large AI applications. If procedural languages are used, however, they can take advantage of the performance offered by conventional scientific computers. These systems will not be addressed in this survey.

The benefits of functional programming are already well established, (Backus, 1978). Since most commercial systems to date support Functional Languages, the discussion will be restricted to these systems. The principles on which these systems are developed are in general equally applicable to other programming paradigms favoured by the AI community, such as Logic (see Chapter 8) and Object-oriented (refer to Chapter 3).

The discussion is presented in two parts. In the first, the sequential execution systems are presented. The role of the von Neumann computer and its suitability to the execution of AI applications, written in functional languages such as Lisp, is discussed. The concept of

Artificial Intelligence in Engineering. Edited by Graham Winstanley
©1991 by John Wiley & Sons Ltd.

the 'semantic gap' is developed, as are the modifications to the von Neumann architecture needed to reduce this gap.

A survey of currently available sequential Lisp platforms is given, with an indication of the likely effect VLSI and RISC processor architectures will have on future systems.

The second part examines the role of VLSI in the continued decline in hardware costs, and leads to the discussion of the development of parallel processing systems. The advantages afforded by parallel processing for AI applications are discussed, together with the obstacles that must be overcome to allow the effective exploitation of these advantages. Several parallel processing systems are detailed. These are presented in increasing levels of parallelism.

Finally, the Connectionist approach to AI processing is explained. Such non von Neumann computational paradigms are likely to have a profound effect on the computer architectures of future AI processing systems.

10.2 SEQUENTIAL IMPLEMENTATIONS

10.2.1 The Semantic Gap

Programming can be defined as the mapping of the abstract software instructions into physical actions on the computer hardware. The extent to which a programming language reflects the architectural features of the computer on which it is run will determine its execution efficiency.

Low-level programming languages, such as assembler languages, directly mirror the architecture of the computer on which they are to be run. Whilst this is very efficient in terms of execution performance, programming in such low-level languages is tedious, error prone and machine specific. To improve programmer productivity, high-level languages such as Fortran, Pascal and Lisp were developed, which provide the programmer with more powerful programming constructs. The high-level language removes the machine-specific features from the programmers' view. Thus, the programmer is presented with a virtual machine model on which to base software production. The use of such high-level languages improves software portability across machines and eases software maintenance. However, these improvements are offset by the need to map the high-level language to the physical machine. The mismatch between the programming language and the physical machine is termed the 'semantic gap'.

It would seem that a conflict arises here between the needs of the programmer, who prefers to work using high-level languages, and the efficient execution of programs. This conflict can be resolved, at least in part, by the automatic translation of high-level languages to machine code. This is the job of the compiler. It is the efficiency with which the compiler can close the semantic gap which largely determines the execution rate. The mechanisms of compiler theory (Aho and Ullman, 1977) are too complex to address here, but for an efficient implementation, the compiler must make use of special architectural features of a particular machine.

The features of two programming paradigms will now be outlined to indicate how the two styles differ, and to determine the performance implications when implemented on a sequential von Neumann computer.

10.2.2 Procedural Programming Paradigm

The von Neumann architecture is based on a global memory of fixed sized, type-less cells, which are used to store both the program's instructions and data. Program execution can be viewed as the sequential application of assignment, and conditional statements to achieve state transitions.

Procedural languages, such as Pascal have been specifically developed in conjunction with the von Neumann architecture. As such, they offer a good compromise between efficiency and ease of use when implemented on a sequential von Neumann computer.

Procedural languages are often strictly typed, which means that a type is associated with every variable, constant and function result. The advantage of this is that memory allocation and type checking can be performed at compile time, rather than at run-time. This means more efficient memory allocation and the ability to detect some errors before run-time, such as incompatible operands to an arithmetic operator. Some procedural languages support the concept of recursion. However the use of recursion, although elegant, may exact a significant performance overhead.

Dynamic memory allocation is also supported in some procedural languages (such as Pascal), whereby a programmer may gain access, via the use of pointers to a user reserved area of memory known as 'heap storage'. However the need to explicitly allocate and de-allocate this memory severely hinders its usefulness.

A number of techniques can be used to improve the execution of programs on the von Neumann architecture. One such technique takes advantage of the sequential control flow and the principle of locality.

The main 'bottleneck' in the von Neumann architecture, is the bus along which data and program instructions must pass on their way to and from the processor and memory. Since it is inefficient to have to wait for data or program instructions to be fetched from main memory, the processor makes use of the fact that a reference to one location in memory can be used to predict the next location needed (i.e. normally followed by another in close proximity to the first). This means that the contents of memory locations can be 'prefetched' from memory, so as to be available in 'caches' when the processor requires them, thus speeding up execution significantly. Of course, there are times when the principle of locality does not hold (such as at the end of a program loop or array accesses) and the prefetched items are not required. In this case there is no choice but to 'suffer' the time delay in accessing the required memory location.

10.2.3 Functional Programming Paradigm

The functional programming paradigm is built upon the idea of combining functions. One of the most widely used functional programming languages is Lisp. As its name implies Lisp programs are concerned with the manipulation of lists. The list is the fundamental data structure in Lisp and is used to represent both programs and data. A list consists of Lisp cells each of which is comprised of two components:

- CAR contains the contents of the Lisp cell;
- CDR contains a pointer to the next Lisp cell in the list

(see Figure 10.1).

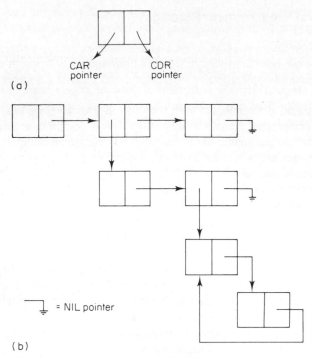

Figure 10.1   (a) A Lisp cell with CAR and CDR pointers; (b) a complex list structure

It is clear from the above that lists are recursive in their structure, may be cyclic and are dynamically alterable. The dynamic nature of lists means Lisp programs require a large heap memory. Because of the complex arrangements of lists, the memory does not display any of the principles of locality associated with procedural language implementations.

Lisp has no type declarations, and all objects (data and programs) are considered to be 'first-class citizens' and therefore available without restrictions. This is one of the most unique features of Lisp. The lack of type declarations and the general availability of objects makes Lisp an ideal language for development work and rapid prototyping, and goes a long way to explaining the popularity of Lisp with the AI community. However this liberal attitude to programming means that no information on types is available at compile time and so all type checking must be done at run-time. Reference should be made to Chapter 3 for a full discussion on the practical software aspects of AI langauges, especially Lisp.

The execution of a Lisp program on a sequential architecture such as the von Neumann machine can be viewed as a collection of dynamically nested, incompleted function calls, the most recent of which is active.

The functional nature of Lisp encourages the programmer to make use of many small functions. However, the overhead associated with the implementation of function calls and returns are more expensive than in lexically scoped languages such as Pascal, since Lisp function evaluation takes place in a 'dynamically bound' context.

10.2.4 Control Flow *versus* Functional Paradigm

It can be seen from the above, that procedural and functional programming languages have very different attributes. Because of the closer mapping between the procedural languages and the underlying von Neumann machine, these tend to lead to efficient implementations of deterministic programs.

Procedural languages have not, to date, proved popular for implementing AI applications. This is due to the fact that they lack the necessary features for the efficient manipulation of symbols. This, coupled with the constraint of static or at best seminal dynamic memory management, precludes their use in the vast majority of applications.

Functional languages provide the facilities for AI development work, but do not map well onto the features of the underlying sequential von Neumann machine. Indeed, some of the architectural features, which are useful to procedural languages, become useless or redundant when programming in a functional language such as Lisp. This large semantic gap leads to inefficient implementation and large memory requirements of Lisp programs on von Neumann machines.

10.2.5 Support for Functional Languages

For efficient execution, Lisp programs require areas of special support, and two areas for concern have been identified. These are: efficient function handling and structured memory.

10.2.5.1 Efficient Function Handling

The functional programming style encourages the recursive use of functions. A typical Lisp program will invoke many functions during its execution. It has already been stated that function calls and returns are relatively expensive operations, but if the implementation does not support function calls efficiently, the execution rate will be compromised and the programmer may be tempted to avoid the use of these costly function calls. This in turn will negate many of the benefits of adopting a functional programming approach. Therefore the efficient implementation of function calls is essential.

Many implementations of functional languages are based on the execution mechanisms of a virtual machine designed to support lambda calculus. Landin (1964) devised one such virtual machine; the S.E.C.D which is stack-based and takes its name from the four principal registers it uses.

Functions can be created dynamically and applied at points remote from their definition. This causes a number of implementation problems. To be implemented correctly, functional languages such as Lisp must obey static binding rules. These state that any free variables in the function body must adopt the values they had when the function was defined, rather than those they may have had when the function was called.

The S.E.C.D machine uses the classical computing technique of employing an environment to provide the binding of values to variables. On a function call the correct environment is pushed onto the stack. Arguments are evaluated on the stack during which time intermediate values may also be added to the stack. An extension of this technique is employed in the S.E.C.D machine, which operates two separate stacks. The S-Stack holds intermediate values, while the D-Dump is used to save the values from the first stack, along with other information, such as the state of the other registers at each function call.

The contents of each list pointed to by the four registers define the state of the machine.

S — Stack, used to reference intermediate results when evaluating expressions.

E — Environment, used to reference the values bound to variables during execution.

C — Control list, used to reference the program being executed.

D — Dump, used to reference a stack which saves the values of other registers when calling a new function.

10.2.5.2 Maintaining the Environment

An important design issue with the Lisp virtual machine, is the choice of variable binding scheme. The way the machine keeps track of the current value of a variable can have a dramatic effect on performance. In current Lisp systems variable binding is based on an associative list of 'name–value' pairs and uses either 'shallow' or 'deep' binding techniques:

- *Deep binding*. The simplest way to implement the environment is by a linear linked list (associative list) of name–value pairs. New items are added at the head of this list on function calls, and deleted from the head on function returns. Whenever a variable is referenced during the evaluation of a function, the list is searched from the head for the first and hence most active instance of that variable name. However, this variable look-up might involve scanning the entire list.

- *Shallow binding*. An alternative scheme, known as 'shallow binding', maintains a table with one entry for each variable name. Each entry contains the current value binding of that variable name.

  Shallow binding alters the variable interrogation problem from one of list search to a simple table look-up. To maintain consistency, the table has to be modified on each function call and return. Bindings in danger of being overwritten by a newer value have to be saved (typically on the binding stack) to be used in restoring the table to its original state on function return.

- *Shallow* versus *deep binding*. In choosing one implementation over the other, there is clearly a trade off between fast function calling and fast variable look-up. For function call and return, shallow binding can be expensive if many parameters are involved, whereas deep binding is trivial and incurs little overhead. For variable look-up with shallow binding, the time is constant, whereas for deep binding variable look-up can be expensive. However, it is possible to consider hybrid implementations, combining the features of both deep and shallow binding.

10.2.5.3 Structured Memory

This is concerned with the management of the dynamic data objects common to AI languages such as Lisp. The main technique for supporting these nested, variable-sized objects is to use Lisp cells. The typeless nature of Lisp together with the dynamic nature of the creation and use of Lisp cells presents three major problems in the implementation of structured memory.

10.2.5.3.1 Self-defining objects.
In typed languages such as Pascal, static data structures such as integers, reals, characters, etc. are mapped onto the physical memory of the system. The memory is a collection of fixed-sized, typeless cells. It is not the memory that defines

the type of contents of any cell, but the instructions that access the memory cell. Since Lisp is a typeless language, the above mechanism cannot be used. However, some way of performing type checking at run-time is required. One way to achieve this is not to associate type with the instructions, but to associate the type of an object with the object itself. This is the concept of self-defining objects.

The type of a object is explicitly stated in a tag field which is stored with the object's value. The tag field specifies the data type of the corresponding object, such as character, string, address, etc. This association of type with the object has a number of advantages:

- In strictly typed languages such as Pascal, operators demand operands of the correct type. Once the programmer has decided on a certain operation, it is the programmer's task to select the operator whose operand's type matches that of the data. The programmer may be faced with the need to explicitly convert the type of the data, such that they match the operator's operand type.

- With self-defining objects, operators can be generic. Since the type of each object can be deduced from its tag field, type dependent actions for a particular operator and the necessary type conversions can be performed automatically.

- This has the advantage of greatly speeding development work, since the programmer is relieved of type declarations and type-dependent operator selection. However, the system still needs to perform type checking, since automatic type conversion cannot resolve all operators' type dependencies, and these have the execution burden of being performed at run-time.

- Tagging objects also has the storage overhead of associating the tag field, and this is directly proportional to the number of types allowed.

- To allow the efficient run-time checking of tags, the fast manipulation and pattern-matching of bit-fields is required, and this is best performed by enhanced instruction sets and special hardware.

10.2.5.3.2 List representation and access.

A large proportion of Lisp execution time is spent searching lists. When lists are represented by the usual pair of pointers in Lisp cells, a good deal of time is spent following pointers and pointer de-referencing. This is inefficient, since the address of the next cell to be accessed only becomes available once the contents of the previous cell have been retrieved from memory. This addressing bottleneck slows down list traversal. What is required, therefore, is a scheme able to exploit the principle of locality, described in Section 10.2.2, by ensuring that adjacent cells in a list can be stored as adjacent memory locations.

One method of achieving this is to employ a technique known as CDR-coding. List cells of a linear list are represented as vectors (arrays) in memory. The CDR pointer of the Lisp cell is assumed, by default, to point to the next element in the vector and can therefore be replaced by a small CDR-code. This CDR-code is used to handle exceptions to the above scheme, such as the CDR pointer of the last Lisp cell in a linear list.

Typically, the CDR-code consists of two bit which may take the following values.

- *CDR-NEXT* corresponds to a Lisp cell at the beginning or middle of a linked list, whose CDR pointer is assumed to point to the next Lisp cell in the linear list.

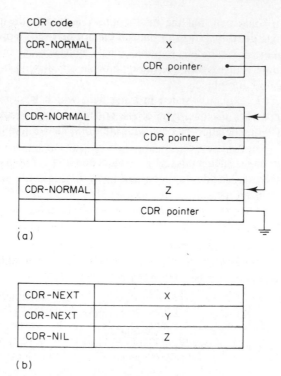

Figure 10.2 (a) Conventional representation; (b) the same structure but implemented using continuous memory locations and CDR-coding

- *CDR-NIL* corresponds to the Lisp cell at the end of the linear list, whose CDR pointer is NIL.
- *CDR-NORMAL* corresponds to the case in which a vector representation is not possible, such as at branches off a linear list. In this case, the system must revert to the previous method and the address bottleneck must be endured.

(refer to Figure 10.2.)

The CDR-coding technique has the advantage of not only reducing the time needed to traverse a list, but also in reducing the storage requirements by approximately half by removing the need to store the CDR pointer.

10.2.5.3.3 Heap maintenance.

Lisp programs execute in an environment that provides support for dynamic memory management. Lisp does not require the programmer to explicitly de-allocate the memory for its return to the heap. Since the heap is of finite size, the implication of the allocation of memory without returning unwanted memory, presents the possibility of the system rapidly using all the heap space available. If this occurs, then the next operation that requires additional memory from the heap cannot be satisfied and the system will abort its execution.

It is the responsibility of the system to manage the allocation and reclamation of lisp cells from and to the heap. The reclamation of redundant Lisp cells is termed 'garbage collection'. The aim of garbage collection is to reclaim, with minimal overhead, those Lisp cells that are no longer accessible via lists. This is achieved by a two-stage process. First the system must identify those items which are no longer needed. Secondly, having identified the candidates for garbage collection, these may then be returned to the heap to await subsequent re-allocation. A number of garbage collection schemes are available and major types are outlined below.

Reference counting This scheme was introduced in 1960, and has since undergone many refinements. The basic mechanism is to maintain a count of the number of references to each object. Whenever the reference count for a object falls to zero, the object is no longer accessible, and it is therefore returned to the heap. Whilst this scheme is very simple, it has a number of drawbacks.

The first involves the storage overhead required to maintain a reference count for every object. The others are associated with the need to update the reference counts each time a reference to an object is added or deleted. Since the garbage collection scheme is continuous, the user 'pays the price' of the garbage collection overhead, even if the application does not require objects to be returned to the heap. In addition, the overheads of maintaining the reference counts are not deterministic, since, when one object's reference count reaches zero, it is removed, which will in turn decrement the reference counts of those objects that it pointed to. If the result of this is that their reference count also reaches zero, then they are garbage collected and the effect can propagate through the system.

For reducing this overhead, a number of modifications have been proposed, which defer the updating of the reference counts until some fixed period has elapsed. Pending updates are held in a table until the periodic update time. This scheme is based on the statistical evidence (Clark and Green, 1977) that in Lisp programs, 97% of objects will not have more than one reference. By postponing the updates, there is a good chance that additional references will be created and deleted in the time slot between updates. This only involves the clearing of the pending update table, rather than the actual updates of the reference counts.

Mark and sweep Again this scheme was first introduced in 1960, and can be viewed as a precursor to the copying collectors described below. As its name suggests, this scheme consists of two phases.

During the 'mark' phase, all accessible (non-garbage) objects are marked. This is achieved by following the pointers from selected 'root' objects and then recursively marking all objects accessible from these roots.

After all the accessible objects have been marked, the entire address space is 'swept', and all 'marked' objects are swept over and left unaltered, except for having their markers reset. Unmarked objects are garbage collected and returned to the heap.

In this scheme, garbage collection is only initiated when there is no heap space left, and a request for more memory is made. During the garbage collection phases, the user's program is suspended.

Copying

The copy garbage collector resembles the Mark and Sweep scheme, in that it uses the same mechanism for identifying garbage objects. Copying garbage collection is the most prevalent scheme amongst modern systems. The address space is divided into two equally sized semispaces: 'oldspace' and 'newspace'. During garbage collection all accessible objects are copied to newspace. Oldspace is then available to be reclaimed as a whole without further inspection. The names of the two semispaces are then swapped, and the system is ready for the next garbage collection phase in the future.

Generation garbage collection

This scheme is based on the observation that 'young objects die young', and results from the premise that most objects are used as temporary storage. Thus, for any new object, the chances are that it will be referenced for a short time and then become garbage. However, if the object survives, it is likely that it will continue to be accessed for some time. The schemes above take little or no account of this fact.

With generation garbage collection, the basic strategy is to divide the address space into several 'generations'. New objects enter the youngest generation. If an object survives garbage collection in any generation a number of times, thus indicating its stability, then it is 'promoted' to the next generation. The number of garbage collections an object survives, can be stored as a count, and the number of garbage collection survivals before promotion can be adjusted as the workload of the system changes. Older generations would be garbage collected less often than younger generations. Each generation can be garbage collected (using copying for instance) without disturbing other generations.

For a full account of garbage collection schemes the reader is encouraged to refer to Cohen (1981).

10.2.6 Lisp Platforms

Having described the characteristics of a functional language such as Lisp, and the mechanisms that may be applied to enhance its execution on a sequential machine, this section outlines the spectrum of sequential machines supporting Lisp. This spectrum ranges from the standard conventional von Neumann architecture, to dedicated workstations such as the Symbolics 3600 series. The recent emergence of Reduced Instruction Set Computers (RISCs) will be examined to indicate their likely effect on the hardware support for Lisp in the future.

10.2.6.1 General Purpose: Conventional von Neumann Machines

Programmers and computer scientists were, until quite recently, 'tied' to the use of multi-user systems. These systems were typically equipped with a single high performance processor (normally 32-bit), a large working memory (several megabytes), and large amounts of backing store on high performance magnetic discs. These computing resources were shared amongst a community of users. The system operated in a time-sharing mode, switching between several user tasks. Subsequently, the resources, that could be dedicated to any one user, were limited and the response of the system was dependent on the processing demands of the other users sharing the machine. This scheme of operation was adopted, mainly because, historically, the hardware of the system was very expensive and time-sharing allows the processor to be fully utilised.

Because of the severe demands that Lisp programs placed on the system in terms of memory, processor time and input/output, this time-shared approach does not represent a satisfactory platform for Lisp development work, unless the user community could be limited in some way. With the continued fall in the cost of computer hardware, it was not long before developers of AI software 'championed' the idea of dedicating a machine of similar power to a multi-user system to a single user. Indeed, it was not just AI that suffered from this situation, other area such as Computer Aided Design (CAD) required large amounts of processing power and a specialised software environment. It was from these ideas that the workstation concept developed.

10.2.6.2 The Workstation

The workstation is typically based around a 32-bit processor, which allows the addressing of large memories and provides effective memory transfers. The first 32-bit microprocessor was produced by Motorola, and it is these devices that found their way into most early workstation designs. Large amounts of working memory are provided, with four megabytes being a standard configuration, but the system allows for expansion to perhaps 16 megabyte. Large amounts of high-speed backing store are provided by Winchester magnetic discs. Most workstations support high-resolution displays, typically 1150 by 900 pixels, and run under an enhanced UNIX type operating system, also supporting a user friendly WIMP (Windows, Icons, Mice and Pull-Down-Menus) environment.

These general purpose workstations have found favour in the AI community. One of the leading manufacturers of general purpose workstations, Sun Microsystems estimate that 6% of their workstations are used solely for AI software development work (Judge, 1987).

10.2.6.3 Personal Computers

Until the recent arrival of 32-bit personal computers (PCs), these machines were not considered feasible development systems. Earlier 8-bit, and to a lesser extent 16-bit designs, had severe address space limitations and lacked flexible pointer manipulation facilities.

The PC market has polarised into the IBM PC and Apple Macintosh camps. The IBM PCs are based on the Intel 80X86 range of processors, while the APPLE machines are based on the Motorola 680X0 processors. Since it was Motorola that produced the first 32-bit microprocessor, it was this range that found favour with the workstation manufacturers. Consequently more development work has been performed on systems based around the Motorola chips than on the Intel range.

Apple developed its PCs, based around a user friendly interface incorporating a WIMP environment. This has resulted in the Macintosh range closely resembling the environments available on workstations. The IBM PC, however, has been largely tied to the MS-DOS operating system, which imposes severe memory-size restrictions.

A number of systems in use today have been delivered on PCs but most required system enhancements such as increased memory capacity and/or the use of co-processor boards, (see Section 10.2.6.5.2).

10.2.6.4 Performance Issues

The users of general purpose machines have a decision to make on which, if any, of the support mechanisms described in Section 10.2.5 their machines are going to support. If they choose to support them, then it must be by software processes, and these will have performance trade-offs as well as benefits. If a particular system is not to support these mechanisms at all, then it may severely limit the type of application. For example, what would be the implications of a system that does not to support garbage collection?

A large working memory would be required, able to absorb the redundant objects and still meet the demand for new memory. This implies that either the size of the problems must be restricted, or the physical memory must be expanded to meet the problem's requirements. The first is impractical for serious development work, and the second becomes very expensive and is limited by the expandability of the system. Thus, the performance penalty of providing software support may seem attractive when compared to the alternative of not supporting them at all.

Some systems also allow the developer to de-activate some of the software support mechanisms, in order to increase the performance of the system. During development, software designed for type checking is activated, to provide a secure environment, at the expense of the software overhead of the run-time checking of type tags associated with objects. When development is complete, the programmer may de-activate the run-time checking of types, and enjoy a performance increase. Whilst this is a useful facility, its injudicious use can cause problems, and it also 'clouds' comparison of performance figures. Performance figures may be quoted for a fully checked implementation, whilst others may appear faster, because the figures were obtained with type checking disabled. The reverse may be true if both implementations were run in a consistent environment.

10.2.6.5 Lisp Machines

Lisp machines, as their name indicates, are specially designed to support the execution of Lisp programs. They may incorporate special hardware, and support the addition of an enhanced instruction set to assist with the manipulation of tags, etc. These enhanced instruction sets reduce the number of instructions required to encode a program, and the amount of translation required by the compiler.

Architectural support for Lisp is nothing new (Pleszkun and Thazhuthaveetil, 1987), and it is interesting to note that the first implementation of Lisp in the late 1950s on the IBM 704 made specific use of that machine's architecture, and these features have survived, albeit in name alone to this day. It is from the architecture of the IBM 704, that the Lisp access primitives CAR and CDR derive their names. The IBM 704 was equipped with a 36-bit data word, which included two 15-bit fields, called the address and the decrement. Special instructions could be used to fetch these fields independently. The names CAR (contents of the address part of the operand) and CDR (contents of the decrement part of the operand), evolved from the two-pointer Lisp cells in these data words.

Later architectural support was also evident in the design of the DEC PDP-6 and PDP-10, which implemented half word and stack-based instructions. This illustrates that the addition of special instructions which exploit the architecture, can reduce the semantic gap, which compromises the efficiency of Lisp running on general purpose machines.

In general Lisp machines are specialised workstations that are highly microprogrammed. They are typically based on the SECD virtual machine described in Section 10.2.5.1. The microprogrammed structure allows key Lisp primitive functions to be implemented in microcode. Lisp machines typically include hardware support for garbage collection and tag handling, so run-time checking does not significantly slow down execution.

Lisp machines offer 'better' performance than that of general purpose workstations, by virtue of the specialised hardware support. However, all this additional hardware has the negative effect of 'pushing up' the price/performance ratio of this type of machine when compared to general purpose workstations.

It is those applications, demanding this extra performance, that the dedicated Lisp machine is aimed, and the poor price/performance must be accepted as there is little alternative. It would appear that there are a large number of applications that justify this extra performance, since it has recently been reported (Johnson and Durham, 1986) that the market for Lisp machines was $200 million (in 1986). One of the most successful Lisp machines to date has been the Symbolics 3600 series which is described below.

10.2.6.5.1 The Symbolics 3600.

The Symbolics 3600 range (Moon, 1985, 1987) has a long and distinguished pedigree. The origins of the design can be traced back to the Lisp machine project of the middle 1970s, at MIT's Artificial Intelligence Laboratory. One of several 'spin-off' companies resulting from that project was Symbolics. Set up in 1980, Symbolics' first commercial machine was the LM-2. This was discontinued in 1983 with the arrival of the 3600 series. The philosophy of the Symbolics company has always been that security should not be sacrificed for the sake of speed. As such, the 3600 range provides full run-time checking of data-types, array subscript bounds and the number of arguments passed to functions. To achieve high performance, these run-time checks are performed by hardware and in parallel with other operations.

The 3600 series is based around a 36-bit word-tagged architecture, and the processor of the 3600s is highly microprogrammed. The microcode is contained in a 112-bit by 8K word microcode memory. The 3600 provides an instruction cache and two large stack buffers. The stack buffers contain the top portion of the Lisp control stack. Instructions are processed in the Instruction Fetch Unit (IFU) by a four-stage instruction pipeline. The IFU decodes instructions, generates microcode addresses and predicts branches. Program loops that fit into the instruction cache operate at full speed with no instruction fetch overhead.

The 3600 series supports a stack-based execution model. Function calls are efficient, due to the stack buffers, which reduce the number of memory accesses required. A given computation is associated with a stack group:

- *Control* contains function nesting information, arguments, local variables and other objects associated with the control or local environment.

- *Binding* records dynamically bound variables.
- *Data* contains stack allocated temporary objects.

Instructions typically read one operand from the current stack buffer, and another from the top of stack and return the result to the top of stack. Many instructions are Lisp functions implemented in hardware or firmware. These include: CAR, CONS, AREF, MEMBER, and EQ.

The 3600 series provides hardware support for 34 data types. The 36-bit words can be in one of two formats: Tagged pointer format or intermediate number format.

- Tagged pointer format:
  (1) 6-bit tag field.
  (2) 2-bit CDR-code.
  (3) 28-bit address.
- Intermediate number format:
  (1) 2-bit tag field.
  (2) 2-bit CDR-code.
  (3) 32-bit immediate numerical data.

It can be seen that both formats include a 2-bit CDR-code (see Section 10.2.5.3.2).

The memory of the 3600 is arranged as 44-bit words. Of these 44-bits, seven are used for double error detection/single error correction checks. This allows the 36-bit words to be accommodated with one spare bit in each word. The 3600 operates hardware support for generation garbage collection (see Section 10.2.5.3.3). The 3600 implements a paged virtual memory system. The 28-bit address of the tagged pointer format allows the system to address 256 million words of memory.

The 32-bit immediate numerical data field of the intermediate number format offers the 3600 compatibility with other systems. Industry standard 32-bit data representations can be directly transferred to and from the Symbolics.

A 'Front End Processor' (FEP) is used to handle the input/output tasks, and leave the Symbolics free to perform the other processing. The 3600 series supports a common Lisp environment. The operating system of the Symbolics 3600 is itself written in Lisp.

It is probably safe to say that Symbolics is currently the market leader in terms of Lisp machines. However, a number of other companies also produce Lisp machines.

10.2.6.5.2 Other Lisp Machines. Other commercial Lisp machine manufacturers include: Xerox, Tektronics, Fujitsu, Lisp Machines Incorporated (LMI) and Texas Instruments (Hwang *et al.*, 1987).

In 1981, Xerox introduced its 1100 series of personal workstations. These machines resulted from the pioneering work at Xerox into integrated workstation technology. These systems support Xerox's Interlisp-D environment.

LMI produced their first machine in 1977, called the CADR. In 1983 it launched its LAMBDA machine, which consisted of a 16-bit multiprocessor system centred around the 'fast bus' developed at Texas Instruments, called NuBus.

Fujitsu entered the market in 1983 with their ALPHA system. However, the ALPHA system was designed to operate as a backend processor to one of its large computers, and as such did not offer the general availability of other Lisp machines.

In 1984, Tektronics launched their low cost 4400 series. These machines were built around conventional 32-bit microprocessors from Motorola. In the same year Texas Instruments launched their Explorer range (Bosshart, 1987).

Texas Instruments began their research programme into AI processing in 1978. Texas Instruments has invested heavily in this programme, and has acquired an interest in LMI. Like the LAMBDA, the Explorer range was built around the NuBus. Originally the 32-bit microcoded processor was built on two boards. Texas Instruments used its semiconductor fabrication experience to design and produce a 32-bit VLSI chip, that contained over a half million transistors. Billed as the world's first AI microprocessor, it allowed 60% of the original processor design to be implemented on a single chip.

The Texas Instruments MicroExplorer range also illustrates the benefits of VLSI technology, by using the VLSI chip to implement a system small enough to fit inside an Apple Macintosh II personal computer, but providing sufficient power to act as a 'delivery vehicle' for those systems that do not require the full power of the Explorer range.

10.2.6.6 Take the RISC

Recent developments in processor design and compiler technology have led processor designers away from the so called Complex Instruction Set Computers (CISCs), to the idea of Reduced Instruction Set Computers (RISCs). The RISCs have taken their design philosophy from the fact that most of the large number of instructions that the CISC designs support are not used to a sufficient extent to justify the design effort and overheads in implementing them. The argument from the RISC proponents is to design a processor that can support very efficiently a small number of primitive instructions, which can be combined if need be to perform more complex tasks.

The advantage of the RISC approach is that it gives better performance and leads to simpler processor design. It should be noted that the RISC approach actually increases the semantic gap. The rationale behind this is that the extra performance gained from executing primitive instructions more than compensates for the extra overheads of translation to close the semantic gap using software.

10.2.6.6.1 A RISC for Lisp.

Most RISC processors to date have been designs based on the instruction mix measurements of procedural languages, such as Pascal and C, when applied to scientific computing type tasks. However, a number of research groups have been investigating the use of RISC chips as an efficient execution vehicle for Lisp (Hill, 1986). Because of the stack-based nature of most RISC designs, attention has been focussed on the type of modifications to this design which would produce the best performance for Lisp programs.

One such project has been underway since the mid-1980s at Berkeley, is known as SPUR (Symbolic Processing Using a Risc), and involves modifications to the Berkeley RISC-II processor. These extensions include hardware tags, special condition branches for tag checking, parallel type-checking on CAR and CDR operations, support for integer arithmetic with run-time type-checking and support for the generation garbage collection scheme. Initial performance figures have proved promising but research is still continuing to ascertain the most effective ways to optimise performance.

10.2.6.6.2 Other RISCs. Perhaps the most significant effect of the availability of fast general purpose RISC processors is that of a dramatic reduction in the price/performance ratios of systems. The workstation market is currently being 'flooded' with 32-bit RISC processor based systems. Recently, Apollo has announced a 64-bit RISC processor based workstation, known as the PRISM (Wilson, 1988). This has involved a re-evaluation of the market from the manufacturers of personal computers (PCs), workstations and specialised systems.

Because of the increased performance of the personal computers now available, made possible with the use of 32-bit microprocessors, available from large manufacturers (i.e. the same manufacturers who have traditionally supplied the workstation manufacturers), the workstation manufacturers have to conceive new ways to maintain their performance premium over personal computers. The resultant action has been for workstation manufacturers to produce their own custom-designed (RISC) processors. One example of this approach is the Sun Microsystems SPARC (Scalable Processor ARCHitecture) chip set (Sun, 1988). It is interesting to note that the SPARC supports the use of tags, and as such has moved to provide limited hardware support for Lisp processing.

The increased performance and continued reduction in the price/performance ratio of general purpose workstations, will undermine the performance advantages previously enjoyed by the dedicated Lisp machines. If significant performance increases cannot be found by further modifications to a single processor design, then these manufacturers will have to explore new ways to maintain their performance advantage. One approach which promises to maintain the performance advantage is to exploit the potential of parallel processing.

10.3 PARALLEL IMPLEMENTATIONS

10.3.1 Parallel Processing

We have seen in the previous sections how conventional and modified von Neumann processors can be used to support the execution of functional languages such as Lisp. These systems are all based on uniprocessor designs. If dedicated machines such as the Symbolics 3600 were capable of providing the necessary power to run all applications, AI processing would present less problems.

A major problem arises if we wish to run an application, which even when implemented on the fastest uniprocessor, provides inadequate performance. It would appear possible to speed up the execution of the problem, by employing more than one of these fast uniprocessors, working at the same time to cooperate in the solution of the problem. Such a multiprocessor system, or parallel computer, could be theoretically N times faster than the fastest uniprocessor, where N is the number of uniprocessors used in the solution. This is the principle of 'linear speedup', and it is the performance gain for which all parallel computers strive. It would appear, then, that potentially 'huge' performance increases can be achieved by the application of a large number of processors.

Another aspect to the parallel processing approach is that of improved price/performance ratio. The advent of VLSI has made it possible to produce large numbers of processors at reasonable cost. Although the performance of each one of these processors may be very much less than that of the most efficient uniprocessor available, if a number can be coupled to cooperate in the solution of a problem, the system may display attractive price/performance figures over the specialised uniprocessor.

However, these benefits do not 'come free', and a number of obstacles must be overcome to effectively exploit the potential of parallel processing. Parallel processing demands that the problem be somehow split, in order that the parts can be run on a number of separate processors at the same time. The size of these partitions is termed the 'grain size'. The design of a parallel computer is intimately coupled to the grain size, and the nature of these tasks.

A number of processors cooperating on the solution of a problem implies some form of synchronisation and communication between the individual processors of the system. The way in which the tasks are controlled and the way data are exchanged, relates to the underlying programming paradigm. Both synchronisation and communication are dependent on some form of delivery system, and so the topic of the interconnection scheme of the cooperating processor becomes important.

10.3.1.1 Processor Configurations

The optimal arrangement for cooperating processors would be to have them all directly connected. This would support a direct communications path from one processor to all other processors. Whilst this ideal configuration is achievable for small numbers of processors, it becomes impractical for larger numbers of processors. With large numbers of processors, the only feasible way to connect them is by a network. Such networks are normally highly regular, to allow easy implementation. The aim of the network is to link as many different processors as possible with the minimum number of intervening processors between those that are not directly connected.

One popular processor configuration, based partly on its simplicity, is the mesh. With the mesh network, processors are connected only to their nearest neighbours. This configuration works well when communications are restricted to local exchanges. Fortunately, a large number of problems map well onto the mesh network, by exhibiting only local communications.

An alternative processor configuration is required however, when the need arises for communication with processors that are not neighbours. This non-local communication is handled very poorly by the mesh network. This is illustrated by the use of an example (see Figure 10.3). In a two-dimensional N by N square mesh of processors, each processor connected to its four nearest neighbours, and it can be easily seen that the greatest distance separating any two nodes is experienced by processors located at diagonally opposite corners of the mesh. Since no direct communication path exists, if these two processors wish to communicate, a route must be constructed from the nearest neighbour connections. The shortest number of such connections to connect these two processors is $2(N - 1)$. This proves very inefficient in terms of the use of the available communications bandwidth within the system and the overheads associated with recruiting the intervening nodes to deal with the handling of the communication.

In addition to its poor performance in the handling of non-local communications, the mesh network has other peculiar characteristics. Processors in the middle of the mesh exhibit a regular structure, each like its neighbours being connected to four other processors. However, processors on the edges and corners have only three and two connections respectively. It is necessary to treat these processors as special cases, and this increases the amount of programming required to map a problem onto the architecture.

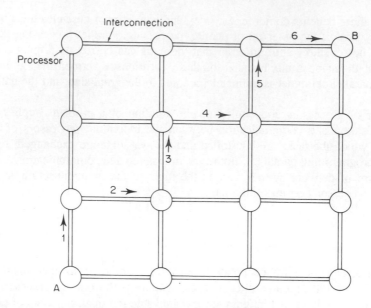

Figure 10.3   A four by four mesh of processors with nearest neighbour connections. If processors A and B wish to communicate the shortest route involves six hops

Many alternative networks to the mesh have been proposed. These include trees, rings, pyramids etc. The binary n-cube or hypercube (Pease, 1977) configuration has been identified as an attractive form for configuring large numbers of processors.

A large number of commercial parallel processors such as the Intel iPSC and the Thinking Machines Connection Machine, use a hypercube topology for their processor's configuration. The concept of a twelve dimension hypercube configuration of processors, used in the Thinking Machines Corporation Connection Machine, seems a daunting concept to comprehend. The hypercube's characteristics are described below.

10.3.1.1.1 What is a hypercube?

Whilst the hypercube configuration appears more complicated than the mesh, the underlying concept of a hypercube is very simple. The number of processors in a hypercube configuration is always a power of two (i.e. 0,1,2,4,8,16,...). This power of two also describes the dimensionality of the hypercube.

From the above, it is clear that a single processor can in fact be considered as a zero-dimensional hypercube configuration, since two to the power of zero (the dimension of the hypercube) is one (the number of processors). A processor configuration based on a one-dimensional hypercube is constructed by duplicating a zero-dimensional hypercube and connecting the processors. Any N-dimensional hypercube can be constructed by taking two N-1 dimension hypercube configurations as connecting the corresponding processors (see Figure 10.4).

The hypercube configuration displays some important characteristics:

- The interconnections are richer than those of the mesh, and the maximum path between processors in an X-dimensional hypercube is X. (Expressed another way, if the number of processors in the hypercube is N then the maximum path length between processors is $\log_2 N$.)

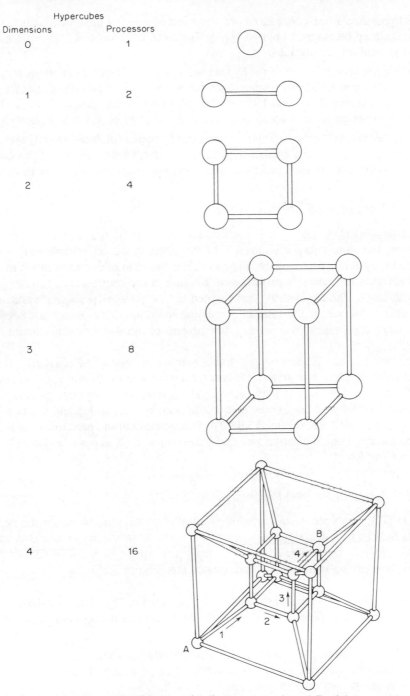

Figure 10.4 In the four-dimensional hypercube if processors A and B wish to communicate the shortest route involves only four hops

- The hypercube's interconnections are a superset of most of the other candidate networks. A mesh may be mapped onto an appropriate sized hypercube, by ignoring a number of the hypercube's connections.

- Another important feature of the hypercube, which has helped promote its use, is the fact that its structure is entirely regular. Each processor in the hypercube has exactly the same number of connections, and looks exactly like every other processor in the hypercube. Thus, there is no need to consider special cases of edge *versus* centre, as with the mesh.

- The hypercube is highly 'scalable'. It has already been explained how any size hypercube can be constructed from smaller hypercubes. To double the number of processors in the system only requires each processor to support one more communications channel.

10.3.1.2 Performance Considerations

Since communication and synchronisation actions must be transferred to the appropriate processor, the time delay associated with the interconnection scheme may significantly degrade the system's performance. In general, it is the ratio of communication to processing that determines the system's performance. Because of the overheads of communication and synchronisation, the theoretical 'linear speed-up' of parallel processor systems is never achievable. However, if the ratio of processing to communication is very large, practical performance can be achieved within a few percent of the theoretical maximum (Genner *et al*., 1988).

If the problem can be decomposed into a number of 'grains' of equal size, the question of load balancing becomes trivial. However, many problems do not 'split' as conveniently as this, and the mapping of these varying sized tasks onto physical processors must be considered carefully, if one processor is not to become overloaded and create a bottleneck to the system, whilst others remain idle. The management of the partitioning of the problem, synchronisation, communication and load balancing will all depend on the type of parallel processor employed.

10.3.1.3 Types of Parallel Processors

Flynn (1972), produced a classification of parallel processors, based on the concept of a stream. A stream is defined as a sequence of items (instructions or data), as executed or operated on by a processor. A processor may have single or multiple data or instruction streams, thus giving rise to four classifications (see Figure 10.5):

- *The SISD*. Single Instruction stream/Single Data stream. This describes the classical serial von Neumann computer. There is one thread of control acting upon a single stream of data.

- *The MIMD*. Multiple Instruction stream/Multiple Data stream. This describes multiprocessor systems, in which each processor possesses its own instruction stream, and acts on its own data stream.

- *The SIMD*. Single Instruction stream/Multiple Data stream. This describes systems that perform the same operation on many data items simultaneously. It describes systems that retain the single thread of control as in SISD systems, but achieve parallel processing by applying the same instruction to a number of data simultaneously.

	Single instruction stream	Multiple instruction stream
Single data stream	SISD (sequential von Neumann computers)	MISD (void)
Multiple data stream	SIMD (processor arrays)	MIMD (multiprocessor computers)

Figure 10.5 The Flynn classification of computer architectures

- *The MISD*. Multiple Instruction stream/Single Instruction stream. This classification appears void, as it implies that a number of different instructions be applied to the same data simultaneously, which does not correspond to any of our computational paradigms.

The MIMD and SIMD type of parallel processors will now be examined more closely.

10.3.1.4 General-purpose Parallel Processors

General purpose parallel processors are those able to support a range of programming languages, and are not built to perform one specific task (that, at least, is the meaning conveyed here). These machines tend to be based around the MIMD principle.

The most obvious way to implement a parallel processor, would be to base the design on replicated von Neumann processors. Such a scheme would be MIMD in nature, since each of the replicated processors could perform instructions from its own instruction stream, and apply these instructions to the contents of its own memory (its data stream). This type of design would afford a number of advantages. The experience of processor design, dating back to the first von Neumann machine can be utilised. The use of VLSI allows a conventional processor to be placed on an integrated circuit. Extensions to the programming style associated with programming a SISD processor can be used to program the machine.

The two major types of MIMD configurations based on replicated von-Neumann architecture are Shared-memory and Message-passing architectures (see Figure 10.6).

10.3.1.4.1 Shared memory. In the shared memory scheme, a number of processors are connected via a switch or bus to a global memory. There are currently a number of successful commercial systems which employ the shared-memory approach to MIMD parallel processing. These range from the very fast supercomputers from Cray Research Systems (Cray, 1984), such as the Cray X-MP, to 'superminis', such as the Sequent Balance

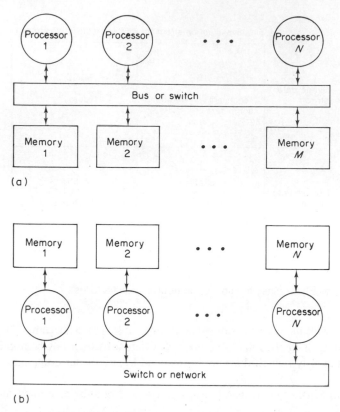

Figure 10.6   (a) Shared-memory systems; (b) message-passing systems

range (Thakkar *et al.*, 1987). The partial success of these systems may be attributed to the fact that they offer an evolutionary path to parallel processing, rather than a more radical departure from conventional uniprocessing systems required by other schemes. Communication and synchronisation is achieved by the use of shared variables within the global memory. This sharing of data requires mechanisms to ensure correct usage and consistency.

Whilst these systems are suitable for a small number of processors, they cannot be extended. This is due to all the processors accessing the same global memory, which results in the need for very high memory bandwidths. This scheme can therefore be seen to exasperate the von Neumann bottleneck between processor and memory. Because of this, performance of the system will fall short of the linear speed-up curve, when additional processors are added. At current memory and switch technology, this cut-off point appears to be at around tens of processors.

10.3.1.4.2 Message passing.   With the message-passing scheme, there is no concept of a single global address space. Instead, each processor has access to its own local memory. Thus, the memory of the system can be considered as truely distributed throughout the system. Since processors have access to only their own local memory, communication and synchronisation are handled by passing messages between processor–memory pairs. The interconnection of the processor–memory pairs may be by a switch or by a network.

10.3.1.4.3 Shared memory versus message passing. The shared-memory approach has the advantage of being an 'evolutionary path' to parallel processing from a uniprocessor system. Whilst shared-memory systems are suitable for large-grain parallel-processing problems, which exploit a few processors, such systems cannot be extended due to the heavy demands made by additional processors on the memory-to-processor bandwidth.

Whilst the message-passing approach may appear a more radical departure from the SISD mode of operation than the shared memory approach, the ease with which these systems can be built, and their extensionability, have resulted in a number of commercial systems based on the message-passing principle. However, the lack of global memory with the message passing scheme, means that the communication of control information and data between processors must be achieved via messages. These messages are conveyed either via a switch or network. If each processor–memory pair were directly connected, this would not present too many problems. However, when more than a few processors are employed, it is impractical to directly connect them.

The problem then arises as to how to deal with messages between non-directly connected processor–memory pairs. Messages must be sent across the switch or network, and directed to the desired destination. When large numbers of processors are involved, it is usual to employ a network. The processor–memory pairs at intermediate points between the source of the message and its final destination have the responsibility of forwarding on the message to other processor–memory pairs along the route. This has two detrimental effects. One is to increase the time taken for the destination processor–memory pair to receive the message, and secondly the intermediate nodes are required to process the message to determine the appropriate action, which incurs a performance overhead on the intermediate processor–memory pairs.

The performance of message-passing systems is determined by the efficiency of the network, since it has implications on the choice of grain size appropriate to the system, and the ratio of processing to communication within that grain size. Interest in message-passing parallel processors has recently been stimulated by advances in VLSI, which make it possible to produce complete processors with local memory on a single integrated circuit. One example of this type of device specifically aimed at the message-passing parallel-processing systems is the Inmos Transputer.

10.3.1.4.4 The Transputer. The Inmos Transputer (Inmos, 1985) is a range of programmable VLSI devices, specifically designed as 'building blocks' for message-passing parallel processors. Often termed 'a-computer-on-a-chip', (see Figure 10.7) a Transputer contains a powerful RISC type processor, several kilobits of single cycle on-chip memory, a memory manager unit to provide access to large amounts of off-chip memory and a number of links by which Transputers to be easily connected together, and along which Transputers can exchange messages.

10.3.1.4.5 Programming considerations. The task of programming such MIMD systems should not be underestimated. The programming paradigm most commonly adopted for these general purpose parallel processors is based on modifications to the sequential control-flow scheme. This requires the programmer to explicitly define the tasks to be implemented on each processor, and the communication and synchronisation control between them. This is achieved by operating system type extensions to sequential languages, in order to make communication and synchronisation primitives available to the programmer.

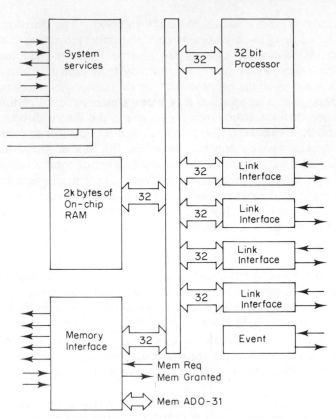

Figure 10.7 T414 transputer system organisation (source: IMS T414 Transputer Reference Manual (Inmos, 1985)). (Reproduced by permission of Inmos Ltd.)

10.3.1.4.6 Commercial message-passing parallel processors. To date, there have been several message-passing MIMD parallel-processor systems available. With one notable exception, most have been aimed squarely at the scientific 'number crunching' applications. The Intel iPSC has always made its possible application in the AI environment clear. The Intel iPSC is described in detail below.

10.3.1.4.7 The Intel iPSC. The Intel iPSC (Personal Supercomputer) (Intel, 1988) emerged from the Cosmic Cube research project, conducted at the Californian Institute of Technology in the early 1980s. The Cosmic Cube project (Seitz, 1985), investigated the feasibility of using a relatively small number of powerful processors, operating in message-passing MIMD mode, and connected in a hypercube configuration. The Intel Corporation took an early interest in the project and donated a number of its 8-bit microprocessor chips to assist in the construction of prototypes. The first prototype was completed in 1983, and consisted of 64 processing nodes, containing the 8086/8087 microprocessor and numerical co-processor, together with 256 Kbyte of random access memory. This gave each processor node roughly the same computing power as the IBM PC. In addition, each node had six communications links, which allowed the nodes to be connected in a six-dimensional hypercube.

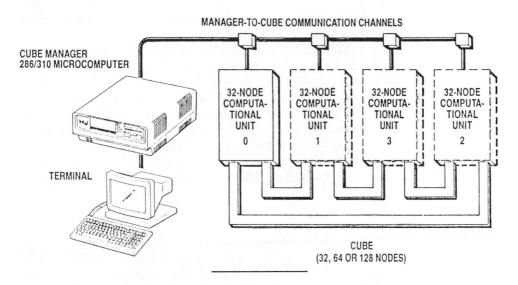

Figure 10.8 iPSC system architecture (source: Intel iPSC System Product Summary). (Reproduced by permission of Intel Corporation.)

Having demonstrated the feasibility of using off-the-shelf components to construct such a system, Intel developed and marketed the system, and in 1985 announced its iPSC range.

The iPSC is comprised of two subsystems: the cube and the cube manager (see Figure 10.8):

The Cube   The cube is constructed as a number of computational units. Each computational unit contains 32 processing nodes. The cube may be constructed from one, two or four computational units allowing the iPSC to offer systems with 32, 64 or 128 processing nodes. These processing nodes are arranged as a five-, six- or seven-dimensional hypercube respectively. Each processing node in the iPSC contains the more advanced 16-bit 80286/80287 microprocessor and numerical co-processor chips. Each node has 512 Kbyte of random access memory. This gives each node an equivalent computational power as the IBM AT PC. A fully configured system will thus have 128 processors with a total distributed memory of 64 Mbyte.

The expandability of the system is limited to a seven-dimensional hypercube configuration, by the need for each node to support a number of communication channels for message passing between processing nodes. Each node has seven communication channels. Each of these are bidirectional and asynchronous. They are implemented as serial Ethernet links operating at 10 Mbit/s. The links are controlled by Intel's 82586 Ethernet Local Area Network controllers. There is also a global Ethernet communication channel which connects all nodes to the cube manager. The circuits for each node are contained on a double extended eurocard.

The commercial viability of the iPSC can be attributed to its price/performance ratio. This was achieved by the use of off-the-shelf VLSI components, which has reduced development costs and at the same time allowed each node to be constructed from a few components. The entire ensemble of 128 processing elements can be accommodated in a cabinet of modest dimensions (2 ft by 2 ft by 5 ft)

Resident on each node, is an elementary operating system called a 'kernel'. This kernel supports the routing of messages between non-adjacent nodes, dynamic message buffer management and the execution of processes within the node.

The Cube Manager

The cube manager supports the programming environment, and servers as the cube's system manager. The cube manager is implemented on an Intel 286/310 multi-user supermicro-computer, with 2 Mbyte of working memory and winchester disc backing memory. Application programs are developed and compiled in the cube manager, and downloaded onto the cube to execute on separate processing nodes.

The store and forward technique, used in the message passing communications between non-adjacent nodes, proved a bottleneck. This was due to the fact that each node in the communications path had to store the message in its memory. The processor at each node along the route was charged with the task of handling the message, and forwarding it on along the appropriate link. During these periods, the processor was not available to continue processing other tasks. Because of this bottleneck, how the problem was to be decomposed, in order to run on the hypercube processors, had to be given very careful consideration, if performance was not to be 'stunted' by these communication overheads.

In late 1987, Intel announced a second generation of hypercube configured systems: the iPSC/2. These machines took advantage of the advances in processor and memory technology from the first generation. Each node's power was now equivalent to a workstation, by the use of the 32-bit 80386 microprocessor. This, together with the 80387 numerical co-processor, and a memory of capacity ranging from 1 to 16 Mbyte made the nodes of the iPSC/2 about four to five times faster than those of the iPSC. These 32-bit processors allowed a much improved software environment to be implemented , as well as being able to maintain software compatibility with the earlier generation.

Perhaps most importantly, the new generation improved the communications bottleneck by providing hardware support to balance communication performance with processing power. This was achieved with the Direct Connect Routing scheme.

Direct connect routing   The Direct Connect Routing (DCR) scheme uses improvements in both hardware and software to shorten message passing times. The store-and-forward method of the iPSC is replaced by a hardware switching system. The hardware switching system is implemented in CMOS programmable gate array logic to keep the component count as low as possible. With the DCR, only the sending and receiving processors are directly involved in the communication. For intervening nodes, the DCR hardware automatically takes care of the forwarding of the messages, leaving the processor free to continue processing the application's tasks.

The DCR is quoted as being able to maintain communications performance between any two nodes within 10% of that which would be experienced if the nodes were directly connected. By making message times roughly insensitive to message distance, the topological properties of the processing nodes and the mapping of the application onto these processing nodes requires less effort to achieve the same performance increases. This is not, however, to say that this task has become totally trivial, since it would be possible by the injudicious placing of tasks on processing nodes, to achieve poor load balance or to generate so many messages that the communications bandwidth of the system cannot cope. Either one of these actions would seriously degrade the system's performance.

Whilst the Intel iPSC range was originally conceived as a general purpose parallel processor for scientific applications, the applicability of the iPSC range for AI development work was greatly enhanced with the porting of Gold Hill Computer's Common Concurrent Lisp (CCLISP) (Intel, 1987), to run on the iPSC system.

CCLISP was the first commercial Lisp for parallel processors. CCLISP provides a powerful environment to allow multiple common Lisp processes to cooperate asynchronously, and to spawn new processes. Within CCLISP, a library of standard functions that cause messages to be sent or received, together with functions for the creation and destruction of processes, are 'callable' from a Common Lisp program. It is for the programmer to explicitly define the behaviour of the parallel processes and their communications. To exploit the CCLISP environment the CCLISP has to be resident on each processing node, and each node requires over 4 Mbyte of memory.

An indication of the importance of these types of parallel processors can be seen by the recent increase in the number of articles written on MIMD message-passing hypercubes, and the introduction of a number of hypercube products by various manufactures. The hypercube configurations are well placed to take advantage of leaps in technology. The next generation machines may see the introduction of a increased number of nodes, many times more powerful than current systems, perhaps utilising custom-made VLSI chips which may contain several nodes per chip. However, on a cautionary note, it is interesting to note that one of the recent manufacturers to enter the market, Floating Point Systems with their T-Series (Judge, 1988) of large dimension hypercubes, has recently withdrawn from the market. This may have been due in part to the relatively small market for such systems at present, currently about $50 million, or to an over-estimation of the amount of parallelism that can be successfully exploited by the MIMD message-passing paradigm.

10.3.1.5 Language-Based Parallel Processors

We have seen in the section on general purpose parallel processors, that the problem of exploiting the potential of parallel processing is not largely associated with the building of such machines, but more with the effective programming of them.

The extent of this problem becomes apparent, when the problems of sequential programmming are examined. In sequential control flow languages such as Pascal, there exists a single thread of control. The need to consider communication and synchronisation between different threads of control is thus redundant. The sequential execution of code has a number of advantages, in that the programmer is assured that when one statement appears below another in a program listing, they will be processed in that order. Thus, there exists an amount of 'assumed synchronisation' and communication within a sequential program. This is not normally considered by the programmer, as it is inherent within the language, and sequential implementation. Even so, the writing of correctly functioning sequential code is, in many cases, a non-trivial task.

It is instructive to consider the additional tasks assigned to the programmer, to allow the exploitation of parallelism, by providing extensions to sequential control flow languages. With the need to determine parallel tasks, their synchronisation and the communication between them, it is not surprising that the construction of correctly functioning code to run on parallel processors, which effectively exploits their potential, is very much more difficult than in the sequential uniprocessor arena. This problem of programming parallel processors is currently providing a limitation on the 'penetration' of such machines in all the computing markets. In order to ensure greater utilisation of parallel processing techniques, new ways to conceal the underlying complexity of the parallel processor must be found. One way to achieve this, is to 'move away' from languages that have a sequential base, and 'look' for more promising programming paradigms. A programming language based on mathematical principles would appear particularly suited.

One of the most attractive features of functional languages is their suitability for programming parallel computers. To explain this valuable asset, it is necessary to review briefly the salient features of functional programming languages and how these allow the exploitation of parallel processing.

Strict functional languages do not allow the use of variables or assignment to variables. Functional programs can be viewed as a nested set of expressions. These expressions allow functions to be applied to arguments. Both functions and arguments can be simple values or subexpressions. The only control structure that is permitted is recursion. The values of expressions are returned by the function and are not assigned to temporary variables. By banishing the assignment statement from the programmer's vocabulary, functional programs are 'side-effect free'. This results in functional programs being referentially transparent. This property ensures that the value of an expression is determined only by its definition, and not by its computational history. In other words, the execution of a function is guaranteed to return the same result (assuming termination), regardless of the order of evaluation.

Referential transparency allows the subexpressions of a function to be evaluated in parallel, without the need to explicitly define the synchronisation and communication between them. Functional languages allow programs to be written, in which the parallelism is implicit in the data dependencies of the program. The synchronisation and communications are handled implicitly within the semantics of the language, thereby relieving the programmer from the need to explicitly specify these lower level tasks. This allows the programmer to work at a 'higher level of abstraction', and hence improve productivity. It is for these reasons that functional languages provide such an attractive basis for exploiting parallel computers.

However the implementation of functional languages present a number of problems. Since referential transparency demands that there may only be one definition of every function within the program, some means must be found of relating a reference to a function with the function's definition. We have already seen in Section 10.2.5.1 how, with the SECD virtual machine, it is possible to employ an environment to form this association. The use of the environment, allows the function body to remain unaltered, and while this scheme may be acceptable in a sequential implementation, an alternative scheme is required if the potential of processing subexpressions within the functional program is to be realised. The reduction model of computation allows this and is described below.

10.3.1.5.1 *Reduction computation.*   Reduction programs are constructed from nested expressions. Execution is based on the recognition of reducible expressions and the transformation of these expressions. This transformation is termed 'rewriting'. Execution proceeds by progressively rewriting expressions, using the appropriate rewrite rule to generate simpler expressions. These simpler expressions have mathematically the same meaning as the original expression, but are reduced. Execution terminates when a rewrite results in a constant expression that cannot be further reduced, representing the result of the execution.

During execution, functions are applied to arguments. The arguments are examined to see if execution is possible. If the arguments are simple values, then execution can continue, and the effect is to rewrite the expression by the returned value. However, if the arguments to a function are themselves subexpressions, then these must be reduced recursively until they return the desired values. This is termed 'demand driven' evaluation (Treleaven *et al.*, 1982), since the requirement for an argument's value triggers the operations that will generate it. Demand driven evaluation of an expression spawns the reduction of one or more subexpressions. The original expression is then blocked awaiting the values to be returned by these subexpressions. When these arrive, the reduction process of the original expression can continue.

Referential transparency makes reductions independent of context and sequencing and so subexpressions may be evaluated in parallel. Two forms of reduction evaluation are possible, differentiated in the way that the arguments are manipulated:

- *String Reduction*   The basis of string reduction is that each expression accessing a particular definition will take and manipulate a separate copy of the definition. This gives the appearance of a string of expressions which during execution, expand and contract as each expression is rewritten.

- *Graph Reduction* The basis of graph reduction is that each expression, accessing a particular definition, will be replaced by a pointer to the body of the definition. This gives the image of an expression not stored as a string, but as a graph, with subexpressions as nodes in this graph.

- *String versus Graph Reduction*

Because of the properties of referential transparency, both string and graph reduction evaluation of an expression will give the same result (assuming termination). However, graph reduction has a number of attractive features when practical implementations of reduction machines are being considered. Moving from string to graph reduction involves being progressively 'lazier' about making copies. One of the advantages of using graph reduction over string reduction is that there is no need to re-evaluate common subexpressions. Graph reduction also allows the manipulation of infinite data structures, as long as only the values of some finite part are demanded.

A good introduction to reduction can be found in Amamiya (1988). A description of the ALICE machine, which uses graph reduction to evaluate functional programs is given in section below.

10.3.1.5.2 ALICE. The ALICE, or Applicative Language Idealised Computing Engine, was proposed by John Darlington and Mike Reeve of Imperial College of Science and Technology, in London. ALICE (Darlington and Reeves, 1981; Eisenbach and Sadler, 1988), represents one of the first parallel computers designed for the parallel execution of functional programs.

The processors in ALICE are called 'reduction agents', and these perform graph reduction on functional programs. By employing multiple reduction agents, ALICE can simultaneously reduce many subexpressions in parallel. The expression graphs are represented by collections of packets, which are stored in a collection of memories known as packet pool segments. Conceptually, the ALICE machine consists of a collection of reduction agents, and packet pool segments, linked by an interconnection network (see Figure 10.9):

Packets Expressions are stored as packets. Packets are fixed sized blocks of data divided into a number of fields. Each packets comprises of three primary fields and three secondary fields (see Figure 10.10). The primary fields are used to represent the nodes in the graph. The secondary fields provide the control information required by the evaluation mechanism.

- *Identifier*—uniquely identifies the packet. This directly corresponds with an address in a packet pool segment.

- *Function*—contains a pointer to the relevant function associated with the node in the code pool.

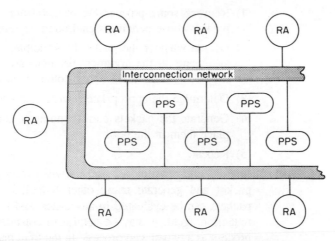

Figure 10.9 An ALICE machine RA (Reduction Agent) PPS (Packet Pool Segment) (Eisenback and Sadler, 1988). (Reproduced by permission of Butterworths & Co.)

Identifier	Function	Argument list	Status	Signal list	Reference count

Figure 10.10 An ALICE packet

- *Argument list*—defines the arguments for the node. May contain literal data or references to other packets.

- *Status*—contains the 'pending signal count', which identifies the number of packets on which this packet is directly dependent. This packet is suspended until all these packets have been evaluated, and the pending signal count is decremented to zero.

- *Signal list*—records the identifiers of the packets which are suspended, awaiting the evaluation of this packet and must be signalled when this packet produces a result.

- *Reference count*—used for garbage collection. Initially contains a count of the signal set. When the reference count is decremented to zero, it is no longer needed, and may be garbage collected. Garbage collection is immediate and takes place concurrently with processing throughout the system. It is necessary to garbage collect in this way because of the high rate of packet growth during graph creation.

Reduction agent operation The reduction agents executes the following sequence of tasks:

(1) Remove some processable packet from the pool, i.e. a packet whose pending signal count is zero.

(2) Decide whether the packet is reducible, i.e. whether the arguments in the argument-list field are constants, or if they contain the identifiers of other packets.

(3) Determine the appropriate rewrite rule based on (2).

(4) Generate the packets corresponding to the 'rewrite' and place them in the pool.

(5) Repeat.

Initially, the rewriting of packets will overwrite the existing packet and generate many other reducible packets. These packets can be evaluated in any order, and by any available reduction agent. It may be helpful to consider the reduction process as a two-staged process. In the first, packets 'demand' values to their arguments. This creates a number of reducible packets which, on completion, send 'control' signals back to the demanding packet, which may then resume.

Some form of communications network is required to allow the distribution of subexpressions, the transmission of definitions (rewrite rules) and the relaying of signals. It is clear that two distinct types of access are made to the packet pool. The first is directed at a packet with a particular identifier, for example to signal completion. The second type of access is to find any processable or empty packet. These two types of accesses are supported by two different communication systems in ALICE. The first mechanism is based on the network allowing each reduction agent access to every packet pool segment. The second mechanism is based on a distribution system consisting of a two channel communications ring, containing the identifiers of processable and empty packets respectively.

A prototype version of ALICE has been constructed. The ALICE machine was one of the first to use the processing power and parallel processing facilities of the Inmos Transputer (see Section 10.3.1.4.4).

Reduction Agents

Each reduction agent is constructed from a diamond of five Transputers (see Figure 10.11.). Two Transputers are dedicated as rewrite units. Each rewrite unit has 64 Kbyte of RAM cache memory, in which the rewrite rules that are currently available to the rewrite unit are stored. A single Transputer provides the definition checker, whose task it is to maintain a look-up table of the rewrite rules in the caches of the two rewrite units. The two remaining Transputers are dedicated to the management of the communication of packets between the reduction agent and the packet pool segments.

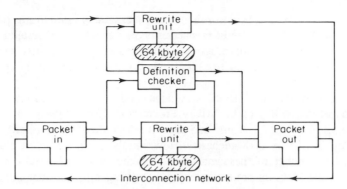

Figure 10.11 A reduction rule (Eisenback and Sadler, 1988). (Reproduced by permission of Butterworths & Co.)

Packet Pool Segments The packet pool is distributed throughout the system in a number of packet pool segments. Each packet pool segment can be accessed independently of any other, and a connection, via the interconnection network, to any reduction agent is possible. The packet pool segments contain 2 Mbyte of memory. Two Transputers are incorporated into each packet pool segment. One Transputer buffers the inputs from the interconnection network and handles the segment memory. The other is dedicated to performing packet transfers onto the interconnection network.

Interconnection Network The interconnection network allows any reduction agent to communicate with any packet pool segment. The design of the network is critical to the overall performance of the system. The network is based on the Delta multistage switch topology. A custom made ECL (Emitter Coupled Logic) cross bar switch integrated circuit, forms the basic building block of the network. ECL technology is used to achieve the 200 Mbit/s communications bandwidth, by reducing the switch latency times.

   The prototype design, which accommodated up to 40 reduction agents and packet pool segments, proved the feasibility of a practical implementation for the parallel execution of functional programs. However, one of the major problems with the design was the interconnection network. Since this is a shared resource between the packet pool segments and the reduction agents, as the number of packet exchanges increases, the network will become saturated, and contention for the network will result. This problem may be exasperated by the 'burst' nature of packet exchanges during rewriting packets. This bottleneck is familiar to all shared memory parallel processors. Other problems are a direct result of the functional programming paradigm. It is sometimes necessary to suppress the parallel execution of subexpressions. This sequential execution mode or 'Lazy evaluation' is flagged by the programmer in the source code.

The future development of ALICE was subsumed within the £15 million Flagship project (Watson *et al.*, 1987) of the Alvey Programme. The Flagship projects draws on the experience of the ALICE team, and also the Dataflow work (Gurd *et al.*, 1985) conducted at the University of Manchester. However the Flagship architecture now bears little resemblance to either of these machines. Together with industrial cooperation with ICL and Plessey, the Flagship project is aimed at creating a parallel computer architecture to support a range of declarative programming paradigms.

Other interesting projects in the reduction execution of functional programs include Grip (Peyton-Jones, 1987) and the wafer scale integration Cobweb project (Anderson *et al.*, 1988).

10.3.1.6 Massively Parallel Computers

An alternative to the schemes described in Sections 10.3.1.4 and 10.3.1.5 is to employ the use of massively parallel computers. With this type of computer, the combined power of a large number of simple processors is harnessed to achieve a significant performance increase, on a limited range of problems, over that achieved by sequential uniprocessors.

The ICL Distributed Array Processor (DAP) (Reddaway, 1978), Goodyear Massively Parallel Processor (MPP) (Batcher, 1980) and the Thinking Machines Corporation Connection Machine (see Section 10.3.1.6.1) are examples of such designs. These machines utilise thousands of 1-bit processors, each of which has access to a small memory. The processors execute in SIMD mode, under the control of a host computer. This fine grain parallelism approach can be very effectively used, because a number of applications exhibit a high degree of data parallelism. In such applications, the same operation is performed on the elements of a large data set. If the application was implemented on a sequential uniprocessor, the operation on each element in the data set would have to be performed one after the other. Execution times could be dramatically improved if the operation could be performed on all the elements of the data set simultaneously. To achieve this, the application must be partitioned in terms of the elements of the data set, and not as we have seen in the previous schemes, in terms of tasks. This can be achieved by the following architecture.

The memory of the host system is partitioned into a large number of elements, each typically a few kilobits in size. Within this memory, the elements of the data set are distributed and stored. This distributed memory can be accessed and modified by the host in the same way as it would access a conventional memory system. Each element of this distributed memory is allocated a simple processor. These processing elements are directly connected to the host, and also connected to a number of other processors. All processors receive the same instruction broadcast from the host, via the direct connections. These instructions allow processors to access their memories, and to perform the instruction on operands contained in the memory, and/or exchange data with their neighbours. The processors operate synchronously in SIMD mode. This eases the effort required in programming such a system, as there remains only one thread of control. The major difficulty is to exploit effectively the data parallelism within the application, and successfully map this onto the existing hardware. Since each processor/memory pair is intimately connected with the host which possesses the only thread of control, these are best considered as intelligent memory cells of the host system, rather than a collection of tiny processors.

There is a trade-off between the complexity of each processor and the number of processors that can be included in the system, if acceptable costs are to be maintained. If a large number of processors are to be used, these must be very much simpler than in the other schemes investigated. The processor's complexity is reduced by the fact that the host supplies the instructions to execute, hence there is no need to supply each processor with an instruction fetch unit. Even so, given the current levels of integration available with VLSI, if several thousand processors are to be included these will be restricted to 1-bit devices.

The use of 1-bit processors may seem counter-productive when compared to the current trend to exploit 32-bit microprocessors. However, many of the operations to be performed on the 'intelligent memory' for a number of applications, involve Boolean operators, and not high precision arithmetic. If more complex operations are required, then these may be performed by the 1-bit processors by combining cycles of operations. This of course carries a performance overhead. However the aggregate performance of such a large number of processors is still capable of delivering performance comparable with fast sequential uniprocessors, working on the principle of 'many hands make light work'.

Finally, the question of how to arrange the large number of 'intelligent memory' cells has to be considered. Connections based on a two-dimensional square mesh, with nearest neighbour connections, were favoured by earlier designs such as the Goodyear MPP, which was designed for low-level image-processing operations (see Chapter 7). In these designs, each cell is allocated a picture element, or pixel. Thus, the two-dimensional mesh allows the direct mapping of the data element set onto the hardware. As such these machines are considered special purpose machines. A more general purpose massively parallel computer is the Connection Machine built by the Thinking Machines Corporation which is described below.

10.3.1.6.1 The Connection Machine.
The Connection Machine (Hillis, 1985; Tucker and Robertson, 1988), represents the 'far end' of the parallel processor spectrum, by its exploitation of VLSI to achieve massive parallelism. The Connection Machine has a very fine grain architecture, incorporating 65 536 1-bit processors, each with 4 Kbit of memory. These processor–memory pairs or processing elements are connected by a novel interconnection network. This network allows processing elements to be connected together to form arbitary patterns. All processing elements operate under the control of a host in SIMD mode.

The Connection Machine design was a result of research work conducted by Danny Hillis, whilst at MIT, in the late 1970s and early 1980s. Hillis was investigating the way computers could be used to simulate common-sense reasoning. A common-sense reasoning system must inspect the relationships between stored concepts and determine the appropriate action. As the number of concepts within the common-sense reasoning system increases, the number of possible relationships among these concepts experiences a combinatorial explosion. When these systems were implemented on a sequential uniprocessor, the system's response degraded rapidly as the system was made 'smarter' (i.e. contained more concepts). This was due to the relationships being examined one at a time by the sequential uniprocessor.

The potential for parallel processing is enormous since all the searches could be performed simultaneously. As the same operation is being conducted throughout the system, but on different stored concepts, the system can operate in SIMD mode.

It was the poor performance of sequential uniprocessors and the recognition of the gains afforded by utilising fine-grain parallel processing that prompted Hillis to design a massively parallel processor aimed specifically at AI applications. By 1983, Hillis had co-founded a company called Thinking Machines, and with funding from DARPA, built a 16 384 processor prototype, by May 1985. In November of the same year, a fully configured demonstration model of the commercial CM-1 system containing 65 536 processors was ready, and the system went to the market in April 1986. The design by now had moved a little more towards a massively parallel processor, aimed at supporting a general class of problems, and not just specifically AI applications.

Whilst some aspects of the design of the Connection Machine are novel, it appears heavily influenced by previous single-bit processor arrays, such as the ICL Distributed Array Processor (DAP), and the Goodyear Massively Parallel Processor (MPP). The Connection Machine's novelty, and where it derives its name, is from the incorporation of a complex switching network to provide a flexible processing element interconnection scheme. The network allows arbitrary connections to be made between processing elements under program control. This allows, for example, arrangements of processing elements to represent concepts in the form of a semantic networks. As a result of this flexibility, the programmer is free to devise an algorithm that is the most appropriate for solving the problem, without having to consider the limitations of the wiring scheme of the underlying architecture.

The CM-1 contains the following system components (see Figure 10.12):

- up to 65 536 processing elements,
- an interprocessor communication network,
- one to four sequencers,
- one to four front-end computer interfaces.

Each processing element within the CM-1, consists of a one-bit Arithmetic Logic Unit (ALU) with associated latches and registers. The registers include an activity flag, which indicates if a certain processor should obey or ignore a broadcast instruction. Each processor can address a 4 Kbit RAM memory. The processors are implemented on a custom CMOS chip. Conservative processor speeds and fabrication technology were employed to ease design and increase reliability. Sixteen processors can be accommodated on each chip. Many more processors could have been packed onto a single chip, had it not been for the room used by the router unit, which will be discussed later.

To connect 65 536 processors in a hypercube configuration, would have taken a 16-dimensional hypercube. The wiring overheads of such a configuration make this impractical. To reduce the wiring complexity to a manageable size, the CM-1 is arranged as a 12-dimensional hypercube, each node containing 16 processors and a router, ie one custom chip. A fully configured CM-1 consists of 4096 custom chips, with 32 MByte of memory, connected by 24 576 wires in a 12-dimensional hypercube topology. It can be appreciated that, even with this construction, there are considerable wiring costs.

Communication along these wires is based on a serial packet switching scheme. A packet routing strategy allows the routes of the packets to be dynamically determined, to effect the most efficient path between processors. The on-chip router handles the message switching and routing tasks for its 16 processing elements, and forwards incoming messages if they are not addressed to that node. Any processor can inject a message into the system, by sending a

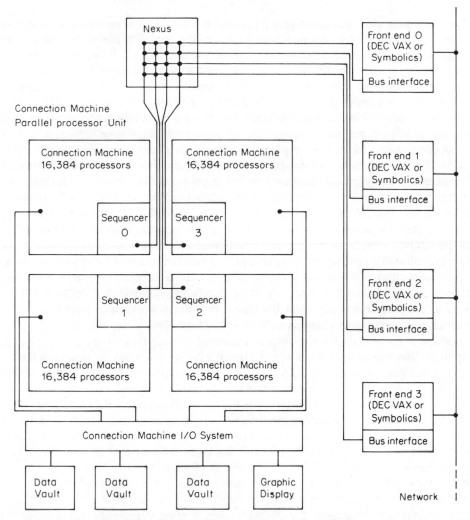

Figure 10.12 The Connection Machine's system organisation (Tucker and Robertson, 1988). (Reproduced by permission of IEEE © 1988 IEEE.)

message to its router. Each message has a data field and a destination address field. The router operates by forwarding messages serially over one of the twelve hypercube communication wires. The router's actions are governed by the destination address fields of the message packet, and the existing message passing activity on the hypercube. The destination address field of the packet can be considered as a relative address of the destination from the source. Contention for communications paths will delay messages, since only one message may be sent across each link in one cycle of operation. However, the hypercube configuration provides a number of efficient communications routes between source and destination pairs. The router selects one of these routes, if they are not already allocated, and forwards the message on, modifying the destination address field as it does so to maintain the correct relative address. The same operation is conducted at intervening nodes. When the message contains a zero destination address field, it has arrived at the desired destination.

This communications scheme makes it possible for the CM-1 to establish many different patterns of communication. An important aspect of the system's use is that these details are hidden from the user. This allows the transparent mapping of the user's problem onto the underlying architecture of the CM-1, without the need for explicit program control. The CM-1 processing elements are controlled from a host machine. Up to four host machines may be connected to the CM-1. These host machines are either a DEC VAX or a Symbolics workstations. Instructions from the hosts are routed through a crossbar switch, called the 'Nexus', to the four microprogrammed sequencers. Each sequencer controls 16 384 processing elements. The Nexus crossbar switch provides a partitioning mechanism to allow the CM-1 to be split into four quadrants. Each quadrant may be under the control of a different host/sequencer pair. Alternatively, a single host can control all the processing elements, by connection via the Nexus to all four sequencers. Instructions from the host are sent to the sequencers. Each 'high-level virtual instruction' from the host will cause the sequencer broadcast a stream of 'nano-instructions' to the processing elements. These nano-instructions control the timing and operation of the processing elements.

The programming environment for the CM-1 is based on Lisp. This reflects the origins of the Connection Machine. Connection Machine Lisp (CMLisp) is an extension of Common Lisp, designed at MIT to support the SIMD mode of operation of the Connection Machine. The programmer is presented with a conventional workstation interface, and with a programming environment supporting CMLISP. The programming style demanded by the Connection Machine exhibits the normal sequential control flow of other von Neumann computers. The only extension is that CMLISP allows operations to be expressed over parallel structures. The communications and hypercube topology of the Connection Machine remain hidden from the programmer in this environment.

The Connection Machine's flexibility has led to its use in many different areas (Waltz, 1987). In April 1987, Thinking Machines launched a modified version of the CM-1, called the CM-2. This machine is software compatible with the CM-1, but included a number of modifications and improvements. The CM-2 has taken advantage of the improvements in memory densities, to increase the memory capacity of the CM-2 to over 16 times that of the CM-1. A more efficient communications strategy, which allows messages to be interleaved, has been implemented, and limited support for floating point arithmetic has been included to improve the performance of the machine, when applied to scientific applications. This marks a further move away from the pure AI origins of the original design. However, over a dozen of these machines were sold before the end of 1987. With each machine costing around $3 million the move made commercial sense.

10.4 FUTURE DIRECTIONS—CONNECTIONISM

Another way to exploit the potential of massive parallelism is by the use of Connectionism. Connectionism represents a completely different computational paradigm to those previously discussed. A connectionist system can be described as a collection of simple computational units, operating asynchronously and in parallel. These computational units, or nodes are arranged in a pattern reminiscent of biological neural networks. Nodes are connected via weighted links, whose values are modified during training to improve the systems performance.

The concept of connectionism is not new. A restricted form of connectionist system, known as Perceptrons (Rosenblatt, 1959), was developed in the late 1950s by Rosenblatt. Most research into these systems was abandoned after the publication of a paper by Marvin Minsky and Seymour Papert (Minsky and Papert, 1969), which proved that the Perceptron model was flawed, and failed to operate when applied to certain classes of important problems. However, a small number of researchers persevered with the idea, and in the early 1980s, a number of research groups had independently extended the Perceptron model to overcome the previous limitations. These systems emerged under many different names; Connectionist (Feldman and Ballard, 1982), Artificial Neural Networks (Hopfield and Tank, 1986), Parallel Distributed Processing (PDP) (Rumelhart and McCelland, 1986) and Boltzmann Machines (Hinton, 1984). Whilst these systems differ in subtle ways, they share the same general principles. The term 'connectionist system' will be used to describe all such systems in this section.

The rationale behind the development of connectionist systems has been to approach AI processing in a radically new way. The success and failure of current AI systems is dependent on the ability to form some compact description of the task. This may take the form of rules for instance, which allow the manipulation of information in a governed way. However, there are many applications, such as speech recognition, and the visual interpretation of images, where attempts to find compact descriptions has only had limited success. The fact that natural systems, such as the human brain appears to handle these applications with ease, has led to a re-evaluation of the processing of these tasks.

One approach is to investigate the processing mechanisms that are present in the brain. The brain is, of course, a very complex organ, which despite recent advances in neurobiology and neurophysiology still remains largely an enigma. However, certain principles regarding its operation can be deduced. One of the brain's basic principles is that it appears to operate in a massively parallel fashion. This is due to the slow speed of the basic processing unit within the brain: the neuron. The neuron's switching time is in the millisecond range (10^{-3} s). Current high speed digital electronic circuits operate in the nanosecond range (10^{-9} s). Thus, digital electronics can operate about a million times faster than neurons. The slow speed of the neuron can be attributed to the fact that natural systems are constrained by the amount of energy they can dissipate, and the conducting media used to 'wire' the circuits. In the case of natural systems, this wiring is performed by conducting 'salty water'. Given the brain's 'wetware' slow operation, it must operate in parallel if it is to achieve the performance observed.

A typical human brain contains between 10^{10} and 10^{11} neurons. Each neuron may be connected to as many as 10 000 other neurons. Thus, the brain may be considered as a highly connected system. The operation of neuron is not yet fully understood, but in simple terms the neuron remains passive until the total activity on its inputs exceeds some threshold value. When this threshold level is exceeded, the neuron 'fires', which sends a wave of electrical activity down all its outputs. These outputs fan out, and form input connections to many other neurons. The arrival of the wave of activity at any one of these other neurons may cause the current activity level associated with its inputs to rise above the threshold level. The activity can therefore be visualised as a wave of activity, which is absorbed or regenerated as it propagates through the neural network.

It is this architecture and behaviour that is mimicked in a connectionist system.

10.4.1 Connectionist Systems

A connectionist system consists of a number of processing nodes, which have a number of inputs and a number of outputs (see Figure 10.13). These nodes are normally arranged in a number of layers, consisting of an input layer, an output layer and a number of intermediate or 'hidden' layers. The number of nodes in the hidden layers are less than those in the input or output layers. The nodes are connected such that the outputs from one layer form the inputs to the layer above. These connections have associated with them a modifiable weight. When a connection is active, the weight associated with that connection can be considered as a input value to the node on the end of the connection. When the sum of these values appearing at a node's input exceeds a certain threshold the node's outputs become active (else they remain inactive).

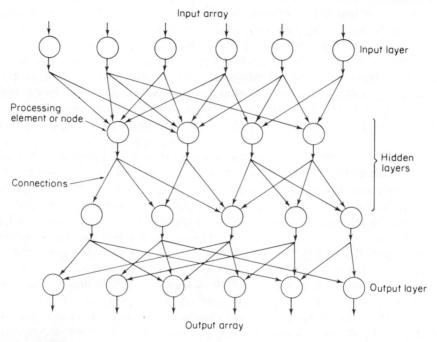

Figure 10.13 A connectionist system

Connectionist systems are specified by their net topology, node characteristics and the training or learning rule used. There is no concept of programming a connectionist system. The connectionist system has to be trained to perform the desired task. This is achieved by the use of a learning rule. These rules specify an initial set of weights, and indicate how the weights should be adapted during training to improve performance. One such rule is the Back Propagation method (Lippmann, 1987). In this scheme, a series of input patterns are applied to the inputs of the connectionist system. Each input pattern has a target output pattern. After the input pattern has been applied, an output pattern will form at the output to the connectionist system. From the differences between these two, an error can be calculated. Then, starting at the output and working backwards (hence the term back propagation), a figure (or gradient) can be calculated for each weight, that indicates how much a particular weight's adjustment will modify the calculated error. These gradients are summed at the end of one cycle of training input patterns, and the weights modified accordingly. The process is repeated until the error for all the patterns is within a tolerance or it becomes apparent that the training set is inconsistent or insufficient.

After training, the connectionist system has developed an internal representation of the training patterns. There is no concept of the information being stored at a specific site within the connectionist system. The internal representation is held in the weights linking the nodes and is totally distributed throughout the system. When trained the connectionist system displays some interesting characteristics. If presented with a novel input pattern, the system will respond with the output associated with the training case, whose internal representation matches closest to that of the new input.

The connectionist system can be considered as a direct implementation of an associative store. It is this feature of the connectionist systems that has most excited AI researchers. Most research to date has been largely on the experimental side, aimed at determining the best network topology and learning rule combination for various tasks (Mehta, 1988). However, most systems at present are merely software simulations of connectionist systems running on conventional sequential computers. The poor performance of the sequential computer, when implementing connectionist systems, limits the size of network that can be realistically investigated. Despite its name, the Connection Machine is not a connectionist machine. However, its highly parallel nature, coupled with its flexible communications strategy makes the Connection Machine an excellent platform for the running of software simulations of large connectionist networks. The true benefit of connectionist systems will, however, only become apparent when hardware specifically designed to support connectionist systems is available. One of the first commercial systems to use connectionist principles was the Wisard system (Aleksander, 1984), developed by Igor Aleksander. The Wisard system was implemented using standard integrated circuits. More specialised chips are under development, and one example is that of Synaptics (Durham, 1987) of San Diego, which has developed a hybrid analogue-and-digital chip to implement a real-time system, to simulate the optical system of a frog to preprocess visual information obtained from a video camera at the end of a robotic arm.

Whatever their underlying construction, it is clear that VLSI fabrication technology will be required to implement the many nodes required. The remaining problem of how to form the interconnections between large numbers of nodes remains unsolved. One possible solution is to use optical devices, which combine the use of lasers and holograms to form the interconnections for such systems (Yaser *et al.*, 1987).

It is clear from the interest that connectionist systems have generated that they will play an important role in AI processing in the future. The reader is encouraged to refer to (Rumelhart and McCelland, 1986) for a good overview of the topic.

BIBLIOGRAPHY

Aho, A.V. and Ullman, J.D. (1977) *Principles of Compiler Design*, Addison-Wesley, Reading, MA.

Aleksander, I. (1984) Wisard: A Radical Step Forward in Image Processing, *Sensor Review*, (July), 210–224.

Amamiya, M. (1988) Data Flow Computing and Parallel Reduction Machine, *Future Generations Computer Systems*, **4**, 53–67.

Anderson, P., Hankin, C.L., Kelly, P.H.J., Osman, P.E. and Shute, M.J. (1988) Cobweb-2: Structured Specification of a Wafer Scale Supercomputer, In: *Parallel Architecture and Languages in Europe, Lecture Notes in Computer Science*, Vol. 258, Eds J.W. de Bakker, A.J. Nijman and P.C. Treleven, Springer Verlag, Berlin, pp 51–67.

Backus, J. (1978) Can Programming be Liberated from the von Neumann Style? A Functional Style and its algebra of Programs, *CACM*. **21**, 613–41.

Batcher, K.E. (1980) Design of a Massively Parallel Processor, *IEEE Trans*. C-29, 836–840.

Bosshart, P.W. (1987) A 553K-Transistor LISP Processor Chip, *ISSCC 87: Digest 1987 Int. Solid State Circuits Conf.*, IEEE, New York, pp 202–3.

Clark, D. and Green, C. (1977) An Empirical Study of List Structures in LISP, *CACM*, **20**, pp 78–86.

Cohen, J. (1981) Garbage Collection of Linked Data Structures, *ACM Computing Surveys*, **13** pp 341–367.

Cray Research Inc. (1984) The CRAY X-MP Series of Computer Systems, Publication MP-2101,Cray Research Inc., Minnesota.

Darlington, J. and Reeves, M. (1981) ALICE: A Multi-Processor Reduction Machine for the Parallel Evaluation of Applicative Languages', *Proc. 1981 ACM Conference on Functional Programming Languages and Computer Architecture*.

Durham, T. (1987) Introducing the Man with Silicon Nerves, *Computing*, 2 July, p 28.

Eisenbach, S. and Sadler, C. (1988) Parallel Architecture for Functional Programming, *Information and Software Technology*, **30**, 355–364.

Feldman, J.A. and Ballard, D.H. (1982) Connectionist Models and Their Properties, *Cognitive Science*, **6**, 205–254.

Flynn, M.J. (1972) Some Computer Organizations and their Effectiveness, *IEEE Trans. Comput.*, C-21, 948–960.

Genner, R.E., Gustafson, J.L. and Montry, R.E. (1988) Hypercube Linear Speedup: Development and Analysis of Scientific Applications programs on a 1024-processor hypercube, SAND 88-0317 Sandia National Laboratories, Albuquerque, February.

Gurd, J.R., Kirkham, C.C. and Watson, I. (1985) The Manchester Prototype Dataflow Computer, *Commun ACM*, **28**, 34–52.

Hill, M. (1986) Design Decisions in SPUR, *Computer*, (Nov), 8–24.

Hillis, W.D. (1985) *The Connection Machine*, MIT Press, Cambridge, MA.

Hinton, G.F. (1984) Boltzmann Machines: Constraint Satisfaction Networks That Learn, Technical Report CMU-CS-84-119, Carnegie-Melllon University, May.

Hopfield, J.J. and Tank, D.W. (1986) Computing with Neural Circuits, *Science*, **233**, 625–633.

Hwang, K., Ghosh, J. and Chowkwanyun, R. (1987) Computer Architectures for Artifical Intelligence Processing, *Computer*, 19–27.

Inmos Ltd. (1985) IMST414 Transputer Reference Manual, Inmos Ltd, Bristol.

Intel Scientific Computers (1988) A Technical Summary of the iPSC/2 Concurrent Supercomputer, Intel Scientific Computers, Beaverton, Oregon, Order Number 280115–001.

Intel Scientific Computers (1987) Concurrent Common LISP, Concurrent Common LISP Data Sheet, Intel Scientific Computers, Beaverton, Oregon, Order Number CC0186–01.

Johnson, T. and Durham, T. (1986) *Parallel Processing: The Challenge of New Computer Architectures*, Ovum Press, p 189.

Judge, P. (1987) AI Delivery, *Systems International*, (May) 37–40.

Judge, P. (1988) No More Hypercubes, *Parallelogram* (8 Nov), p 20.

Landin, P.J. (1964) The Mechanical Evaluation of Expressions, *Computer* **J.6**, 308–320.

Lippmann, R.P. (1987) An Introduction to Computing with Neural Nets, *IEEE ASSP Magazine*, **3**(4), 4–22.

Mehta, A. (1988) Promise of Progess and Probability, *Computer Weekly*, (23 February), p 8.

Minsky, M. and Papert, S. (1969) *Perceptrons*, MIT Press, Cambridge, MA.

Moon, D.A. (1985) Architecture of the Symolics 3600, *Proc. Twelth Int. Symp. on Computer Architecture*, pp76–83.

Moon, D.A. (1987) Symbolics Architecture, *Computer*, 43–52.

Pease, M.C. (1977) The Indirect Binary N-Cube Microprocessor Array, *IEEE Trans. Comput*, **C-26**, pp 458–473.

Peyton-Jones, S.L. (1987) GRIP—a parallel processor for functional languages, *IEE Electronics and Power*, (Oct), 633–636.

Pleszkun, R. and Thazhuthaveetil, M.J. (1987) The Architecture of Lisp Machines, *Computer*, (Mar), 35–44.

Reddaway, S.F. (1979) *The DAP Approach, Infotech State of the Art Report: Supercomputers* Vol. 2, eds. C.R. Jesshope and R.W. Hockney, Infotech Int. Ltd., Maidenhead, pp 311–329.

Rosenblatt, R. (1959) *Principles of Neurodynamics*, Spartan Books.

Rumelhart, D.E. and McCelland, J.L. (1986) *Parallel Distributed Processing: Explorations in the Microstructure of Cognition*, Vols 1 and 2, MIT Press, Cambridge, MA.

Seitz, C.L. (1985) The Cosmic Cube, *Comms ACM*, **28**, 22–33.

Sun Common Lisp 3.0: Technical and Performance Report (1988) Software Products Division, Sun Microsystems Inc., CA, August.

Thakkar, S., Gifford, P. and Fielland, G. (1987) Balance: A Shared-Memory Multiprocessor System, *Proc. Int. Conf. Supercomputing*, St Petersburg, Fla. 93–101.

Treleaven, P.C., Brownbridge, D.R. and Hopkins, P. (1982) Data-Driven and Demand-Driven Computer Architecture, *ACM Computing Surveys*, **14**, 93–143.

Tucker, L.W. and Robertson, G.G. (1988) Architecture and Applications of the Connection Machine, *Computer*, (Aug), 26–38.

Waltz, D.L. (1987) Applications of the Connection Machine, *Computer*, (Jan), 85–97.

Watson, I., Sargent, J., Watson, P., and Woods, V. (1987) Flagship:Computational Models and Machine Architecture, *ICL Technical Journal*, **5**, 555–574.

Wilson, I. (1988) Reduced Instructions: RISC Chips Compared, *Systems International* (Oct), 23–33.

Yaser, S., Abu-Mostafa and Pslatis, D. (1987) Optical Neural Computers, *Scientific American*, **256**, 66–73.

N.J. Avis
Department of Medical Physics and Clinical Engineering
University of Sheffield and Health Authority
Royal Halamshire Hospital
Glossop Road
Sheffield
South Yorkshire S10 2JF

Index

Bold entries denote headings